T0174911

Practical Sampling Techniques for
INFRARED ANALYSIS

Practical Sampling Techniques for INFRARED ANALYSIS

Edited by
PATRICIA B. COLEMAN

CRC Press
Taylor & Francis Group
Boca Raton London New York

CRC Press is an imprint of the
Taylor & Francis Group, an **informa** business

CRC Press
Taylor & Francis Group
6000 Broken Sound Parkway NW, Suite 300
Boca Raton, FL 33487-2742

© 1993 by Taylor & Francis Group, LLC
CRC Press is an imprint of Taylor & Francis Group, an Informa business

First issued in paperback 2019

No claim to original U.S. Government works

ISBN 13: 978-0-367-44976-6 (pbk)
ISBN 13: 978-0-8493-4203-5 (hbk)

This book contains information obtained from authentic and highly regarded sources. Reasonable efforts have been made to publish reliable data and information, but the author and publisher cannot assume responsibility for the validity of all materials or the consequences of their use. The authors and publishers have attempted to trace the copyright holders of all material reproduced in this publication and apologize to copyright holders if permission to publish in this form has not been obtained. If any copyright material has not been acknowledged please write and let us know so we may rectify in any future reprint.

Except as permitted under U.S. Copyright Law, no part of this book may be reprinted, reproduced, transmitted, or utilized in any form by any electronic, mechanical, or other means, now known or hereafter invented, including photocopying, microfilming, and recording, or in any information storage or retrieval system, without written permission from the publishers.

For permission to photocopy or use material electronically from this work, please access www.copyright.com (http://www.copyright.com/) or contact the Copyright Clearance Center, Inc. (CCC), 222 Rosewood Drive, Danvers, MA 01923, 978-750-8400. CCC is a not-for-profit organization that provides licenses and registration for a variety of users. For organizations that have been granted a photocopy license by the CCC, a separate system of payment has been arranged.

Trademark Notice: Product or corporate names may be trademarks or registered trademarks, and are used only for identification and explanation without intent to infringe.

Visit the Taylor & Francis Web site at
http://www.taylorandfrancis.com

and the CRC Press Web site at
http://www.crcpress.com

PREFACE

Patricia B. Coleman

I. THE PURPOSE OF THIS BOOK

This book is intended to serve as a practical guide for sample handling for routine analysis in the mid-infrared region using commercially available instrumentation and accessories. (Over the years, the definition of mid-infrared has changed, mainly based on the available instrumentation. Forty years ago, a mid-infrared spectrum was obtained with a dispersive infrared spectrophotometer that covered the range 4000 to 200 cm^{-1}. In the last 20 years, as FT-IR spectrometers with potassium bromide (KBr) based beamsplitters became more prevalent, the definition of mid-infrared has changed to mean the range 4000 to 400 cm^{-1} due to the KBr cut-off. This book covers sampling techniques for the mid-infrared range defined as 4000 to 400 cm^{-1}.)

In the course of assembling this book, it became apparent that techniques considered routine in one laboratory are rarely used in others. For example, in one laboratory solid samples are always run using a diamond anvil cell, even though numerous alternative techniques are available. It is hoped that this book will provide the spectroscopist, the chemist, and/ or the technician the background needed to consider widely diverse sampling techniques.

II. HOW AND WHY THE BOOK WAS WRITTEN

Three years ago (after an ASTM meeting held in conjunction with FACSS '90), a lively discussion ensued about the problems and misuses of sampling techniques in infrared spectroscopy. All agreed that too often a new, more expensive technique is used, rather than the "old standby", just because it is "new" and not because it neccessarily represents a better way to analyze the sample. Dissolving a sample in solvent and examining it in a sealed cell by transmission, or grinding the sample and forming a KBr pellet are techniques that have fallen into disfavor because they are perceived as "old-fashioned." A common misconception is that the mere presence of an FT-IR in the lab, requires newer sampling techniques expressly designed for that instrumentation. Another serious misconception is that the analyst no longer has to worry about carefully preparing the sample, because the FT-IR's computer can be used to manipulate the data in order to "correct" for sloppiness in sample preparation. An example was mentioned of an "expert" witness who testified in court that the spectrum in question could not be identified relative to a published reference spectrum (even though they appeared identical) because the reference had been obtained on a dispersive instrument, and the unknown had been obtained on an FT-IR! It is a sad state.

The concept of this book was inspired by that discussion. The chapters are written by experts in the field who utilize the various sampling techniques on a regular basis. (Some of the material in this book was presented in a day-long symposium organized by the editor at FACSS '91.)

Because of the many sampling techniques available for a particular sample, it is not possible to provide a unique recipe for the infrared analysis. For example, a cured polymer sample could be cryoground, diluted with KBr, and analyzed either as a KBr pellet or by diffuse reflectance. Which of these methods is better? This choice depends on the sample type, its chemical and physical properties, the equipment available, and the skill of the analyst. If the polymer sample is changed as a result of the pressure used to make a KBr pellet, then diffuse reflectance would be a better choice. If the laboratory does not have a high quality (e.g., Carver) press for making KBr pellets, then again diffuse reflectance would be a better choice. If the particular model of diffuse reflectance accessory is not easily aligned or the analyst is not skilled in aligning it, then a KBr pellet would represent a better choice.

On the other hand, if the cured polymer is too hard to cyrogrind, perhaps a small chip of it could be broken off and a spectrum acquired using PAS. Likewise, if a diamond anvil cell was available, the chip could be analyzed in that cell either using a beam condenser or an infrared microscope. If only qualitative information was needed, the sample could be scraped against a piece of carborundum paper and a diffuse reflectance spectrum could be obtained.

Clearly the "proper" technique depends on the information needed (qualitative or quantitative), sample form (solid, liquid, or gas), sample availability (is it micro analysis?), the sampling preparation that would change or destroy the sample, the equipment and/or accessory available, and the skill of the anaylst. Many of the sampling techniques require accessories — diffuse reflectance, ATR, PAS, GC/FT-IR, diamond anvil cells, beam condensers, infrared microscopes — and not all laboratories have access to all of these. The analyst has to decide which technique is better suited for the samples and problems that arrive in his or her laboratory.

III. ABOUT THE BOOK

Chapter 1, 'Theoretical Considerations', by this editor is a brief look at what factors determine the nature of an infrared spectrum. After reviewing the remainder of the book, it was felt that a short summary would help understanding.

Chapter 2, 'Sample Handling in Infrared Spectroscopy — An Overview', by W. D. (Paul) Perkins covers basic techniques for analyzing solids and liquids such as KBr pellets and solution cells. Based on his years of practical experience in analyzing samples by infrared and also organizing and running numerous training courses, Perkins presents this material in a clear, pedagogically-sound fashion.

The Chapters (3, 4, and 5) on internal reflectance (IRS), diffuse reflectance (DRIFT), and photoacoustic spectroscopy (PAS), respectively, outline the theory behind each technique, specify the accessories needed to perform the technique, and illustrate their application through examples. Chapter 3, 'Optimization of Data Recorded by Internal Reflectance Spectroscopy,' by Senja and David Compton covers the analysis of liquids and solids, both organic and inorganic, using IRS. Chapter 4, 'Diffuse Reflectance Infrared Spectroscopy: Sampling Techniques for Qualitative/Quantitative Analysis of Solids,' by Scott Culler discusses different methods for solids analysis by DRIFT and the use of an internal reference material for quantitative DRIFT. Chapter 5, 'A Practical Guide to FT-IR Photoacoustic Spectroscopy,' by John McClelland et al. details the use of PAS for macro and micro sample analysis.

Chapter 6, 'The Versatile Sampling Methods of Infrared Microspectroscopy,' by Patricia Lang and Lisa Richwine includes discussion of the various undesirable optical effects that can occur with infrared microscopes and how to avoid them. The chapter also elaborates on the use of the infrared microscopy for transmission, reflectance, emission, grazing angle, internal reflection, and even chromatographic effluent analysis. The analysis of chromatographic effluent is further discussed in Chapter 7, 'General Methods of Sample Preparation for the Infrared Hyphenated Techniques,' by Gregory McClure. A number of different sampling devices that can be used for gas chromatographic sample introduction are described.

Quantitative aspects are covered in great detail in Chapter 8, 'Quantitative Analysis — Avoiding Common Pitfall,' by Senja and David Compton. This chapter also discusses the selection and preparation of standards, as well as a review of virtually all of the sampling techniques discussed in other chapters. Several examples of quantitative analysis are also included in Chapter 4 an 5.

Chapters 9 and 10 include numerous examples of laboratory experiences with a wide variety of samples and the resultant artifacts that can cause problems with infrared analysis. Chapter 9, 'Errors in Spectral Interpretation Caused by Sample-Preparation Artifacts and FT-IR Hardware Problems,' by Robert Williams describes silicon contamination and problems due to sample crystallinity. Also described are errors that can be introduced into the spectrum by incorrect use of the computer software and by FT-IR hardware anomalies. Chapter 10, 'Sampling Techniques for Vibrational Analysis in the Environmental and Forensic Laboratories,' by Kathryn and Victor Kalasinsky covers a number of "typical" samples with which the analyst must deal. These analyses include transmission, reflectance, microscopy, and chromatographic techniques.

In Chapter 11, 'Infrared Sampling Techniques Training,' by Thomas Porro and Silvio Pattacini, the annual training courses available for infrared analysis are summarized. Included are sample handling courses and those offered by the various manufacturers of FT-IR instrumentation.

IV. SAFETY ISSUES

Several of the materials that are commonly used in conjunction with infrared analysis can pose a variety of safety hazards (e.g., KRS5, carbon tetrachloride, carbon disulfide, etc.) While some mention is made about handling of these materials in the individual chapters, these discussions are not meant to be inclusive. Detailed information on safety procedures for handling chemicals can be found in the Material Safety Data Sheet (MSDS) which is readily available in chemistry libraries or through a laboratory safety engineer.

V. ACKNOWLEDGMENTS

Thanks to Clara Craver (Craver Consulting), David Compton (Sonoco), Rachael Barbour (BP America), Roy Cain (Firestone), Kathryn Kalasinsky (Division of Forensic Toxicology, Office of the Armed Forces Medical Examiner), and Richard Duerst (3M Company) for suggesting scientists to participate in writing this book. Special appreciation is expressed to Christina Martin of CRC Press for her support and assistance.

CONTRIBUTORS

Patricia B. Coleman, Ph.D.
Ford Motor Company
Analytical Sciences Department
Dearborn, Michigan

David A.C. Compton, Ph.D.
Bio-Rad
Digilab Division
Cambridge, Massachusetts

Senja V. Compton, M.S.
Bio-Rad
Digilab Division
Cambridge, Massachusetts

Scott R. Culler, Ph.D.
Abrasive Systems Division
3M Corporation
St. Paul, Minnesota

R.W. Jones, Ph.D.
Ames Laboratory
Iowa State University
Ames, Iowa

Kathryn S. Kalasinsky, Ph.D.
Division of Forensic Toxicology
Armed Forces Med. Exam.
Armed Forces Inst. Pathology
Washington, D.C.

Victor F. Kalasinsky, Ph.D.
Department of Environmental and
  Toxicologic Pathology
Armed Forces Inst. Pathology
Washington, D.C.

Patricia L. Lang, Ph.D.
Department of Chemistry
Ball State University
Muncie, Indiana

S. Luo, Ph.D.
Ames Laboratory
Iowa State University
Ames, Iowa

J.F. McClelland, Ph.D.
Ames Laboratory
Iowa State University
MTEC Photoacoustics, Inc.
Ames, Iowa

Gregory L. McClure, Ph.D.
Organic Analysis Division
The Perkin-Elmer Corporation
Norwalk, Connecticut

Silvio C. Patacini
The Perkin-Elmer Corporation
Norwalk, Connecticut

W.D. Perkins, Ph.D.
The Perkin-Elmer Corporation
San Jose, California

Thomas J. Porro
The Perkin-Elmer Corporation
Norwalk, Connecticut

Lisa J. Richwine
Department of Chemistry
Ball State Unviersity
Muncie, Indiana

L.M. Seaverson, Ph.D.
MTEC Photoacoustics, Inc.
Ames, Iowa

Robert C. Williams, Ph.D.
The B.F.Goodrich Company
Research & Development Center
Brecksville, Ohio

DEDICATION

This book is dedicated to my husband, David Coleman, and my daughter, Jennifer Roush, for their continuing help and encouragement.

TABLE OF CONTENTS

Chapter 1
Theoretical Considerations 1
Patricia B. Coleman

Chapter 2
Sample Handling in Infrared Spectroscopy — An Overview 11
W. D. Perkins

Chapter 3
Optimization of Data by Internal Reflectance Spectroscopy 55
Senja V. Compton and David A. C. Compton

Chapter 4
Diffuse Reflectance Infrared Spectroscopy: Sampling Techniques for
Qualitative/Quantitative Analysis of Solids 93
Scott R. Culler

Chapter 5
A Practical Guide to FT-IR Photoacoustic Spectroscopy 107
J. F. McClelland, R. W. Jones, S. Luo, and L. M. Seaverson

Chapter 6
The Versatile Sampling Methods of Infrared Microspectroscopy 145
Patricia L. Lang and Lisa J. Richwine

Chapter 7
General Methods of Sample Preparation for the Infrared Hyphenated Techniques 165
Gregory L. McClure

Chapter 8
Quantitative Analysis — Avoiding Common Pitfalls 217
Senja V. Compton and David A. C. Compton

Chapter 9
Errors in Spectral Interpretation Caused by Sample-Preparation Artifacts
and FT-IR Hardware Problems 255
Robert C. Williams

Chapter 10
Sampling Techniques for Infrared Analysis in the Environmental and
Forensic Laboratories 275
Kathryn S. Kalasinsky and Victor F. Kalasinsky

Chapter 11
Infrared Sampling Techniques Training 289
Thomas J. Porro and Silvio C. Pattacini

Index 295

Chapter 1

THEORETICAL CONSIDERATIONS

Patricia B. Coleman

CONTENTS

I. Introduction 2
II. General Concepts 2
III. Group Frequencies 7
References 10

0-8493-4203-1/93/$0.00+$.50
© 1993 by CRC Press Inc.

I. INTRODUCTION

Although the main purpose of this book is to describe proper sample handling techniques for infrared analysis, a cursory understanding of how an infrared spectrum is generated is useful. This chapter covers infrared theory in a limited fashion, since numerous texts cover this material in great detail.[1-5]

By understanding the theory, it is possible to see how the infrared spectra of complex molecules can be interpreted. It then becomes possible to see how the infrared spectrum of a material can be different depending on the state of the molecule (i.e., solid, liquid, or gas). This ultimately speaks to experimental realities associated with sample handling techniques. When a sample is dissolved in solvent and the spectrum is run in a solution cell, the solvent that is used may hydrogen bond with the sample and cause bands in the spectrum to shift. When a sample is ground, mixed with potassium bromide (KBr) and pressed to form a pellet, there can be shifts in the band frequencies due to reaction between the sample and the alkyl halide which can occur from the pressure used in making the pellet. If the same sample is ground and mixed with KBr, but the spectrum is run by diffuse reflectance, the band frequencies may not be shifted in the same manner that they were in the KBr pellet. When a sample is analyzed using gas chromatography/Fourier Transform Infrared (GC/FT-IR) the resulting spectrum is that of the vapor phase of the compound. That spectrum, with its sharp narrow lines, looks very different than the condensed phase spectrum of the same material. Not only the physical state of the sample, but also the manner in which the sample is diluted, the accessory used to obtain the spectrum, and the care that is employed in the sample preparation can have large effects on the characteristics and quality of the resulting infrared spectrum.

II. GENERAL CONCEPTS

Spectroscopy is the study of the interaction of electromagnetic radiation with matter. Electromagnetic radiation can be divided into different regions of energy that correspond to different spectroscopic techniques (see Figure 1). The region from 4000 to 400 cm^{-1} is considered the mid-infrared region where vibrational and rotational bands are observed. Sampling techniques for the mid-infrared region are covered in this book.

Classical theory described electromagnetic radiation as a continuous wave, as shown in Figure 2. This wave can be described in terms of its frequency or its wavelength which are related to each other in terms of the speed of light, as shown in Equation 1. Classical theory explained properties of electromagnetic radiation, such as refraction, diffraction, and other optical phenomena. Quantum mechanical theory described electromagnetic radiation as particles with energy that can only occur at discrete levels. This theory can be used to explain absorption and emission of electromagnetic radiation. The discrete particles (called photons) have no resting mass and energy equal to hv where h is Planck's constant, as shown in Equation 2. Most properties of radiant energy can be explained in terms of this dual wave-particle behavior of electromagnetic radiation. Equation 2 shows that energy is proportional to frequency, while Equation 3, which is derived from Equations 1 and 2, shows that it is inversely proportional to wavelength. Thus, the shorter the wavelength the higher the energy.

$$\lambda v = c \tag{1}$$

$$E = hv \tag{2}$$

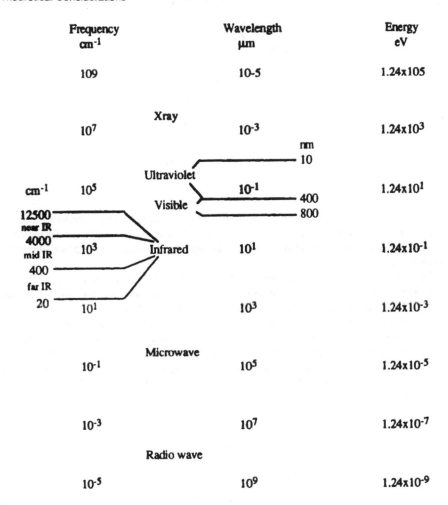

Figure 1. Electromagnetic spectrum.

$$E = \frac{hc}{\lambda} \tag{3}$$

$$E_{total} = E_{electr} + E_{vib} + E_{rot} \tag{4}$$

The total energy of a molecule is due to electronic, vibrational, and rotational contributions, as described by Equation 4. The electronic component is due to state-to-state electron transitions. These processes usually generate ultraviolet and/or visible spectra. For example, the 589 nm line of sodium in a flame is due to the first excited state of sodium as the electron goes from a $3p^1$ state to a $3s^1$ state. Vibrational energy is due to changes in the internuclear distance of two or more atoms. This process may generate an infrared spectrum. Rotational energy is due to changes in the molecule as it rotates around its center of mass. This process often results

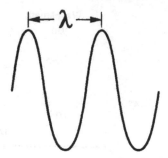

Figure 2. Continuous wave.

in fine structure superimposed on vibrational bands; it is usually only observed for small, gaseous molecules in the infrared, the far-infrared, or the Raman. For larger, condensed phase molecules the fine structure from the rotational bands is broadened to the point where it is generally not observed.

The manner in which molecules absorb electromagnetic radiation can be explained in terms of an analogy that takes into consideration both the classical and the quantum mechanical models. The molecule can be viewed as made up of springs connected to balls to represent the bonds that connect the atoms. The springs obey Hooke's law for a harmonic oscillator, that is, the force that the spring exerts on the ball is proportional to the displacement from the equilibrium state. The simplest diatomic molecule with an internal vibrational frequency can be modeled as two balls, of mass m_1 and m_2, connected by a spring. The oscillation frequency can be calculated using Equation 5, where k = force constant and m_r = the reduced mass of the atom. Reduced mass is calculated from the individual atomic masses, m_1 and m_2 using Equation 6.

$$v = \frac{1}{2\pi} \sqrt{\frac{k}{m_r}} \tag{5}$$

$$m_r = \frac{m_1 m_2}{m_1 + m_2} \tag{6}$$

If mass m_1 is small, such as for a hydrogen atom attached to heavier atom such as carbon, the reduced mass in Equation 6 is approximately equal to m_1, the mass of hydrogen, and the frequency becomes that of a point oscillator. The hydrogen atom behaves as if it were a ball connected by a spring to an immovable wall, rather than two balls that are connected by a spring. This approximation is useful for predicting frequencies of hydrogen stretching vibrations.

If there are no losses due to friction, the harmonic oscillator will remain in motion with constant energy. This energy is the sum of the kinetic and potential energy of the system. As the balls move on the spring, the relative amount of potential and kinetic energy changes depending on their specific location. Assuming that the model obeys Hooke's law, the potential energy of the molecule is given by Equation 7, a parabolic function.

$$PE = \frac{1}{2} k x^2 \tag{7}$$

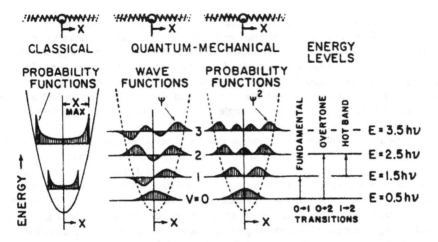

Figure 3. Representation of the classical and quantum mechanical oscillators. (From Colthup, N.B., Daly, L.H., and Wiberley, S.E., *Introduction to Infrared and Raman Spectroscopy*, Academic Press, New York, 1975. With permission.)

When the atoms are at the equilibrium state, the kinetic energy of the molecule is zero. As an atom moves to displacement x_{max} the total energy is equal to the potential energy. Since the total energy is dependent only on the force constant and the displacement, a continuum of energy levels are allowed, as shown in Equation 8. This is the distinguishing mark of the classical harmonic oscillator model. The energy and probability function for the classical harmonic oscillator is shown in the left part of Figure 3.

$$E = \frac{h}{2\pi}\sqrt{\frac{k}{m_r}} \tag{8}$$

Quantum mechanics explains the presence of discrete energy levels. When this consideration is added, energy can be expressed in terms of the harmonic oscillator model, as shown in Equation 9, or in terms of frequency, Equation 10. The principal quantum number is n. The energy can only have discrete levels of positive half-integer multiples of hv. The lowest energy possible for a quantum mechanical oscillator is 1/2 hv. Even at 0 K the molecule possesses this amount of vibrational energy. If the molecule could lose all of the vibrational energy, both the equilibrium position and the momentum would simultaneously be known, and this would violate Heißenberg's uncertainty principle, one of the fundamental axioms of quantum mechanics.

$$E = \left(n + \frac{1}{2}\right)\frac{h}{2\pi}\sqrt{\frac{k}{m_r}} \qquad n = 0,1,2,3... \tag{9}$$

$$E = \left(n + \frac{1}{2}\right)hv \qquad n = 0,1,2,3... \tag{10}$$

The wave-particle duality of matter requires that there be a wavefunction Ψ to specify the amplitude of the wave as a function of location. The Schrödinger equation provides a generalized wavefunction that is applicable to spectroscopic problems. Born provided the physical interpretation for the amplitude of the wavefunction. The square of amplitude Ψ^2 is the probability of finding the particle at a given location. Figure 3 shows the energy levels and probability functions of both the classical and the quantum mechanical harmonic oscillator for the simplest case of a single mass on a spring that is only free to move horizontally.

The harmonic oscillator model predicts that energy levels are equally spaced. Using this model, absorption of infrared radiation would result only when the molecule is excited from one vibrational state to the next highest state, and emission would occur only when the molecule decays from one state to the next lowest state. This is the selection rule for the harmonic oscillator. The frequency corresponding to this excitation ($\Delta v = 1$) is called the fundamental or normal mode. Based on the Boltzmann distribution, only the ground state would be highly populated at room temperature. Therefore, the only allowed transition would be from the ground state to the first excited state (v_0 to v_1). However, molecules are not strictly harmonic oscillators. When anharmonicity is invoked, resultant energy levels are not equally spaced. The higher the energy level, the more closely spaced they are, thus making it easier to attain the higher levels. The change in vibrational level Δv can now be greater than one, for example, v_0 to v_3 or v_1 to v_3, etc. These weaker absorptions (typically an order of magnitude weaker) are called overtones and are found at integer multiples of the fundamental frequency. For example, the fundamental of the carbonyl in ethyl formate is at 1750 cm^{-1}, and an overtone of the carbonyl band is observed at 3500 cm^{-1}. Real molecules, due to their anharmonicity, can also have combination and difference bands in their spectra. A combination band occurs when a particular frequency excites two fundamental modes. For example, a frequency v1 + v2 could excite fundamentals at v1 and v2. A difference band occurs at a frequency that is the difference of two fundamentals. [Note: the harmonic oscillator model does not predict overtones or combination bands, and therefore does not allow for the existance of the near-infrared spectral region. Indeed, the majority of spectral bands observed in the near infrared are overtones or combinations of hydrogen stretching vibrations or combinations of hydrogen stretches and other vibrations.]

Two types of molecular vibrations correspond to the normal mode of the molecule: stretching and bending. Stretching is rhythmical movement along the bond axis and can be symmetric or antisymmetric. Two balls attached by a spring can move simultaneously away from the center of the spring or they can move in towards the center. Either of these motions is termed a symmetric stretch. Two balls attached by the spring could move simultaneously to the right or to the left. Either of these motions is termed an antisymmetric stretch. Bending vibrations arise from a change in bond angle between two atoms or movement of a group of atoms, relative to the remainder of the molecule. The terms scissoring, wagging, rocking, and twisting are often used to describe these motions. Stretching and bending vibrations are shown in Figure 4.

In order for a molecule to absorb infrared radiation, several conditions must be met. The frequency of the infrared source radiation must be identical to one of the transitions between the discrete energy levels of the molecule, and there must be a change in dipole moment. For example, when both oxygen atoms in carbon dioxide (which is linear) move symmetrically in or out, there is no change in the dipole moment, and no infrared radiation is absorbed. The symmetric stretch is therefore infrared inactive. However, when the oxygens both move either to the left or to the right, the antisymmetric stretch, there is a change in dipole moment, and infrared radiation is absorbed at 2350 cm^{-1}.

Stretching

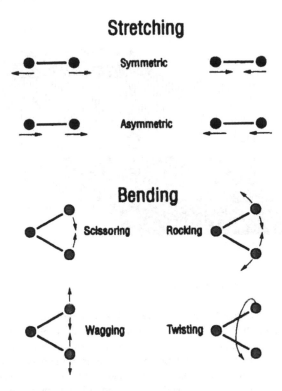

Figure 4. Typical stretching and bending modes.

III. GROUP FREQUENCIES

The position of a molecule in three dimensional space can be described in terms of the x, y, and z coordinate of each atom in the molecule. For a molecule of n atoms, there are 3n distinct coordinates to describe its shape, position, and orientation defining 3n possible degrees of freedom or movement. Any other motion of the molecule is some combination of these. Any translation, rotation, or vibration of the molecule can be described in terms of changes within this coordinate system. Most movements of the molecule are combinations of these three. In considering molecular vibrations, pure translation or rotation is ignored. Three of the degrees of freedom represent translations of the entire molecule along each coordinate axis. There is also one rotation along each axis. In a nonlinear molecule, this represents three additional degrees of freedom. In a linear molecule, this represents two degrees of freedom, because there is no rotation about the molecular axis. All other molecular motions include a vibrational component. For a nonlinear molecule, there can be a maximum of $3n-6$ degrees of freedom, and therefore $3n-6$ fundamental modes. For a linear molecule there can be a maximum of $3n-5$ fundamental modes. The linear triatomic molecule of carbon dioxide has four fundamental modes. These consist of the symmetric stretch, at 1388 cm^{-1}, the antisymmetric stretch, at 2350 cm^{-1}, and two bending modes. The bending modes are a degenerate pair, as shown in Figure 5, which means they both occur at the same frequency. In the case of carbon dioxide this is at 667 cm^{-1}. Even though there are four fundamental modes, only two characteristic bands are seem in the infrared spectrum. As stated earlier, the symmetric stretch does not involve a change in dipole moment and is therefore infrared inactive. However, since the

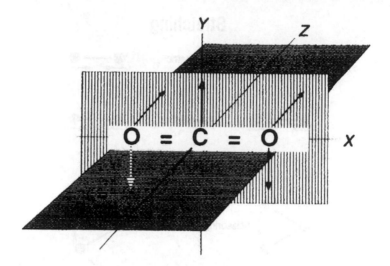

Figure 5. Degenerate bending modes of carbon dioxide.

symmetric stretch is polarizable it is Raman active. For molecules that have a center of inversion (e.g., carbon dioxide), the rule of mutual exclusion states that any mode cannot be *both* infrared active and Raman active. For small symmetrical molecules, Group Theory can also be used to determine fundamental modes and to classify them as infrared and/or Raman active. For example, nitromethane has 7 atoms and would therefore have 15 fundamental modes. Of these there is a symmetric and an antisymmetric stretch of the nitro group. In this example both the symmetric and the antisymmetric stretch cause a change in the dipole moment and are therefore infrared active. The symmetric stretch occurs at 1350 cm^{-1} and the antisymmetric stretch occurs at 1550 cm^{-1}. In general the antisymmetric stretch occurs at a higher frequency than the symmetric stretch.

For large polyatomic molecules, numerous fundamental modes can exist. However, when spectra of large molecules containing diverse functional groups are studied, it is observed that the infrared absorption frequency of a particular functional group is surprisingly constant. For example, if a series of compounds which all contain an -OH group are examined, the infrared spectra universally show bands near 3500 and 1650 cm^{-1} corresponding to the -OH stretching and bending modes, respectively. Even though most other atoms in the molecule also vibrate, their effect on the -OH constituent is relatively small. The vibrational frequency is largely determined by the force constant of the -OH bond. The force constant depends on the electronic nature of the vibrating bond, which may be perturbed by neighboring atoms. For example, a series of compounds containing a C=O group have a characteristic frequency in the range of 1600 to 1800 cm^{-1}. The *exact* location of the C=O frequency varies depending on the atoms attached to the carbonyl. Electron donating groups, electron withdrawing groups, resonance effects, and hydrogen bonding all cause the force constant of the C=O bond to vary and therefore the frequency of the carbonyl absorption to change. Typically, resonance weakens the strength of the C=O, causing absorption at a lower frequency. For example, the carbonyl of a ketone absorbs at 1715 cm^{-1}, whereas the carbonyl of an α,β-unsaturated ketone occurs at 1675 cm^{-1}. Table 1 summarizes typical absorption frequencies for a variety of carbonyl compounds. Carboxylic acids absorb at a lower frequency than ketones due to intermolecular hydrogen bonding. In amides, the resonance effect of the lone pair of electrons on the nitrogen predominates, whereas in esters

TABLE 1. Typical Carbonyl Frequency in Wavenumbers

Anhydride I	Acid chloride	Anhydride II	Ester	Aldehyde	Ketone	Acid	Amide
1810	1800	1760	1735	1725	1715	1710	1690

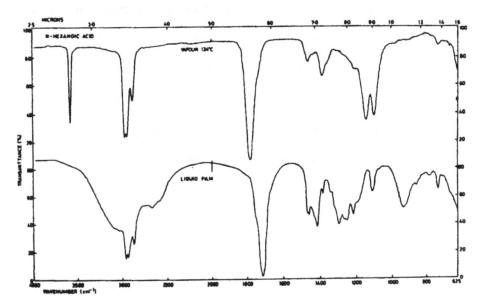

Figure 6. Spectrum of *n*-hexanoic acid in vapor phase and as liquid film. (From *Infrared Vapour Spectra*, Heyden and Sons. With permission.)

the electron withdrawing effect is stronger. Ketones absorb at a slightly lower frequency than aldehydes because in comparison to the hydrogen atom of the aldehyde, the second alkyl group on the ketone donates electrons to the carbonyl and thus weakens the bond. The α chlorine in the acid chloride has a considerable electron withdrawing affect on C=O. The chlorine draws electrons to itself increasing the π bond nature of the C=O bond. This increases the C=O bond force constant causing higher band frequencies. For example, absorption due to the carbonyl in acetyl chloride is at 1805 cm^{-1} while that due to acetone is at 1712 cm^{-1}. The electronegative oxygens in anhydrides cause a strong electron withdrawing effect on the carbonyl. Also, there are two anhydride carbonyl frequencies due to the symmetric and antisymmetric stretching vibrations. These are but a few examples of group frequencies. Several books have been written about the infrared band assignments of various group frequencies. Among these are the classic text by Bellamy[6] and the recent handbook by Lin-Vien et al.[7]

It is evident that the nature and quality of the observed spectrum will depend on the chemical (e.g., solid, liquid or gas) and physical (e.g., pathlength, turbidity, surface condition, etc.) properties of the sample. For example, the spectrum of water in a 15-μm cell reveals several regions that are not totally obliterated by water. The same sample in a 15-mm cell is unusable due to the totally absorbing water. Figure 6 shows the spectra of *n*-hexanoic acid in the vapor phase and as a liquid film. It is quite apparent that there are very few similarities in the spectra. Every aspect of the sample itself, as well as instrumental factors, are important in collecting and interpreting infrared spectra.

REFERENCES

1. Colthup, N.B., Daly, L.H., and Wiberley, S.E., *Introduction to Infrared and Raman Spectroscopy*, Academic Press, New York, 1975.
2. Potts, W., *Chemical Infrared Spectroscopy*, Vol. 1, Techniques, John Wiley & Sons, New York, 1963, chap. 1, 2, and 8.
3. Alpert, N.L., Keiser, W.E., and Szymanski, H.A., *IR-Theory and Practice of Infrared Spectroscopy*, Plenum Press, New York, 1973, chap. 4 and 5.
4. *Physical Methods in Organic Chemistry*, Schwarz, J.C.P., Ed., Oliver and Boyd Ltd., Edinburgh, 1964, chap. 1.
5. Hair, M.L., *Infrared Spectroscopy in Surface Chemistry*, Marcel Dekker Inc., New York, 1967, chap. 2.
6. Bellamy, L., *The Infrared Spectra of Complex Molecules*, Chapman and Hall, 1975.
7. Lin-Vien, D., Colthup, N.B., Fateley, W.G., and Grasselli, J.G., *Handbook of Infrared and Raman Characteristic Frequencies of Organic Molecules*, Academic Press, San Diego, 1991.

Chapter 2

SAMPLE HANDLING IN INFRARED SPECTROSCOPY — AN OVERVIEW

W. D. Perkins

CONTENTS

I. Introduction 12

II. Defining a "Good" Spectrum 12

III. Computer Data Processing 14

IV. Gas Phase Sampling 17
    A. Sample Manipulation and Spectral Characteristics Using Short Path Cells 17
    B. Trace Analysis Using Long Path Cells 20

V. Liquids and Solutions 21
    A. Demountable Cells and Sealed Cells 21
    B. Choosing Thickness and Window Material 23
    C. Aqueous Solutions 26

VI. Solid Sampling Techniques 27
    A. Mulls 27
    B. Pressed Pellets 30
    C. Cast Films 33
    D. Pyrolysis 35
    E. Photoacoustic Spectroscopy 36
    F. Reflection Techniques 37
        1. Specular Reflectance 37
        2. Diffuse Reflection 39
        3. Internal Reflection 44

VII. Microsampling 47

VIII. Conclusion 49

References 52

0-8493-4203-1/93/$0.00+$.50
© 1993 by CRC Press Inc.

I. INTRODUCTION

Infrared Spectroscopy is a particularly useful analytical technique because of its enormous versatility. Spectra can be obtained, often nondestructively, on samples in all three states of matter — gases, liquids, and solids. For a given sample, there will usually be at least two or three, and sometimes as many as four or five, different sampling techniques that can be used in obtaining the spectrum, thus permitting the spectroscopist a choice that may be dictated by available accessory equipment, personal preference, or the detailed nature of that particular sample. In this chapter, we will present a broad overview of the various techniques for sample handling. Subsequent chapters in this book will deal in detail with a number of specific procedures.

II. DEFINING A "GOOD" SPECTRUM

The quality of the information that can be derived from a spectrum is directly related to the quality of the spectrum itself. The goal is to generate good spectra, but what are the standards of goodness? One rather pragmatic definition is that a good spectrum is one that tells you what you want to know. Is the cell clean or dirty? Is the sample thickness appropriate? There is much to be said for this definition, and sometimes the analyst spends excessive time in generating a reference quality spectrum when a less elegant result would serve his needs just as well. But a definition as simple as this one fails to set even minimal standards and can encourage poor practices which soon degenerate into routinely poor performance. Something better is needed.

At the other extreme of the quality spectrum are the Coblentz Society class I, II, and III specifications for spectra.[1] They define the highest quality of reference spectra, but the time needed to generate spectra meeting these standards is usually more than the average laboratory can afford to devote.

Somewhere in between there lies a set of guidelines that will produce reliable, reproducible, and readable information that can be generated within a reasonable time, even in a busy laboratory.

The first consideration is the intensity of the bands in the spectrum. Intensity is related to sample thickness (concentration of sample in solution, concentration of sample in potassium bromide [KBr] pellets, etc.) which should be chosen so that the strongest band in the spectrum has an intensity in the range of 5 to 15% transmittance (T). These are not absolute limits, never to be transgressed, but rather are sensible guidelines. If the strongest band has, for example, a transmission of only 50%, then many of the weaker bands will be observed only with difficulty or not at all. Ordinate expansion is not always a remedy for this situation since the noise in the spectrum will be expanded along with the signal; signal-to-noise ratio remains unchanged. It is preferable to rescan a thicker sample.

If the sample is so thick that the strongest bands have transmittances approaching zero, a photometric accuracy problem can arise. This is particularly the case in optical null-type double-beam dispersive instruments where photometric accuracy degrades rapidly as the transmission goes below the 5% level. In the extreme case of very thick samples, the strongest bands may actually become totally absorbing over wavenumber ranges as wide as two or three hundred wavenumbers, a condition known as "bottoming out". Refer to Figure 1 which shows the spectrum of m-xylene run at thicknesses of 0.025, 0.1, and 0.2 mm. The carbon-hydrogen stretching modes near 3000 cm⁻¹ are well-defined at 0.025 mm. With the thicker cells, not only are the exact frequencies unreadable, but the fact that there are multiple bands can no longer be seen.

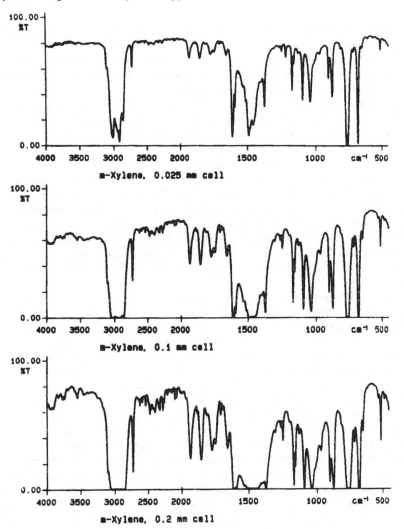

FIGURE 1. Spectra of *m*-xylene run at thicknesses of 0.025, 0.1, and 0.2 mm.

Dispersive instruments usually plot spectra directly on the recorder chart paper during the scan. To ensure maximum readability of the chart, the 100% control should be set so that the highest point in the spectrum lies somewhere between 95 and 100% T. Fourier Transform Infrared (FT-IR) spectrometers usually ratio sample scans to a stored background and then present the result as a transmission spectrum on a cathode ray tube (CRT) screen. Hard copy, if desired, can be sent to a printer or to a plotter. One should choose a transmission range such that the spectrum fills the ordinate of the plotter paper.

Another characteristic of a good spectrum is a relatively flat baseline. Sometimes the use of an accessory (internal reflection, for example) will introduce slope into the baseline. With double beam instruments one can put a second identical accessory in the reference beam, but this doubles the accessory costs, and it is usually difficult to align the two accessories identically in both beams. With FT-IR instruments these problems can usually be avoided by running the background with the accessory in place and then installing the sample and making the sample scan.

More often a sloping baseline will be the result of poor sample preparation — using fogged windows or inadequate sample grinding in the case of mulls or pellets. In these instances, a new sample preparation is usually the best procedure.

The signal-to-noise level in a spectrum will determine the ability to detect the presence of weak bands, and it will also have a profound effect on the cosmetic appearance of the spectrum. In quantitative analyses, the noise level will also limit the ultimate photometric accuracy and hence the analytical accuracy.

Signal-to-noise level is an inherent quality of the spectrometer, but one that will be degraded when we are forced to work under low energy conditions such as a vignetted sample beam or an energy inefficient sampling technique. With dispersive systems, signal-to-noise ratio (SNR) can be improved by using longer time constants at the expense of often much longer scan times. In an FT-IR instrument, the SNR can be improved by averaging a larger numbers of scans. In the dispersive case, a reasonable level of SNR would probably be in the range of 100:1 to 200:1 for most routine spectra. Because of the higher energy throughputs available in even the lower priced FT-IR spectrometers, one can easily attain SNRs of 500:1 to 1000:1.

Another artifact that detracts from the appearance of an infrared spectrum is atmospheric uncompensation. Unless an accessory that lengthens the optical path is used in the sample beam, this is seldom a problem in double beam instruments which usually have chopping speeds of 10 to 15 cycles per second. In the case of FT-IR where the sample scan is ratioed against a stored background that may have been taken minutes or perhaps even a few hours earlier, uncompensation is more likely to occur. Water vapor has a band in the 3600 to 3700 cm^{-1} region and another centered near 1600 cm^{-1} while carbon dioxide (CO_2) has a strong band at 2350 cm^{-1} and a much weaker one at 667 cm^{-1}. At 4 cm^{-1} resolution the rotational structure of the water vapor is readily observed while the 2350 cm^{-1} CO_2 band appears as a doublet. If atmospheric water vapor and CO_2 are present in the sample compartment at a greater level when the sample is scanned than they were when the background was taken, these bands will appear in the ratioed spectrum going downscale (see Figure 2). If the same bands go upscale, it indicates higher levels of these gases during the background scan than during the sample scan. Since humidity and CO_2 levels in the laboratory can change during the day, uncompensation in the ratioed spectrum will be observed from time to time. It can usually be eliminated by running a new background and then re-scanning the sample. It can also be eliminated by purging the optical path with a dry inert gas, but if the sample compartment purge is broken to introduce a new sample, the next scan must be delayed until the purge level has been re-established.

III. COMPUTER DATA PROCESSING

When a spectrum does not meet the criteria just described, there are usually two choices: re-run the sample using corrected technique or turn to the computer to artificially correct the appearance of the spectrum. FT-IR spectrometers invariably have a large collection of data processing routines built into their software. Among the more frequently used ones are smoothing to reduce noise, flattening to level out sloping baselines and to correct for baseline curvature, and ordinate expansion. Ordinate expansion should always be done on spectra in an absorbance format so as to maintain the relative intensities of the bands during the expansion.

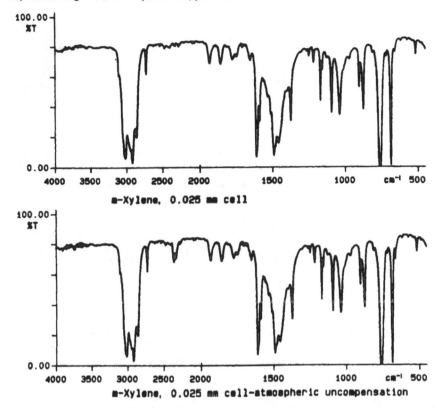

FIGURE 2.   Spectra of *m*-xylene free of uncompensation (top) and with atmospheric uncompensation (bottom). Note the presence of water vapor (3600 to 3700 cm^{-1}) and carbon dioxide (2350 cm^{-1}) in the bottom spectrum.

Properly used, these computer techniques can enhance the cosmetic appearance of a spectrum, improve its readability, and facilitate comparison with reference spectra, which are usually run according to standards of good practice. Improperly used, computer processing can become little more than a crutch used to conceal sloppy sample preparation by "sweeping the bad results under the rug". It is important to remember that data processing means changing the original data. Because of this, it is desirable when copying spectral data to a disk to always save the original, unprocessed spectrum even if processed spectra will be used for all further work. If quantitative analysis is being carried out, it is preferable to work with the original data. Although most quantitative techniques tend to be forgiving if preparation and processing steps are performed in scrupulously repeatable fashion, for both standards and samples, one cannot be certain that a computer algorithm will always treat a series of spectra in exactly the same way each time.

One data processing step that deserves further discussion is that of digital smoothing, frequently done using the Golay-Savitzky algorithm,[2] sometimes also called moving-point polynominal smoothing. It is generally used to reduce the background noise level in a spectrum and thus to artificially enhance the SNR. The degree of smoothing can be varied by changing the number of points in the moving point function (gate width). Sometimes overlooked is the fact that the algorithm does arithmetic, not chemistry or physics, and that it is incapable of distinguishing between a noise pulse and a sharp, weak band. It is possible to oversmooth, and

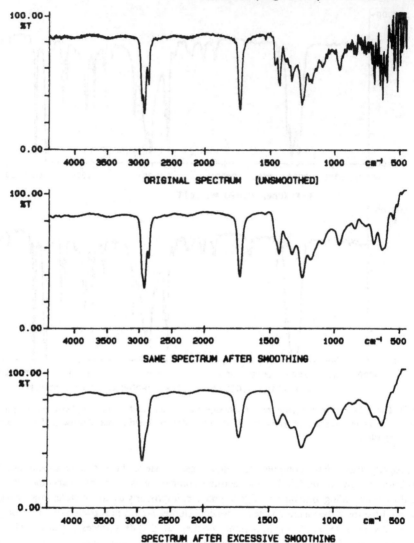

FIGURE 3. Smoothing will reduce noise, but as the level of smoothing is increased, spectral information can be lost. These three spectra are shown unsmoothed (top), after moderate smoothing (center), and after excessive smoothing (bottom). Note the progressive loss of detail in the absorptions around 2928 cm⁻¹ and 1435 cm⁻¹ and also the broadening and loss of apparent intensity in the band at 1733 cm⁻¹.

as the gatewidth is increased, weak sidebands can turn into shoulders and shoulders can disappear. Strong, sharp bands will lose intensity and become broader. In other words, excessive smoothing will lose data (see Figure 3).

The alternative to smoothing is spectral averaging. Noise will be reduced in proportion to the square root of the number of scans averaged. Furthermore, only the noise will be reduced — none of the real data will be lost. Clearly then, averaging is preferable to smoothing. With dispersive instruments connected to computers, where a single scan can take upwards of 5 min, the trade-off between purer data and longer times for data acquisition may still favor judicious use of smoothing or some combination of averaging followed by smoothing. With FT-IR spectrometers, which have higher throughputs and faster scan times, averaging is the preferable technique. Nevertheless, smoothing still has valid applications. If previous work is being re-examined, the data may be recovered from disk storage long after the original sample is gone. If the sample is changing rapidly with time, it may not be possible to average enough scans to attain the desired noise level without sacrificing the time-resolution that motivated the study in the first place. And, finally, if one is working with a difference spectrum, smoothing is immediate and rapid; averaging, on the other hand, requires that the two spectra being subtracted must both be rescanned over longer time periods. Samples are not always still available.

IV. GAS PHASE SAMPLING

A. Sample Manipulation and Spectral Characteristics Using Short Path Cells

The density of a sample in the vapor phase is much less than that of the same sample in its condensed phase. So, in order to get a spectrum of suitable intensity, it is necessary to resort to cells with much longer pathlengths than are used for liquids and solutions. A typical short path gas cell has a pathlength of 5 or 10 cm and is usually fitted with 1 or 2 stopcocks for introducing and removing the gaseous sample. Glass-bodied cells are most common and should not be used at pressures greater than 1 atmosphere absolute. In fact, a pure gas will often need to be run at subatmospheric pressure in order to prevent the bands from bottoming out. The intensity of the spectrum is readily controlled by varying the pressure inside the cell, thus eliminating the need for a series of cells of various pathlengths. The cell is best filled using some type of a vacuum manifold system that includes a manometer for measuring pressures.[3,4] A glass manifold is preferred so as to avoid attack by corrosive gases. A cold trap should be used to prevent contamination of the vacuum system, and the stopcocks should be lubricated with a low-volatility hydrocarbon grease. Silicone greases should never be used since silicones are very strong infrared absorbers and have a tendency to creep. Metal manifolds, which are less easily broken, may also be used when it is known that none of the gases employed will be corrosive to the metal. Filling the cell by flowing a gas stream through it is not good practice because it is difficult to be sure that all of the old gases in the cell have been swept out and replaced with the new sample. The geometry of the gas cell and the locations of the filling ports are important considerations. Grab samples can be taken by first evacuating the cell and then opening the stopcock to the environment being sampled.

Another point to be considered when doing gas phase spectroscopy is the difference between the vapor phase spectrum and the corresponding condensed phase spectrum. In the vapor phase, the molecules are free to rotate; transitions between the various rotational levels are superimposed on the pure vibrational transition, leading to a spectrum with considerable structure. Individual sharp rotational lines occur at closely spaced frequencies both below (P branch) and above (R branch) the frequency of the pure vibrational transition. For some vibrations, a sharp Q branch will also occur between the P and R branches; in other vibrations this feature is absent. The appearance of a typical vapor phase spectrum is shown in Figure 4 which depicts the C–O–H band of methanol run at three different resolutions using a dispersive spectrophotometer. The bottom scan was run with the narrowest slit program available, corresponding to a resolution of about 0.5 cm^{-1}. Rotational lines in the P branch are well-resolved; those in the R branch, where the lines are closer together, are less well-resolved but

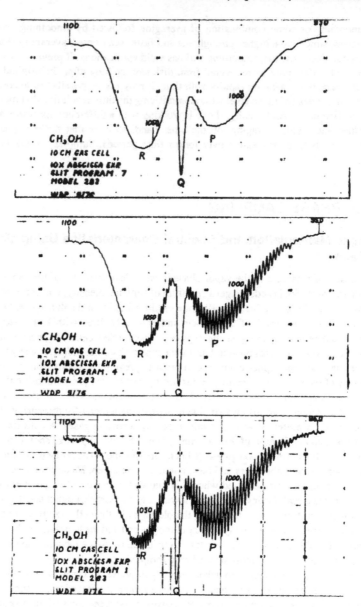

FIGURE 4.   Spectra of the C–O–H band in vapor phase methanol run at three different resolutions.

discernible. The center scan was made with a resolution of approximately 1 cm⁻¹ which
considerably reduced the extent of the splitting in the P branch and almost eliminated the
structure in the R branch. In the top scan, run with the widest slit program, the resolution is
so poor (a little over 2 cm⁻¹), that only the band envelope is seen; none of the rotational
structure is resolved.

The separation between the rotational lines varies inversely with moment of inertia; the
heavier the molecule, the closer the spacing. Hence, the appearance of a vapor phase spectrum

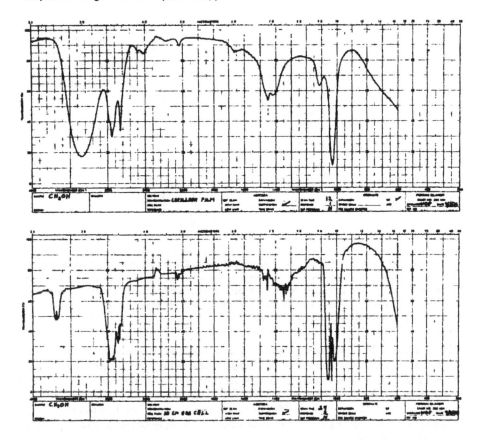

FIGURE 5. Spectra of methanol run as a capillary film (top) and in the vapor phase (bottom).

will depend strongly on the resolution of the spectrometer. At a typical resolution of 4 cm^{-1} usually only the band envelope for all but the very lightest molecules is observed.

In going from the vapor phase to the condensed phase, the molecules are much closer together, and collisions between molecules and with the container wall occur in the same time frame as molecular rotation. Band shapes are smeared out, and the general appearance of the spectrum is significantly changed. Figure 5 compares the liquid phase spectrum of methyl alcohol (top) with that of the vapor phase (bottom). Note the shape of the C–O–H band at about 1030 cm^{-1}. Notice also, the appearance of the O–H stretching vibration which is a strong, broad band occurring at about 3350 cm^{-1} in the liquid spectrum, but which appears in the vapor phase spectrum as a weaker but much sharper band at about 3700 cm^{-1} with a typical gas phase contour. The explanation for this difference is, of course, hydrogen bonding which is very strong in the liquid phase. In the vapor phase, the molecules are too far apart to hydrogen bond and the free O–H frequency is observed. Because of differences like these, vapor phase libraries should be used when making comparisons with spectra taken as gases.

A good spectrum of a pure gas can usually be obtained in a 10 cm pathlength cell, often at subatmospheric pressure. In a mixture of gases, the intensity of each species will depend on the partial pressure of that component. Intensities can be enhanced (pressure broadening) by increasing the total pressure in the cell, sometimes done by introducing an inert gas such as dry nitrogen.[3]

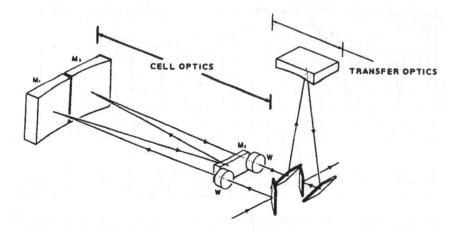

FIGURE 6. Optical diagram of a typical multiple-pass gas cell. Transfer optics located in the spectrometer sample compartment direct the beam through a window (W) into the cell. As shown, the beam makes four traversals before exiting the cell, thus resulting in an optical pathlength four times the physical length of the cell. By varying the dihedral angle between mirrors M1 and M3, the number of passes can be increased to give even longer optical pathlengths.

With dispersive spectrometers, components in a gaseous mixture can be detected and determined quantitatively in the 1 to 20% range. The limiting factor is the attainable SNR. Much higher SNRs can be achieved in FT-IR spectrometers. With room-temperature deuterated triglycine sulfate (DTGS) detectors and signal averaging, detection limits can be reduced by two or more orders of magnitude. Mercury-Cadmium-Telluride (MCT) detectors offer little advantage with short-path gas cells since the detector is usually saturated at the high energy throughput of the cell.

B. Trace Analysis Using Long Path Cells

The detection limits attainable with short path cells are not low enough to meet today's needs for environmental analyses where regulations require measurements in the parts per million and in the parts per billion range. In order to make these measurements, cells with pathlengths of several meters are needed. Commercially available cells have been manufactured with pathlengths ranging from 1 m to as much as 100 m. These very long optical paths are achieved by multiple passing the infrared beam between two spherical mirrors contained inside the cell (see Figure 6). In the original design by John White,[5] the beam made four traversals before leaving the cell. In a later modification, one of the mirrors was split, and by varying the dihedral angle between the two halves of the split mirror, additional traversals can be made. One of the most widely used cells is a "20 m" cell in which the pathlength can be varied from 0.75 to 20.25 m in 1.5 m increments using an external adjustment. The cell is metal bodied with a teflon-coated interior and can be pressurized and heated. A variety of other cells with both fixed and variable paths is also available. Factors in choosing a cell include pathlength needed, cell volume, available windows, construction (glass or metal body), and, of course, cost.

With a 20-m cell and a dispersive instrument, detection limits for most gases will fall in the 1 to 50 ppm range. The actual limit for a given gas will depend on the absorptivity for that

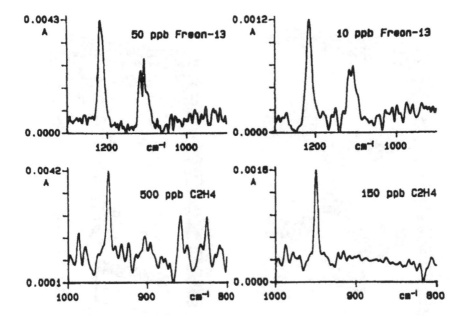

FIGURE 7. Vapor phase spectra in the parts per billion range: (A) Freon-13, 50 ppb, DTGS detector; (B) Freon-13, 10 ppb, wide-band MCT detector; (C) Ethylene, 500 ppb, DTGS detector; (D) Ethylene, 150 ppb, wide-band MCT detector.

specific molecule. With an FT-IR spectrometer, using a DTGS detector, concentrations of a few hundred parts per billion can be obtained with a few minutes of spectral averaging.

Detection limits can be lowered still further by using a liquid nitrogen cooled MCT detector. The full benefit expected from a narrow band MCT will probably not be realized because the energy through a multiple pass cell is usually so high that the narrow band detector is saturated. The beam must then be attenuated, resulting in the throwing away of a part of the signal-to-noise enhancement that motivated going to the narrow band detector in the first place. A wide band MCT detector, on the other hand, will usually require little or no attenuation of the beam, and while its sensitivity is less than that of a narrow band detector, it is still more sensitive than a DTGS detector. With an MCT detector, detection limits in the range of 10 to 100 ppb can be reached for many gases. Figure 7 shows spectra of Freon-13 and ethylene run in a 20-m gas cell in the parts per billion range using both a DTGS detector and a wide band MCT detector.

V. LIQUIDS AND SOLUTIONS

A. Demountable Cells and Sealed Cells

The accessories most often used to obtain spectra of liquids and solutions are demountable cells and sealed cells (Figure 8). The demountable cell consists of a pair of infrared transmitting windows with a spacer (usually lead or teflon) placed between them to establish sample thickness. As its name implies, the cell must be taken apart and reassembled each time it is used. Disassembly allows access to the windows for easy cleaning or for repolishing, when

FIGURE 8. A demountable cell is shown assembled on the left. The cell on the right is a sealed-amalgam cell. It is filled and emptied through the Luer-Lok syringe fittings on the front plate of the cell.

they become fogged or react chemically with the sample. Cell thickness can be changed readily by changing the spacer; thus a single cell mount and one pair of windows can provide a full range of thicknesses. The demountable cell can be used with viscous liquids which would be difficult to get in and out of a sealed cell, and with smears, greases, and mulls. On the other hand, it can be difficult to use the demountable cell with highly volatile liquids which may evaporate partially during assembly, leaving air bubbles in the cell. Volatile liquids may also escape from the cell after filling if the cell is not clamped firmly together during assembly. Finally, the exact thickness of a demountable cell is determined primarily by the spacer chosen, but will vary slightly depending on how tightly the assembly is clamped together. This is not a problem for qualitative analysis, but the inability to precisely reproduce thickness makes the cell unsuitable for single band quantitative analysis.

The sealed cell differs from the demountable cell in two major respects. First, during factory assembly, the lead spacer is amalgamated with mercury, thus creating a very tight positive seal and hence enabling the cell to contain even the most volatile of liquids. Secondly, the filling and cleaning of the cell are done through a pair of hypodermic syringe fittings which connect to the cavity inside the cell. There are both advantages and disadvantages to the amalgam sealed cell. Since it cannot be taken apart by the user, one cannot change the spacer to change thickness — a separate cell is needed for each thickness. Greater care must be exercised regarding samples introduced into the cell. If the sample reacts with the window material, the damage is permanent. (Fragments of broken windows should be saved and can be used as test plates to determine sample reactivity.) If the cell is very thin (0.015 to 0.025 mm) and/or the sample is very viscous, the cell can be difficult to fill and even more tedious to clean. On the other hand, the tightness of the amalgam seal and the use of the syringe to fill the cell allow the transfer and running of the most volatile samples. Using a syringe simplifies the filling procedure and helps to prevent the formation of air bubbles in the cell. Finally, since the cell is never disassembled, the thickness remains unchanged throughout its lifetime. Interference fringes in the spectrum of an air-filled cell allow a highly precise calculation of cell thickness, and the measurement can be repeated whenever thickness must be verified. Because of its constant reproducible path length, the sealed cell is the cell of choice for quantitative analysis.

B. Choosing Thickness and Window Material

In choosing a cell, one must select a thickness and a window material. Spacers are available ranging from 0.015 mm to as much as 1.0 mm thick. The thickness should be selected such that the strongest bands in the spectrum will have a transmission in the range of 5 to 15% T. With most pure organic liquids, this condition can be met using thicknesses of 0.015 to 0.025 mm. Sometimes, however, a few of the strongest bands may bottom-out even in an 0.015 mm cell. A thinner cell is dictated, but spacers of less than 0.015 mm are difficult to fabricate and are unavailable commercially. In this instance, the sample can be run as a capillary film. A demountable cell is used and a few drops of the liquid are placed on the bottom window. The top window is put in place without the use of a spacer, the two windows are clamped into the mount, and the sample scanned. The exact thickness of a capillary film sample cannot be measured, but it will always be thinner than 0.015 mm. Highly volatile samples will occasionally evaporate during the scan, especially if the windows are not polished flat, but the technique will produce a spectrum of suitable intensity for an overwhelming majority of the samples encountered.

Frequently, it is necessary to scan the spectrum of a solution, especially during quantitative procedures. At a typical concentration of 5 to 10%, a cell thickness of 0.1 to 0.2 mm will usually produce a good quality spectrum. With solutions, the cell thickness is less critical than for pure compounds since the concentration of the solute can be adjusted to keep the absorbance in the proper intensity range. In quantitative methods, the intensity of a single weak band must sometimes be measured, perhaps a band coming from a trace component. In these instances, a thicker cell may be chosen so that the band being measured has a suitable intensity over the range of analyte concentration with the result that other bands in the spectrum, which are not of interest, may be totally absorbing.

Finally, it should be remembered that the spectrum of a mixture contains all the bands of all the components with intensities proportional to their concentrations. In a solution, the solvent bands will predominate and will often interfere with the bands of the solute. Solvents with wide spectral regions that are free of absorption should be chosen whenever possible. Two such solvents that meet this need and complement each other are carbon tetrachloride and carbon disulfide. Carbon tetrachloride has only one weak and somewhat broad band (near 1550 cm^{-1}) in the range between 4000 and 1350 cm^{-1}, and carbon disulfide has only one even weaker band below 1350 cm^{-1}. If solutions of identical concentration are prepared with each solvent and then the carbon tetrachloride spectrum from 4000 to 1350 cm^{-1} is merged with the carbon disulfide spectrum below 1350 cm^{-1}, the resulting spectrum will, with the two minor exceptions noted, contain only the bands of the solute. This procedure is described as running the sample as a "split solution".

Care must be taken when handling these and other solvents. Both carbon tetrachloride and carbon disulfide are toxic and should be used in a well ventilated hood. Carbon disulfide is

TABLE 1.   Properties of Infrared Window Materials

Window Material	Transmission Range μM	Transmission Range cm⁻¹	Refractive index at 1000 cm⁻¹	Solubility G/100 G H$_2$O @ 20°C
Sodium chloride, *NaCl*	0.25–16	40,000–625	1.49	36.0
Potassium bromide, *KBr*	0.25–26	40,000–385	1.52	65.2
Potassium chloride, *KCl*	0.25–20	40,000–500	1.46	34.7
Cesium iodide, *CsI*	0.30–50	33,000–200	1.74	160.0 (at 61°C)
Fused silica, *SiO$_2$*	0.20–4	50,000–2,500	1.42 (at 3333 cm⁻¹)	Insoluble
Calcium fluoride, *CaF$_2$*	0.20–9	50,000–1,100	1.39 (at 2000 cm⁻¹)	1.51×10^{-3}
Barium fluoride, *BaF$_2$*	0.20–13	50,000–770	1.42	0.12 (at 25°C)
Thallium bromide-iodide, KRS-5	0.60–40	16,600–250	2.37	$<4.76 \times 10^{-2}$
Silver bromide, *AgBr*	0.50–35	20,000–285	2.2	12×10^{-6}
Zinc sulfide, *ZnS* (*Irtran-2*)	1.0–14	10,000–715	2.20	Insoluble
Zinc selenide, *ZnSe* (*Irtran-4*)	1.0–19.5	10,000–515	2.41	Insoluble
Polyethylene (high density)	16–333	625–30	1.54 (at 5000 cm⁻¹)	Insoluble

flammable and should be kept away from all sources of ignition including steam baths. Moreover, carbon disulfide reacts with primary and secondary amines and must not be used as a solvent for these compounds.

A bewildering number of window materials are available for infrared spectroscopy. Some of the more common ones are listed in Table 1. Major factors in selecting a window material are cost, transmission range, and reactivity with the sample. Sodium chloride (NaCl) and potassium bromide (KBr) are by far the most widely used and also have the advantage of being least expensive. They both transmit well throughout the mid-infrared all the way to their low frequency transmission cut-offs. NaCl stops transmitting at approximately 600 cm⁻¹ and KBr at approximately 400 cm⁻¹. Cesium iodide (CsI) will transmit down to 200 cm⁻¹, but costs four to five times as much as NaCl or KBr. KBr is probably the best all around choice since its transmission range matches that of the KBr substrate beamsplitters used in most FT-IR spectrometers. Users of dispersive instruments may wish to select a window material with a transmission range that matches the scanning range of their spectrophotometer, but even with instruments that scan to 200 cm⁻¹ the cost of CsI windows may not be justified unless the data between 400 and 200 cm⁻¹ are of critical importance.

All three of these window materials are water soluble and should not be used with samples containing even traces of moisture. Moisture will cause the windows to fog and this in turn will result in spectra with sloping baselines. (Fogged windows can be repolished to correct this condition.)

Wet samples, then, require the use of water insoluble windows, and a number of such windows are available. Unfortunately, they tend to be more expensive and usually have shortened transmission ranges. Fused silica (SiO$_2$), usually considered to be a UV-VIS window material, will transmit to about 2500 cm⁻¹. This, however, is far enough to permit observation of the C–H stretching frequencies around 3000 cm⁻¹, and fused silica cells are widely used in a method for determining petroleum hydrocarbons in soil and water. Calcium fluoride (CaF_2) transmits to about 1000 cm⁻¹ and barium fluoride (BaF_2) to about 800 cm⁻¹. Both of these materials are water insoluble and can be used when a wider scanning range is required, but neither will allow observation of the full fingerprint region of the spectrum. Silver bromide (AgBr) is also water insoluble, and it has a cutoff of 285 cm⁻¹. Its transmission range is ideal, but it has other properties that may restrict its use. It cannot be amalgamated and hence cannot be used in sealed cells. Physically, it is a very soft material and easily scratched. Scratching leads to scattering with attendant loss of transmittance and sloping baselines. Repolishing is not feasible. AgBr is also photosensitive and will darken irreversibly on exposure to intense ultraviolet light, but with reasonable care this is not as much a problem as some of its other properties. Finally, it is more reactive chemically than most of the other window materials listed in Table 1. Despite this long list of negatives, AgBr is still widely used when wet samples are involved. One very important application is as a window for cast films made from aqueous emulsions.

Yet another category of water insoluble windows which also have high resistance to chemical attack, which are relatively hard, and which have excellent thermal properties are the Irtrans™. Their generic names are zinc sulfide (ZnS) (Irtran-2™, Cleartran™) and zinc selenide (ZnSe) (Irtran-4™). ZnS can be used to about 715 cm⁻¹ and ZnSe to 515 cm⁻¹. Both of these materials, as well as AgBr, have relatively high refractive indices (2.2 to 2.4). Because of this, interference fringes are often observed even with a filled cell because of the mismatch between index of refraction of the window and that of the sample. (Most organic compounds have indices of refraction in the range 1.4 to 1.6.) This is illustrated rather vividly in Figure 9 which shows spectra of cyclohexane in an 0.025 mm NaCl cell (top) and in an 0.025 mm ZnS cell (bottom). The fringes in the lower spectrum make it impossible to observe some of the very weak bands and also present a problem in constructing a baseline for making net absorbance measurements. Notice also the lower overall transmittance of the ZnS cell. This is due to the higher reflectance losses (associated with the higher index of refraction) encountered as the sample beam enters and leaves the cell. The effect is noticeable but does not present problems.

From the standpoint of chemical attack, there is no perfect universal window material. The alkali halides are water soluble. Calcium floride and BaF_2 should not be used with ammonium salts. ZnS and ZnSe probably have the highest chemical resistance but even they can be attacked by strong acids and especially by strong bases. If the reactivity of a sample is unknown, it may be advisable to place a drop of it on the corner of a loose window (or a window fragment) in order to determine whether it will attack that material. This can be very important if the sample is going to be run in a sealed cell.

None of the windows described above can be used in the far infrared (generally defined as the region below 200 cm⁻¹). Not many materials transmit in this region. Diamond has usable transmission but is prohibitively expensive. Naturally occurring crystal quartz, when cut along the proper optic axis, begins transmitting at around 200 cm⁻¹ and can be used into the far infrared, but it is not readily available from accessory vendors. The most commonly used window for the far infrared is also one of the least expensive — polyethylene. Between 625 and 30 cm⁻¹ it has only one somewhat broad and not too intense absorption at about 70 cm⁻¹. It cannot be amalgamated, and at reasonable window thicknesses it lacks rigidity, thus making it difficult to precisely control cell thickness. This disadvantage is at least partially offset by the fact that absorptions in the far infrared tend to be less intense than those in the mid-infrared range. Thicker cells are usually used in the far infrared region, thus making minor thickness variations more acceptable.

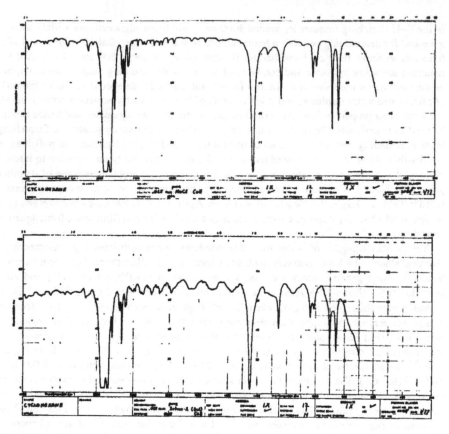

FIGURE 9. Spectra of cyclohexane run in an 0.025 mm-thick cell with sodium chloride windows (top) and with zinc sulfide windows (bottom).

C. Aqueous Solutions

Aqueous solutions have always been difficult to run by infrared spectroscopy. See Figure 10, which shows superimposed spectra of water and of an aqueous solution of water-soluble aspirin. Liquid water has a very strong, broad band which, even in an 0.015 mm cell, is totally absorbing in the 3700 to 3100 cm^{-1} range. There is a weaker band around 2000 cm^{-1} and another strong band at 1640 cm^{-1}. Finally, at about 800 cm^{-1}, water stops transmitting altogether. Observation is limited to the 3000 to 800 cm^{-1} region, with strong interference in the 1640 cm^{-1} region. Another restriction is that the window material used must be water insoluble. (Barium fluoride, whose 800 cm^{-1} transmission cutoff matches the 800 cm^{-1} cutoff of liquid water is an excellent choice for a window material).

Solute bands will often lie on the shoulders of the very broad water bands, and spectral readability can be enhanced by subtracting the water spectrum from that of the solution. Spectral subtraction is meaningless in regions of total absorption and usually produces somewhat noisy difference spectra in regions where the solvent bands approach zero transmittance. This is precisely the situation at 1640 cm^{-1} in aqueous solutions. Two remedies are available, and both should be utilized whenever possible. By averaging a larger number of scans than usual, the SNRs of the original spectra can be improved. The difference spectrum will then also

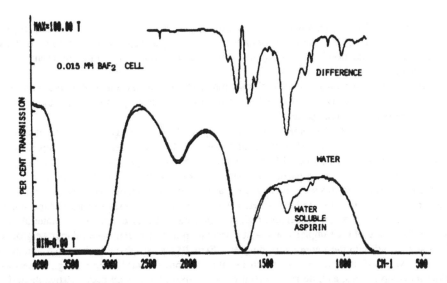

FIGURE 10.   Spectra of water and of an aqueous solution of water-soluble aspirin (superimposed). Run in an 0.015 mm-thick barium fluoride cell. The computer-calculated difference spectrum is shown above.

exhibit a higher SNR. The quality of the difference spectrum can also be improved if a cell thickness that maximizes the transmission of the 1640 cm^{-1} water band is selected. The data shown in Figure 10 were taken using a nominal 0.015-mm thick BaF_2 sealed cell. Commercial cells are sold to a nominal thickness value; the actual thickness can easily vary by as much as ±10% depending on the amalgamation process for that particular cell. Hence, a 0.015 mm cell may have an actual thickness ranging from 0.0135 to 0.0165 mm. If several cells are available for use, interference fringe patterns run on the air-filled cells can be used to measure the actual thicknesses, and the thinnest cell can then be selected for the aqueous work. A thickness of 0.010 to 0.012 mm would probably be ideal, but cells this thin are not commercially available and would be increasingly difficult to fill and to clean. Internal reflection techniques now offer an attractive alternative to transmission spectroscopy for obtaining spectra from aqueous solutions. These procedures will be described later.

VI. SOLID SAMPLING TECHNIQUES

A. Mulls

One of the earliest ways of obtaining an infrared spectrum of a solid sample was to dissolve it in a suitable solvent and then run a transmission spectrum of the solution. Even today, with a considerably wider choice of techniques available, much is to be said for the solution method.[6] In a dilute solution using a nonpolar solvent, the molecular environment will be common and relatively independent of the molecular species involved. Spectra-structure correlation frequencies are most accurate for dilute solutions. Hydrogen bonding will be reduced or eliminated, and crystal lattice effects will be nonexistent. In mixtures, interactions between components will be minimized. In quantitative analysis, an important additional benefit is the ability to use sealed cells which bring the advantage of constant and reproducible pathlength.

On the other hand, a nonpolar liquid that is transparent across the entire mid-infrared region and which will dissolve all samples does not exist. Running as a split solution requires making up two separate solutions and running two separate spectra. Solubility of the solid sample can be an additional problem.

As a consequence, most solid samples are run by other methods, and numerous alternatives do exist. Probably the two most widely used methods are the preparation of a Nujol™ mull and the pressing of a KBr pellet.

In preparing a mull, the sample must first be ground to an average particle size of at least one fifth of the shortest wavelength of light that is to be passed through the sample. For a spectrum that begins at 4000 cm⁻¹ (2.5 μm), this is 0.5 μm. Larger particles, approaching the wavelength of the light being used, will scatter (scattering is proportional to the fourth power of the frequency), and this will result in spectra with sloping baselines, such as seen in Figure 13 for an inadequately ground KBr pellet. The grinding must be done in a hard surface mortar such as agate or alumina; porcelain or glass should never be used since these materials are softer than some of the samples that may be encountered, and mortar material may be ground into and mixed with the sample. Grinding of a 20 to 30 mg sample will take 3 to 5 min even for relatively soft samples and as much as 10 to 15 min for hard samples (e.g., minerals). Smaller sample sizes, as little as 5 to 10 mg, can be used, but greater manipulative dexterity will be required to fully remove the mull from the mortar. Very dygroscopic samples may need to be prepared in a dry box.

Grinding is complete when the sample is evenly distributed over the surface of the mortar and when it takes on a smooth, glossy appearance. A caked or clumpy look indicates inadequate grinding. At this point, a very small drop of the mulling agent (usually Nujol™) is added to the mortar and grinding is continued until the sample becomes suspended in the mulling agent and forms a smooth, creamy paste with a consistency similar to that of cold cream, vaseline, or toothpaste. If the preparation goes to dryness, another tiny drop of mulling agent can be added and the mixing continued until the desired consistency is reached. A common mistake is to add too much mulling agent and "drown" the sample. The ultimate spectrum then becomes that of the mulling agent with few or no sample bands.

The mull is then scraped down from the sides of the mortar and from the tip of the pestle using a rubber policeman and transferred to the central portion of an infrared window. Next, a second window (no spacers) is placed on top of the first, and the two are pressed together until the daub of mull spreads out to a thin translucent film. The assembly is now clamped into a demountable cell mount and the spectrum is run.

The most commonly used mulling agent is a clear white mineral oil sold under the trade name of Nujol™. The spectrum of Nujol™ (shown in Figure 11) exhibits C–H stretching frequencies in the region of 3000 to 2800 cm⁻¹ and deformations at 1460 and 1375 cm⁻¹. A weaker methylene rocking vibration occurs at 720 cm⁻¹. These bands will all be observed in the spectrum of the mull and will overlap any corresponding carbon-hydrogen frequencies in the sample. In most cases, this is not a serious problem for identification since all of the other sample bands are free from interference by the Nujol™ . If detailed information is needed at these wavelengths, a different mulling agent — one containing no C–H bonds — must be used. Fluorolube™ (a perfluorinated hydrocarbon) serves this purpose. Its spectrum is also shown in Figure 11, and it will be noted that it has no appreciable absorptions between 4000 and 1350 cm⁻¹. If a full-range spectrum, completely free of any mulling agent bands, is required, it can be obtained by merging spectra of a Fluorolube™ mull (4000 to 1350 cm⁻¹) and a Nujol™ mull (below 1350 cm⁻¹) to give a "split-mull" spectrum. Care must be taken to match the thicknesses of the two mulls; this can be done with computer processing by normalizing on a band free of interference in both spectra. Figure 12 shows spectra of stearic acid run both as a Fluorolube™ mull and as a Nujol™ mull.

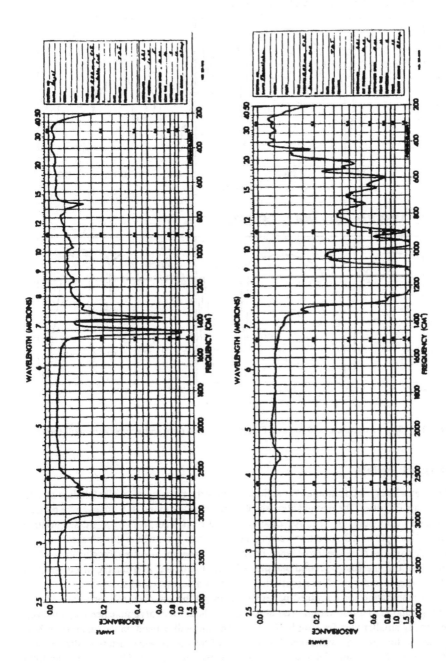

FIGURE 11. Spectra of Nujol™ (top) and Fluorolube™ (bottom). Both are commonly used as mulling agents.

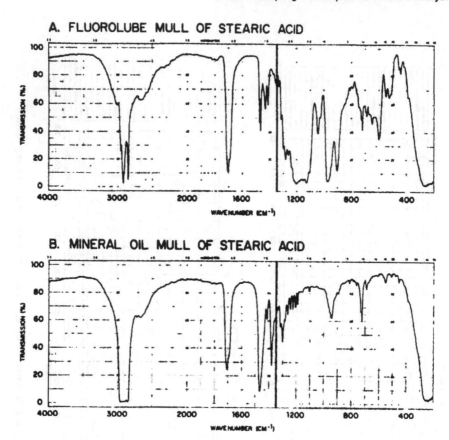

FIGURE 12.   Spectra of stearic acid run as a Fluorolube™ mull (top) and as a Nujol™ mull (bottom).

B. Pressed Pellets

The other widely used technique for solid samples is the pressing of an alkali halide pellet,[7] most often a KBr pellet. The method consists of mixing finely ground (0.5 μm average particle size) sample with a pure, dry spectroscopic grade of KBr powder, usually at a concentration of about 1% sample in KBr, and transferring the mixture to a die where it is pressed until the alkali halide particles coalesce into a clear, hopefully transparent, — often translucent — disk. As with the mull, the finely ground sample is being distributed in a matrix of comparable refractive index in order to reduce scattering by the sample particulates.

Sample preparation for a pressed pellet begins much the same as that for a mull — the sample is ground in a hard surface mortar until glossy. The amount of sample needed, however, is significantly smaller than for a mull. The exact amount will depend on the quantity of KBr to be used and that, in turn, will depend on the size of the die. A wide variety of die sets and presses are available commercially, and the vendor's instructions should be followed carefully if good pellets are to be prepared. One widely used die set presses a 13-mm diameter pellet. When using this device, 1 to 3 mg of sample are mixed with 300 mg of KBr. The amount of KBr is important, but not critically so. Too little KBr will produce a thin and fragile disk that is easily broken during handling and transfer to the spectrometer. (Very thin KBr disks may even exhibit interference fringes in their spectra). If an excess of KBr is used, the pellet may be so thick that it does not coalesce into a good transparent disk.

A number of other die set designs allow the fabrication of pellets with diameters of 7 to 8 mm, and some even as small as 3 mm and 1 mm. Correspondingly smaller amounts of sample and of KBr are used with these devices. In choosing a die, consider such factors as cost and the availability of a press (if needed). Will the pellet be formed in a metal collar or holder, and how will this affect its storability? Can the pellet be saved while a new pellet is being prepared, or must it be destroyed before the next pellet can be made? Different dies will offer different options.

If many pellets are made each day, or if the samples that must be run are very hard materials and difficult to grind, it may be desirable to consider a vibrating ball mill. These devices are very efficient for grinding and for mixing. Hard grinding, that may take 5 to 10 min in a mortar, can often be done effortlessly in less than a minute in the stainless steel capsule used in a vibrating mill. If such a device is to be used, it must be remembered that KBr is usually much softer than most samples, and in a mill the KBr will be ground preferentially to a much smaller particle size than the sample. This in itself is not a disadvantage, but smaller particle size means greater surface area and an increased ability to adsorb water vapor from the atmosphere. For this reason it is recommended that when a vibrating mill is used to grind the sample, the mixing with KBr be done by hand in a mortar. Very hygroscopic samples may need to be prepared in a dry box.

It is important that the sample be well mixed with the KBr. Whether the sample is ground initially in the mortar, as previously described, or whether an appropriate quantity of sample, already ground, is transferred to a clean mortar for mixing, the KBr should be added a little at a time, mixing thoroughly between each addition. The amount of KBr can be doubled each time so that with each addition roughly equal amounts of fresh KBr and of homogeneous KBr-sample mixture are being mixed. These amounts can be estimated by eye, but it should take at least five or six additions to incorporate all of the KBr into the mixture.

The mixture is now transferred to the die. Before pressing, rotate the plunger of the die several times to level out the powder inside. If the mixture is heaped high on one side, that part of the pellet will be well-pressed, but the other side may not receive enough force to make the KBr coalesce into a good, uniform pellet. Then press the pellet according to the recommendations of the vendor who provided the die. These recommendations will vary depending on pellet size, amount of force, and when a hydraulic press is employed, the calibration of the pressure gauge. Regardless of the appearance of the pellet, transfer it to a sample holder and scan the spectrum. Even if it is not a good pellet, the spectrum will tell you how to proceed next.

A good spectrum should have a relatively flat baseline, and the strongest bands should have intensities in the range of 5 to 15% T. If the baseline has a pronounced upward slope, this is an indication of scattering (scattering is proportional to the fourth power of frequency and hence is more pronounced at the shorter wavelengths), most likely caused by not grinding the sample finely enough. Figure 13 depicts a well-ground sample, acetanilide (top), and a poorly ground sample, diphenylacetylene (bottom). If the bands are too weak, it means that the concentration of sample in the KBr was too small. The proper remedy is to start all over again and this time to increase the sample-KBr ratio. If the intensity is only moderately weak, it may be possible to use computer processing to conceal the problem. An ordinate expansion will increase the intensities, but it will also expand the noise along with any uncompensation or other artifacts. Rescanning the pellet with greater signal averaging will reduce the noise level, even in the expanded spectrum, but unfortunately will do nothing for uncompensation. If, on the other hand, some of the stronger bands are totally absorbing (bottom out), the pellet can

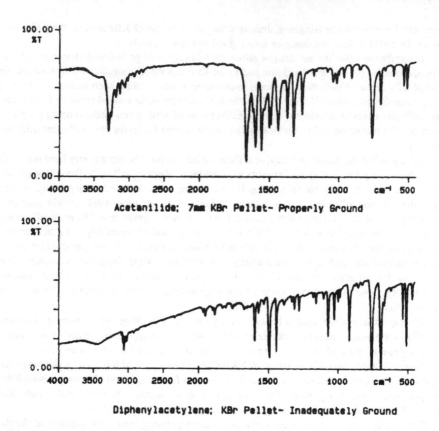

Acetanilide: 7mm KBr Pellet- Properly Ground

Diphenylacetylene: KBr Pellet- Inadequately Ground

FIGURE 13. Spectra run as KBr pellets of acetanilide (top) and diphenylacetylene (bottom). The badly sloping baseline in the diphenylacetylene spectrum is caused by scattering and is a result of not grinding the sample finely enough. A similar effect will be observed in the spectra of inadequately ground mulls.

be broken into several smaller fragments. One of the fragments can then be repulverized in a clean mortar; fresh KBr can be added to further dilute the mixture; and after mixing, a new pellet can be pressed. The amount of dilution with fresh KBr will depend on how much the original concentration was too strong, but a factor of less than five or ten fold will not usually produce a very profound change in the intensities of totally absorbing bands.

Atmospheric uncompensation was discussed in an earlier section of this chapter. Most of the time it can be eliminated by running a fresh background and then rescanning the sample. When running KBr pellets that are smaller than the diameter of the spectrometer beam, a significant improvement can often be obtained by running the background scan through an empty pellet holder so that the beam is vignetted equally for both the background and sample scans.

Solids are run more frequently as KBr pellets than as mulls. Properly carried out, both techniques will produce equally good quality spectra. The preference for pellets probably stems from the fact that KBr has no absorptions of its own over its entire transmission range. With just a single preparation, all the bands observed belong to the sample. On the other hand, KBr is a very hygroscopic material, and it is extremely difficult to keep the powder perfectly dry. Any time that a band is observed around 3300 cm⁻¹ in the spectrum of a KBr pellet, one needs to ask if it belongs to the sample or if the KBr has picked up moisture. Figure 14 shows the spectrum of a "pure" KBr pellet made with "wet" KBr. Both the O–H stretching vibration at 3300 to 3400 cm^{-1} and the weaker O–H bending mode at 1640 cm^{-1} are observed. Absorption in the O–H region due to "wet" KBr can also be seen in the diphenlyacetylene spectrum in Figure 13. This problem does not occur in the spectra of mulls.

KBr can also undergo ion exchange with certain types of samples.[8] For example, the spectrum of a KBr pellet prepared from an amine hydrochloride will usually exhibit some of the bands of the corresponding amine hydrobromide. Inorganic salts can also undergo anion exchange with KBr. Finally, there is the question of polymorphism. The forces involved in grinding, and especially in the pressing of a powdered solid, can convert one polymorphic form into another. The possibility of this happening is probably a little greater for a pellet than for a mull, although it can occur with either type of preparation.[9] While KBr is most frequently used for making pellets, other materials, such as CsBr, CsI, etc. have been used as well. The procedures for pressing these materials are different and the recommendations of the vendor who provided the die should be followed.

C. Cast Films

A completely different technique for obtaining the spectrum of a solid sample is to prepare it as a cast film. This is a procedure that is widely used for, but not limited to, obtaining spectra of polymers. The sample is first dissolved in a moderate to highly volatile solvent. The concentration is not critical. A few drops of the solution are then placed on the central portion of an infrared window, and the solution is allowed to spread out over the surface of the window. The solvent evaporates, leaving behind a very thin film of the sample adhering to the window. The single window is now placed in a demountable cell mount and a spectrum taken. A second window should never be placed on top of the first. To begin with, it serves no useful purpose. More importantly, if all of the solvent has not yet evaporated, a second window will prevent further evaporation and contaminate the spectrum with solvent bands. And if the sample should be tacky, it can sometimes result in cementing the two windows together.

In examining the spectrum of a cast film, first check to be sure that all of the solvent has evaporated. One must always be familiar with the spectrum of the pure solvent. If the spectrum of the cast film contains a band that matches the frequency of the strongest band in the solvent, it suggests that the solvent may not have fully evaporated, but it is also possible that the sample itself has an absorption at this frequency. If the two or three strongest bands of the solvent all appear to be present in the spectrum of the cast film, the solvent has almost certainly not evaporated, and more time should be allowed before running the final spectrum. Gentle heating is sometimes used to hasten the evaporation, but this is a dangerous practice for two reasons. Crystalline window materials, especially NaCl and KBr, have poor heat conductivity and when heated too rapidly will undergo stress and may fracture along their natural cleavage planes. (Save broken window fragments to use as test pieces for possibly reactive samples.) Worse, some samples may volatilize or decompose, even under gentle heat. If it is difficult to get a good cast film with one solvent, try other solvents. So long as it does not react with the sample, the chemical composition of the solvent is unimportant. Its volatility and the ability to dissolve the sample are the important considerations. Among a long list of possibly useful solvents are such compounds as chloroform, carbon tetrachloride, carbon disulfide, hexane, methylene chloride, tetrahydrofurane, xylenes, etc. Care must be taken when handling these and any solvents. If you are not familiar with the safety procedures for handling a particular solvent, check the Material Safety Data (MSD) sheet available for that solvent.

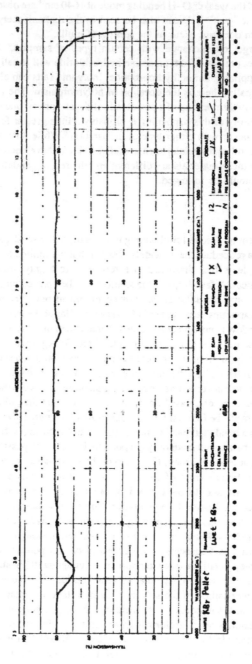

FIGURE 14.   Spectrum of a pellet pressed from KBr (no sample) which has picked up moisture. Note the O–H stretching vibration around 3400 cm⁻¹ and the O–H bending mode near 1640 cm⁻¹.

There is only limited control over the thickness of a cast film. The more concentrated the solution, and the more drops placed on the window, the thicker the film. Yet thickness is important in producing a good spectrum (strongest bands in the 5 to 15% T range). The process of getting the right thickness is one of trial and error, but it is a process that usually converges rapidly. Run a spectrum immediately after casting the first film. If the spectrum is too strong (bottoming bands), wash the film from the window and cast a new film using a smaller quantity of the solution or using a more dilute solution. More often the problem will be a film that is too thin. Sometimes this can be remedied by adding several drops of the solution on top of the existing film and letting it dry, thus building up the thickness. If this is unsuccessful, the window should be cleaned and a fresh film cast using either a more concentrated solution or, if possible, a larger quantity of the solution.

The cast film technique is used very extensively to obtain the spectra of polymeric samples. If the polymer is thermoplastic, another widely used method is to run the sample as a hot, pressed film. A small chip of the polymer can be heated to the softening point and then pressed between sheets of teflon™ or aluminum foil until a film thin enough to give good transmission is formed. In a more controlled experiment, the film can be pressed between heated platens in a Carver press using spacers made from a machinist's thickness gauge to control the sample thickness. A thickness of 25 to 30 μm will usually yield a good quality spectrum.

D. Pyrolysis

Yet another technique, also popular in polymer spectroscopy, is pyrolysis.[10] In the simplest procedure, approximately 100 to 300 mg of the polymer are placed in a 4 in. hard glass test tube and the tube is held in the flame of a Bunsen burner. The sample undergoes a dry destructive distillation or decompostion. Some of the lighter boiling pyrolysis products will issue from the mouth of the test tube and vent to the surroundings where they will be lost. (The pyrolysis should be carried out in a hood.)

The higher boiling pyrolyzates will condense on the cooler surface near the mouth of the test tube, often as a brownish liquid, which can be scraped from the test tube with a microspatula and transferred to an infrared window (or micro window) and then run as a smear.

In a refinement to this procedure, the test tube can be connected to an evacuated 10-cm gas cell, and the gaseous pyrolysis products can be collected and their spectra examined in the vapor phase.[11] In either case (liquid phase or vapor phase), it must be remembered that the spectra obtained are not those of the original sample but rather those of its pyrolysis products. When identifying unknowns, a reference library of pyrolyzates must be consulted. Reference spectra of pyrolyzates will sometimes be found as appendices in books on the infrared spectroscopy and identification of polymers and are also available in commercial collections of reference spectra. The best libraries will be those made in the using laboratory where pyrolysis conditions will be repeatable for both library generation and for the examination of unknowns.

Most polymers will produce liquid pyrolyzates which can be used subsequently for the identification of the basic polymer. Occasionally (polyvinyl chloride, for example), the pyrolysis products will be entirely gaseous, and the technique will be less useful. Pyrolysis finds its greatest utility when dealing with heavily carbon-filled polymers. Carbon black is highly absorbing in the infrared, and when it is present most of the sampling techniques previously described will result in spectra ranging from poor to useless. In pyrolysis, the carbon is reduced to carbon monoxide and/or carbon dioxide which escape with any other gaseous pyrolyzates, thus avoiding this problem.

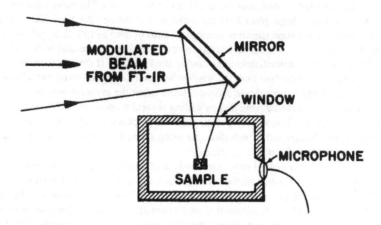

FIGURE 15.   Schematic diagram of a photoacoustic detector.

Despite the simplicity of the pyrolysis process just described, and the limited control over temperature and heating times, the results are surprisingly reproducible. Commercially available accessories, while expensive, will allow greater control over the pyrolysis conditions (temperature and time) and will even provide for direct and automatic deposition of the liquid pyrolyzates on infrared transmitting windows or on internal reflection elements. They can be very useful in laboratories which make extensive use of the pyrolysis technique.

E. Photoacoustic Spectroscopy

Another interesting technique that can be used to obtain the spectrum of a solid sample is photoacoustic spectroscopy (PAS).[12] While not new, this procedure saw only limited use until it was combined with FT-IR spectroscopy. In photoacoustic spectroscopy the sample is placed in a small sample cup which fits into a sealed chamber within the PAS accessory. The chamber is filled with helium (its high thermal conductivity produces the highest sensitivity) or, if helium is not available, nitrogen may be used. The PAS accessory is placed in the FT-IR sample compartment and a mirror directs the incoming beam through a window onto the sample. (See Figure 15.) The heating of the sample by the modulated FT-IR beam heats the gas layer above the sample and causes it to expand and contract with the same modulation frequency, thus generating a sound wave. This acoustic wave is then detected by a very sensitive microphone within the sealed helium-filled chamber, and the signal is fed back into the FT-IR detector electronics.

The resulting spectrum is usually presented in an absorbance format, i.e., with the bands pointing upward. Sample spectra are usually ratioed against a background scan made on a totally absorbing material such as carbon black. Frequencies in the PAS spectrum are, of course, the same as in transmission, but there are differences in relative intensities which depend on wavelength and modulation frequency. These are not easily correctable, but generally present no difficulty in identification or other qualitative applications. Figure 16 shows PAS spectra of a pulverized medium volatility coal (top) and of an extruded polymer pellet (bottom).

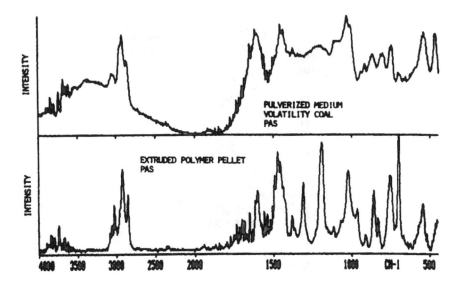

FIGURE 16.   Photoacoustic spectra of a pulverized medium volatility coal (top) and of an extruded polymer pellet (bottom).

The major advantage of PAS is the almost total lack of sample preparation. Samples are run neat, as received, and can have either smooth or rough surfaces. The only requirement is that they be small enough to fit into the sample cup. They can have virtually any shape or geometry and do not need to be ground or diluted. Disadvantages are the cost of the PAS accessory and that noise levels for comparable data collection times tend to be somewhat higher than for alternative techniques.

Penetration depth into the sample is a function of the thermal conductivity and heat capacity of the sample and also varies inversely with the modulation frequency of the FT-IR spectrometer. Modulation frequency is, in turn, the product of OPD (optical path difference) velocity multiplied by wavenumber. At a constant mirror velocity, the penetration depth varies with wavenumber, being 10 times greater at 400 cm⁻¹ than at 4000 cm⁻¹. This accounts for the relative intensity differences between photoacoustic spectra and transmission spectra. The dependence of penetration depth on mirror velocity also makes it possible to do depth profiling experiments — the slower the mirror velocity, the greater the penetration into the sample.

F. Reflection Techniques

1. Specular Reflectance
A final group of techniques applicable to solid sampling are the reflectance methods — specular, diffuse, and internal. Specular reflectance is defined as reflection in which the angle of incidence on the sample is exactly equal to the angle of reflection. Most accessories are designed to direct the beam upwards and reflect it from a sample placed face down on a horizontal supporting stage. (See Figure 17.) Near-normal angles of incidence are most common. In order to produce a good spectrum, the surface of the sample must be smooth and flat. The amount of light reflected is usually very low, sometimes only a few per cent, thus making this technique more useable with FT-IR than with dispersive instruments. The re-

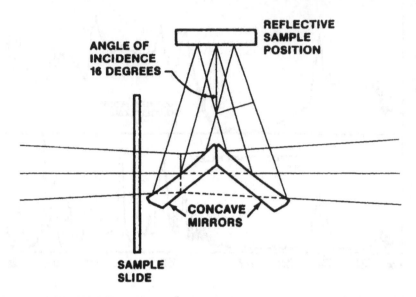

Figure 17. Optical diagram of a specular reflectance accessory.

flected intensity also depends in detail upon the index of refraction, and since the refractive index of a material goes through an anomaly in the region of an absorption band, the true specular reflectance spectrum qualitatively resembles the dispersion in refractive index. The result (see Figure 18) is a spectrum that shows strong features at the absorption frequencies, but which is not easily interpreted. A mathematical process known as the Kramers-Kronig transformation[13-15] can convert the specular reflectance spectrum into a transmission spectrum. The transform works well only if the spectrum is purely specular (no diffuse reflection) and if the dispersion is low at both the beginning and the end of the range transformed. A high SNR will also improve the quality of the transformed spectrum.

Specular reflection accessories are also very widely used for another technique which is often referred to as specular reflection but which is actually a form of transmission through an absorbing film on a reflecting substrate. In the current literature, this method is now becoming known as reflection-absorption (R-A) spectroscopy. A typical sample might be the polymer coating usually applied to the inside surface of a food or beverage tin. If the sample is placed, coated side down, on the stage of a specular reflection accessory, the sample beam penetrates the coating once, reflects from the metal substrate, and passes a second time through the coating before ultimately reaching the detector. The result is a transmission spectrum with an intensity commensurate with that of a sample two times as thick as the actual coating. Figure 19 shows a spectrum obtained in this manner from the coating on a chewing gum wrapper. There is no control over sample thickness, and the technique works well with coatings in the 0.2 to 20 μm range. Thicker coatings usually produce spectra in which many of the bands show total absorption, and thinner coatings result in very weak spectra. Thinner coatings can still be studied if the path through the coating is increased by going to grazing angles of incidence where the slanted path through the coating is longer, and further improvement can then be obtained by using infrared radiation polarized with the electric vector parallel to the plane of reflection as defined by the incident and reflected rays.[16, 17] Monolayers with thicknesses of approximately 1 nm have been studied in this way.

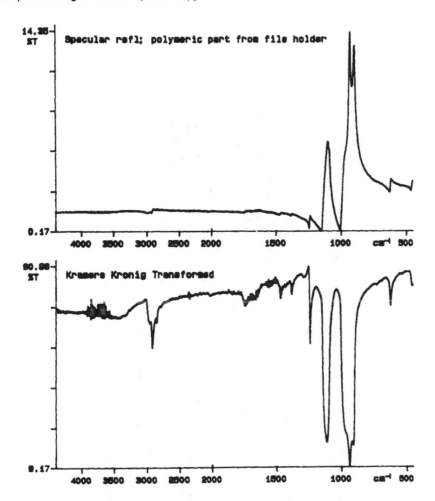

FIGURE 18. The top spectrum was obtained by specular reflection from a polymer block. Note the relatively low reflectance from the sample. Shown below is the Kramers-Kronig transform of the reflectance spectrum.

2. Diffuse Reflection

Diffuse reflection is, by definition, that process in which the angle of reflection is different from the angle of incidence. In the visible- and near-infrared regions of the spectrum, diffuse reflectance measurements have been made for many years using integrating spheres. There have been some attempts to use these devices in the mid-infrared, but the very low levels of reflectivity and the energy limitations of dispersive instrumentation have combined to make the technique unattractive.

The higher throughputs of FT-IR spectrometers have now made infrared diffuse reflectance feasible, if not commonplace. A variety of commercial accessories are available based on an original design described by Fuller and Griffiths.[18-20] The sample is usually ground and diluted with a nonabsorbing material such as KBr or potassium chloride (KCl). The mixture is placed

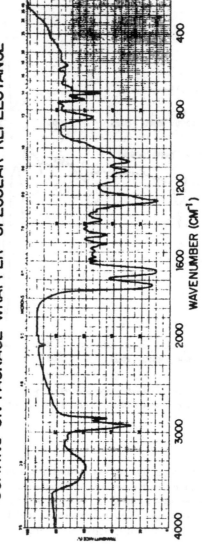

FIGURE 19. The reflection-absorption spectrum from the coating on a chewing gum wrapper. Obtained using a specular reflection accessory.

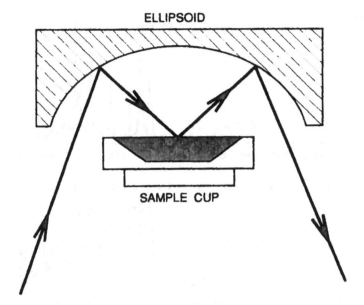

FIGURE 20. Sample configuration used for diffuse reflectance.

in a small cup about 13 mm in diameter and 3 to 4 mm deep. (Micro cups can be used when sample size is limited.) The sample beam is focused onto the surface of the powdered mixture. (See Figure 20.) The ensuing process is a complex combination of reflection, absorption, and scattering, but the result is that energy is reflected diffusely through a 2π solid angle above the surface of the powder in the cup. A large aperture optic collects this diffusely reflected beam over as wide a solid angle as possible, and it is then transferred on to the detector. Spectra are ratioed against a background scan made on a KBr-filled sample cup. The procedure is often referred to as DRIFTS, which is an acronym for Diffuse Reflectance using Infrared Fourier Transform Spectrometry.

The resulting spectrum looks very much like a transmission spectrum of the same sample, except that the relative intensities of the bands will be different in the two spectra. Furthermore, if diffuse reflectance spectra are run on a series of mixtures, the spectra converted to an absorbance format, and a Beer's law plot is attempted, using a peak from one of the components whose concentration is varying, the resulting plot will be non-linear. These problems can be overcome by first performing a Kubelka-Munk (K-M) transform on the diffuse reflectance spectrum.[21] This is illustrated by the three spectra in Figure 21. The top spectrum is that of an aspirin tablet prepared as a conventional KBr pellet; the center spectrum is from the same tablet but run by diffuse reflectance; and the bottom spectrum is the DRIFTS spectrum after performing a K-M transform. The first two spectra are plotted in absorbance. The ordinate for the bottom spectrum is in Kubella-Munk Units. It is conventional to plot K-M transformed spectra in an "absorbance" format with the bands pointing upwards.

The best spectra will be obtained if the sample is first ground (10 to 20 μm average particle size is sufficient) and then mixed with the nonabsorbing diluent. This procedure *must* be followed if the K-M transform is to be carried out and/or if quantitative work is to be attempted. An appropriate concentration of sample in diluent is 1 to 5%. Lower concentrations present no problems but will result in weaker spectra which may then need to be ordinate

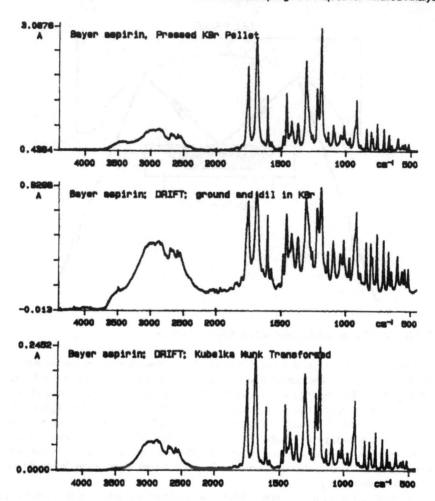

FIGURE 21. Top: spectrum of aspirin run as a pressed KBr pellet. Center: spectrum of aspirin run by diffuse reflection. Bottom: diffuse reflection spectrum of aspirin after Kubelka-Munk transform.

expanded. Concentrations greater than 5% should be strenuously avoided. High sample concentrations, with their correspondingly larger number of sample particulates in the mixture, will often result in specular contributions to the reflected beam which will manifest themselves as distortions in the band shapes.

The recommendations just given should be followed in order to ensure good quality spectra. Sometimes, however, good data can also be created even when these recommendations are ignored. Figure 22 is the spectrum of an ABS polymer, ground to 60 mesh, and run neat. It shows excellent detail. Figure 23 illustrates a novel and very useful technique first described by Spragg.[22] The sample was a cured epoxy resin, too hard and brittle to be easily ground for making a KBr pellet or for the recommended DRIFT procedure, and not readily soluble thus preventing preparation as a cast film, and thermosetting which precluded making a hot pressed film — a really intractable sample. The spectrum was obtained easily and quickly by abrading a chunk of the resin onto a piece of emery paper and then, with the abraded powder

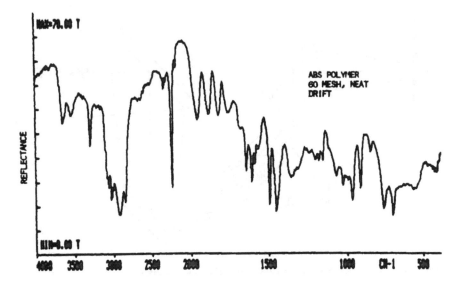

FIGURE 22.   Diffuse reflection spectrum of an ABS polymer, 60-mesh particle size, run neat.

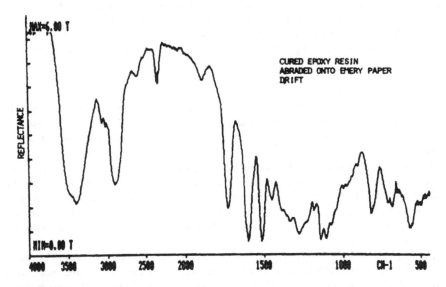

FIGURE 23.   Diffuse reflection spectrum of a cured epoxy resin run in place on the surface of a piece of emery paper onto which it had been abraded.

left on the paper, cutting out a piece of the emery paper just the right size to fit in the DRIFT sample cup. The spectrum was then run vs. a KBr background.

There are several advantages in using the DRIFT technique, perhaps the most important being the ease of sample preparation. Note that the particle size of the ground sample for DRIFT can be much coarser than that needed for preparing a KBr pellet or a mull, where particle sizes need to be 0.5 μm or less to avoid scattering and sloping baselines. Since the

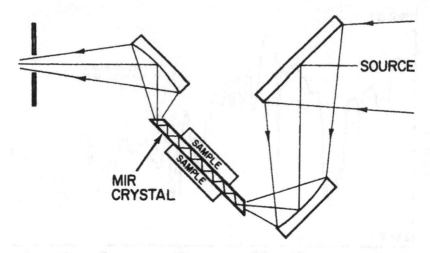

FIGURE 24.   Optical diagram of a typical internal reflectance accessory.

sample is ground more gently, and pressing is not involved at all, polymorphic changes induced by sample preparation are very unlikely. Disadvantages are that relative band intensities in the raw spectra differ from those of the corresponding transmission spectrum, and that the amount of energy diffusely reflected is very low, often only a few percent and sometimes even less. In order to achieve a good SNR, longer data collection times (more scans averaged) are required.

3. Internal Reflection

The third form of reflectance spectroscopy is internal reflectance, also described as ATR (Attenuated Total Reflection), FMIR (Frustrated Multiple Internal Reflection), or just plain MIR (Multiple Internal Reflection). Only the acronyms are different; all refer to the same technique.

The phenomenon of attenuated total reflection in the visible region of the spectrum was reported by Isaac Newton,[23, 24] but it remained until the early 1960s before it came into use as a sampling technique for infrared spectroscopy.[25-27] The theory of Internal Reflection Spectroscopy (IRS) and a number of its applications are discussed in Chapter 3, and are also documented in detail by some of the earlier publications by Harrick.[28, 29] The treatment here will be relatively brief.

Figure 24 shows the optical diagram of a typical internal reflection accessory. The central component is an Internal Reflection Element (IRE) which is usually an infrared transmitting crystal with a high refractive index. Unless the index of refraction of the IRE exceeds that of the sample, total internal reflection cannot take place.

The sample is placed in close optical contact with one or both faces of the crystal. The intensity of the internal reflection spectrum will depend on a number of factors (see Chapter 3), but the two most important considerations are the area of contact between the sample and the IRE, and the proximity of contact between the two. If the sample is placed in contact with the full area of both sides of the crystal, the spectrum will have maximum intensity. If the spectrum obtained in this way is too intense, the sample can be removed from one side of the crystal and left in contact with only the other side. If needed, further intensity reductions can be achieved by using an even smaller sample area so that only a part of one side of the crystal is in contact with the sample.

It is more likely, however, that the spectrum will be too weak rather than too strong. Good optical contact between the sample and the crystal is required. Even a very small air gap between the surface of the sample and the face of the crystal will reduce spectral intensity. If the sample is a liquid or a grease, it can be smeared directly on the crystal and good contact will be achieved. Solid samples will need to be pressed into close contact with the crystal, and most internal reflection accessories will include some type of clamping device to facilitate this. If the spectrum is too weak and the sample is already in contact with the full available area of both sides of the IRE, the next step is to clamp it more tightly to achieve better contact.

Most solid samples can be run by internal reflection, but some sample types lend themselves to the technique better than others. Finely powdered solids can be run but if they are hard they may scratch the face of the crystal; with soft crystals such as KRS-5 (thallium bromoiodide), they may even become imbedded in the crystal. Other techniques may be preferable. Polymer films, which are frequently too thick to run by direct transmission, are ideal candidates for internal reflection since they are usually soft enough to press into good contact with the crystal. Blocks of solid material, if they have a smooth surface to press against the crystal, make good samples provided they are not too bulky to fit into the clamping device. Fibers, fabrics, coatings on paper products, and foams usually make excellent samples. Cast films can be deposited onto an IRE, and since penetration into the sample is so small, the film thickness is less critical than for transmission spectroscopy.

Various crystalline materials have been used as IREs (see Chapter 3). Of these, KRS-5 and ZnSe probably find the greatest use. Since KRS-5 transmits to about 285 cm^{-1}, it can be used over the full scanning range of most FT-IR spectrometers. Its softness makes it easier to get good contact with the sample, but this can also be a disadvantage because it is more easily scratched, leading to scattering and reduced transmission. While not hygroscopic, KRS-5 will, over a period of time, be etched by water. If wet samples or aqueous solutions are to be examined, ZnSe or germanium (Ge) are better crystal choices. Normal handling of these IREs does not create a health hazard. Powdered or granular KRS-5, which may be produced during grinding or polishing, is dangerous since it can be ingested or absorbed through breaks in the skin.

There is one rather specific but important application for which germanium is the crystal of choice. High levels of carbon filling in an elastomer or polymer present problems in the infrared because carbon black is such a strong absorber. This is true for internal reflection spectroscopy just as it is for other sampling techniques. It has been found, however, that using a Ge crystal will often produce useful spectra even from carbon-filled materials. The relatively high refractive index of Ge (4.0) results in a much shallower penetration depth than when ZnSe or KRS-5 are used. Figure 25 shows the internal reflection spectra of a carbon-filled rubber "O" ring run with a KRS-5 crystal (top) and with a Ge crystal (bottom).

Even though internal reflection can be used with both liquid and solid samples, its early uses seemed to ignore liquid sampling. Most of the commercially available accessories addressed solid sampling applications and treated liquid sampling as an afterthought. During the 1980s, this situation changed, and recent years have seen the development of a number of easy-to-use accessories designed especially for liquid samples. These have been marketed under various trade names such as Circle Cell™, Horizontal ATR™, Tunnel Cell™, Dipper Probe™, etc. They come in a variety of configurations including low-volume batch-type devices and flow-through cells. Most use zinc selenide elements, probably because it is a harder crystal than KRS-5 and perhaps in anticipation that many liquids may include an

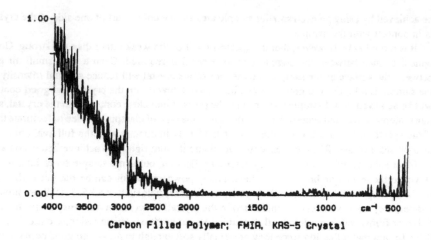

Carbon Filled Polymer; FMIR, KRS-5 Crystal

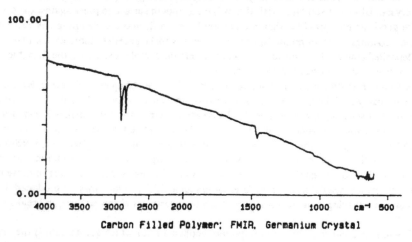

Carbon Filled Polymer; FMIR, Germanium Crystal

FIGURE 25.   Internal reflection spectra of a heavily carbon-filled rubber "O" ring run using a KRS-5 crystal (top) and a germanium crystal (bottom).

aqueous component. The potential applications to high-lab throughput quality control applications and to process analysis are obvious.

Finally, it must be remembered that internal reflection is a surface phenomenon. Only the surface of the sample interacts with the internally reflected beam. This effect is demonstrated by the spectra in Figure 26. The sample was a cigarette package wrapper. One side of the wrapper was clamped against an IRE and the top spectrum was run. Comparison with a spectral library established that it was the spectrum of nitrocellulose. The wrapper was removed from the accessory and the other side of the wrapper was now clamped into contact with the crystal. The center spectrum, completely different, was obtained. It was identified as cellophane. The bottom spectrum was run on the wrapper by transmission. It is obvious that the sample is too thick to run by transmission, but this spectrum matches the cellophane spectrum and we can now see that the wrapper is cellophane with a nitrocellulose coating on one side (probably to prevent moisture penetration). Note that the nitrocellulose spectrum is completely free of any cellophane bands, making identification easy; neither of the other two spectra contain any indication of the presence of nitrocellulose.

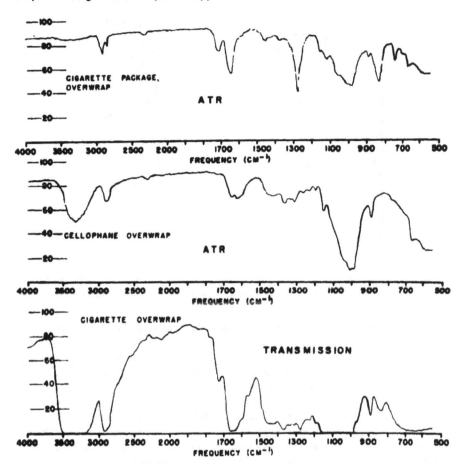

FIGURE 26. Spectra taken of a cigarette package wrapper. Top: internal reflection spectrum with one side of the wrapper in contact with the crystal. Center: internal reflection spectrum with the opposite side of the wrapper pressed against the crystal. Bottom: transmission spectrum of the wrapper.

VII. MICROSAMPLING

It does not usually take a large amount of sample in order to obtain a good infrared spectrum. In many of the techniques just described, a few milligrams or less are more than adequate. But there are exceptions, such as trace evidence in the forensic community, contaminants related to clean room operations, and certain types of failure analysis when even a few milligrams are not available. Sometimes the problem is strictly related to the amount of sample — it just is not enough to permit use of the conventional techniques. Other times this problem may be compounded with an inability to isolate and transfer a minute sample. This is the domain of microsampling.

In microsampling, the concern is with the area of the sample presented to the infrared beam. The thickness of a sample, in order to give a spectrum of good intensity, will need to be the same whether the technique is macro or micro. (Beer's law states that intensity, or absorbance, is proportional to sample thickness.)

When it comes to a good spectrum, one of the important criteria is SNR. Noise, in today's instrumentation is usually limited by the characteristics of the detector. Signal, of course,

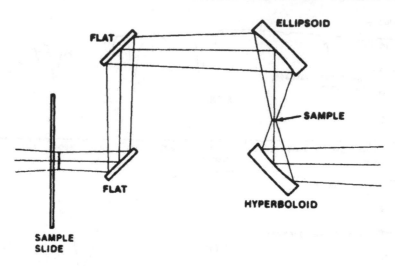

FIGURE 27.   Optical diagram for a typical infrared beam condenser. Note the reduction of the beam size at the sample point, thus permitting full beam energy to pass through smaller samples.

relates to the total energy available from the sample beam. Consider a beam, 6 mm in diameter, passing through the sample compartment. If the sample is 6 mm in diameter or larger, then all of the available energy falls on the sample and the SNR is maximized. On the other hand, suppose that the sample has been prepared as a 1.5 mm diameter KBr disk. The area of the beam that strikes the sample is now 1/16 of its former value, and the signal is now 1/16 of the full beam energy. But the noise, which originates in the detector, is unchanged, and so the SNR is reduced by a factor of 16. Weak absorptions now become increasingly difficult to distinguish from background noise. Repetitive scanning and signal averaging can reduce the noise level, but the reduction goes as the square root of the number of scans, and each factor of two in noise reduction requires four times as many scans as previously.

A more efficient way to maintain SNR is to use a beam condenser. This is an accessory that fits into the sample compartment and which will reduce the size of the image at the sample point to a fraction of the value that it had in the empty compartment. Figure 27 shows an optical diagram for a 5X beam condenser. The top ellipsoidal mirror reduces the beam size fivefold and then the lower hyperboloid re-expands the beam so that it looks the same (convergence and focus) to the rest of the spectrometer as if the beam condenser were not present. Returning to the previous example, if the normal beam size is 6 mm, then all of the energy passing through the 6-mm focal point will now be passing through an image (at the sampling position of the beam condenser) a little more than 1 mm diameter, and now if the 1.5 mm diameter pellet is placed in the beam condenser, once again the full beams energy is impinging on the sample and the SNR is restored to its prior value.

Micro KBr pellets have been a favorite technique for microsampling for many years. It is usually easier to transfer a minute amount of a solid than a liquid, and the ultimate transfer of the microsample from its point of origin until it is in the sample beam of the instrument is usually the most challenging aspect of microsampling. Many of the conventional macro techniques can be scaled down for micro work, and smears, cast films, and micro scale pyrolysis, done in a melting point tube with a match as the source of heat, have all been employed. Here, more than anywhere else, skill and art and technique assume greater importance than hardware and accessories.

One accessory worthy of special mention is the miniature diamond anvil cell (Figure 28). The windows are Type II optical quality diamond, usually about 1 mm across. The sample, which may be a paint chip, a fiber, or a chip of polymer, is placed on the bottom anvil. The cell is assembled and the sample is squeezed between the two diamond anvils until it is thin enough for a transmission measurement. Figure 29 is the spectrum of an automobile paint chip squeezed in the diamond anvil cell and run using a 5X beam condenser.

Infrared microspectroscopy is in some ways an extension and refinement of the beam condenser, but in addition to somewhat higher levels of magnification, it also brings with it a manipulative capability that makes it ideal for even the smallest of microsamples. Figure 30 is a schematic of a typical infrared microscope. Reduction of the image size at the sample position and subsequent reexpansion is done between two in-line mirror assemblies known as Cassegrain optics, chosen for their superior imagery. An optical conjugate of the sample position, magnified for more convenient observation, provides a location for a remote aperture with four independently adjustable blades. By sliding a mirror in and out of position, the sample may be viewed either with the eyes through a binocular microscope or left in the path of the infrared beam for the collection of an infrared spectrum. The manipulative advantages are that a sample may be viewed under magnification and positioned in the field of view. An adjustable aperture can then be used to isolate only that part of the sample whose spectrum is desired, and finally, that spectrum can then be scanned. Sample size is usually diffraction-limited. Samples as large as 20 to 30 μm across can be routinely scanned in a matter of seconds with good SNRs. Figure 31 is the spectrum of a single strand of a blue fiber squeezed flat in a miniature diamond anvil cell and run using an FT-IR microscope. The apertured area producing the spectrum is approximately 20 x 100 μm. Laminates, such as plastic containers or multi-layered paint chips can be examined by infrared microspectroscopy if they are first embedded and then microtomed. Figure 32 shows spectra of the four layers found in an automotive paint chip prepared in this manner.

VIII. CONCLUSION

Infrared spectroscopy is a powerful technique for analyzing samples; both qualitatively and quantitatively, inorganic as well as organic, and can be applied to samples in all three states of matter. Because of this wide diversity, a large variety of sample preparation techniques have been developed over the years as spectroscopists have striven to produce better quality spectra with less effort and in less time. With the commercial availability of FT-IR spectroscopy has come the development of additional sampling methods that are now feasible because of the increased energy throughputs of FT-IR instruments.

Sometimes the ease and simplicity of these new techniques have encouraged spectroscopists to select the easiest technique rather than the best technique. The purpose of this chapter has been to describe the widest possible variety of IR sample preparation methods so that the user can intelligently select the method best suited for his particular sample, consistent with the infrared spectrophotometer and the accessories that are available in the laboratory. The frequent references throughout the chapter to dispersive instrumentation are a tacit acknowledgment that while FT-IR is steadily displacing dispersive IR in the workplace, there are nevertheless many double beam dispersive instruments still in use.

Particular emphasis has been given to the more traditional sampling methods that are still widely used but described mostly in older literature. The newer sampling techniques, which are also discussed in this overview, are described in greater detail in subsequent chapters of this book.

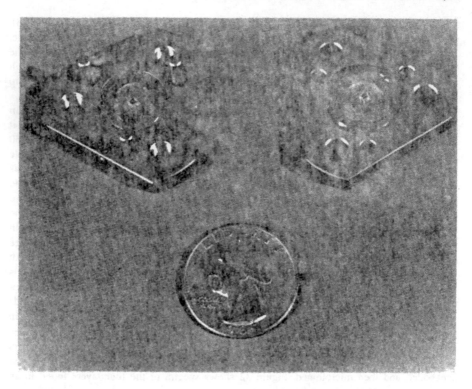

FIGURE 28. Photograph of a miniature diamond anvil cell, disassembled. A 25¢ coin is shown to give a size perspective. The diamond windows are approximately 1 mm across.

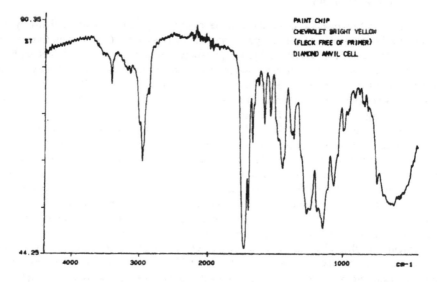

FIGURE 29. Spectrum of an automotive paint chip run in a miniature diamond anvil cell using a 5X beam condenser.

Optical path — transmittance infrared.

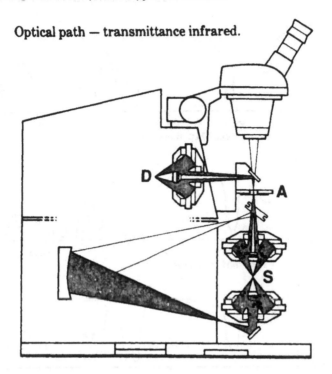

FIGURE 30. Optical diagram of an infrared microscope used in the transmission mode. The sample (S) is located between a pair of Cassegrain optics. An optical conjugate of the sample is formed at the remote aperturing stage (A) and the image is then refocused onto the detector target (D).

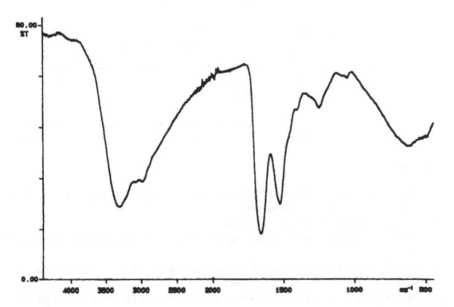

FIGURE 31. Spectrum of a single fiber using a miniature diamond anvil cell and an infrared microscope. The apertured area was approximately 20 X 100 μm.

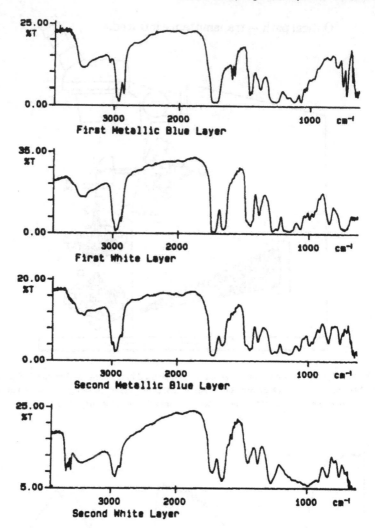

FIGURE 32. Spectra from a multi-layered paint chip run in transmission using an FT-IR microscope. The sample was embedded and then microtomed to a thickness that gave bands of proper intensity.

REFERENCES

1. *The Coblentz Society Specifications for Evaluation of Research Quality Analytical Infrared Spectra (Class III), Anal. Chem.,* 47, 945A, 1975.
2. Savitzky, A. and Golay, M. J. E., *Anal. Chem.* 36, 1627, 1964.
3. Zeller, M. V., *Infrared Methods in Air Analysis,* The Perkin-Elmer Corporation, Norwalk, CT, 1975, chap. 1.
4. Potts, W. J., Jr., *Chemical Infrared Spectroscopy,* John Wiley & Sons, New York, 1963, 127.
5. White, J. U., *J. Opt. Soc. Am.,* 32, 285, 1942.
6. Potts, W. J., Jr., *Chemical Infrared Spectroscopy,* John Wiley & Sons, New York, 1963, 93.
7. Stimson, M.M. and O'Donnell, M.J., *J.Am. Chem. Soc.,* 74, 1805, 1952.

8. Nyquist, R. A. and Kagel, R. O., *Infrared Spectra of Inorganic Compounds*, Academic Press, New York, 1971, 2.
9. Potts, W. J., Jr., *Chemical Infrared Spectroscopy*, John Wiley & Sons, New York, 1963, 139.
10. Harms, D. L., *Anal. Chem.*, 25, 1140, 1953.
11. Pattacini, S. C., *Perkin-Elmer Infrared Bulletin No. 52*; October 1975.
12. Vidrine, D. W., *Fourier Transform Infrared Spectroscopy, Vol. 3*, Ferraro, J. R. and Basile, L. J., Eds., Academic Press, New York, 1982, chap. 4.
13. Robinson, T. S., *Proc. Phys. Soc. London Ser. B.*, 65, 910, 1952.
14. Robinson, T. S. and Price, W. C., *Proc. Phys. Soc. Sondon Ser. B.*, 66, 969, 1953.
15. Roessler, D. M., *Br. J. Appl. Phys.*, 17, 1313, 1966.
16. Greenler, R. G., *J. Chem. Phys.*, 44, 310, 1966.
17. Greenler, R. G., *J. Chem. Phys.*, 50, 1963, 1969.
18. Fuller, M. P. and Griffiths, P. R., *Anal. Chem.*, 50, 1906, 1978.
19. Fuller, M. P. and Griffiths, P. R., *Am. Lab.*, 10(10), 69, 1978.
20. Fuller, M. P. and Griffiths, P. R., *Appl. Spectros.*, 34, 533, 1980.
21. Kubelka, P. and Munk, F., *Z. Tech. Phys.*, 12, 593, 1931.
22. Spragg, R. A., *Appl. Spectros.*, 38, 604, 1984.
23. Newton, I., *Optiks*, Book III, Part 1, Query 29, Dover Publications, New York, 1952.
24. Newton, I., *Principia Phil. Nat.*, Crawford, R.T., Ed, University of California Press, Berkeley, 1947, Book I, Prop. 96.
25. Harrick, J. J., *Phys. Rev. Lett.*, 4, 224, 1960.
26. Harrick, N. J., *J. Phys. Chem.*, 64, 1110, 1960.
27. Fahrenfort, J., *Spectrochim. Acta*, 17, 698, 1961.
28. Harrick, J. J., *Ann. N.Y. Acad. Sci.*, 101, 928, 1963.
29. Harrick, N. J., *Internal Reflection Spectroscopy*, Interscience, New York, 1967.

Chapter 3

OPTIMIZATION OF DATA RECORDED BY INTERNAL REFLECTANCE SPECTROSCOPY

Senja V. Compton and David A. C. Compton

CONTENTS

I. Introduction 56
II. Nomenclature 56
III. Theory 56
IV. Internal Reflection Elements (IREs) 62
    A. IRE Properties 62
        1. KRS-5 62
        2. Silicon and Germanium 63
        3. Zinc Selenide and Cadmium Telluride 64
    B. IRE Cleaning 65
    C. IRE Contamination 65
    D. Cushioning Pressure Pads 66
V. Instrumentation 68
    A. Experimental Design 68
    B. Stray Light in Alignment 68
    C. Liquid Sample IRAs 68
        1. Description 68
        2. Aqueous Solutions 69
        3. Experimental Procedure 69
    D. Solid Sample IRAs 73
        1. Description 73
        2. Horizontal IRAs 73
        3. Short Pathlength Accessories 74
        4. Sample Types 75
            a. Fibers 75
            b. Large or Irregular Shapes 75
            c. Films 76
            d. Powders 77
            e. Surface Treatments 77
            f. Semisolids 79
    E. Experimental Procedure 79
VI. Results and Discussion 80
    A. Surface Examination 80
    B. Solute Identification 80
    C. Quick, Repetitive Analysis 81
    D. Multilayer Samples 81
    E. Reaction Monitoring 83
    F. Black, Carbon-Filled Materials 83

0-8493-4203-1/93/$0.00+$.50
© 1993 by CRC Press Inc.

G. Trapped Air Bubbles 85
H. Inadequate Contact 85
I. Total Absorption 86
   1. Partial IRE Coverage 86
   2. IRE Selection 88
   3. Accessory Choice 88
   4. Angle of Incidence (θ) 89
J. Background Selection 89
VII. Conclusions 91
Acknowledgments 91
References 92

I. INTRODUCTION

IRS (Internal Reflection Spectroscopy) is a quick and easy nondestructive sampling technique for obtaining the infrared spectrum of a material's surface or of material which is either too thick, or strongly absorbing, to be analyzed by more traditional transmission methods. Samples examined by IRS generally require minimal, or no, sample preparation. Achieving optimal spectral results, however, requires some understanding of the physical principles involved. The purpose of this chapter is to furnish a practical guide for data acquisition, and therefore, the reader is referred to the literature for a more rigorous discussion of the theory and descriptions for the more advanced methods of analysis.[1,2]

II. NOMENCLATURE

Internal Reflection Spectroscopy (IRS) — The technique of recording optical spectra by placing a sample material in contact with a transparent medium of greater refractive index and measuring the reflectance (single or multiple) from the interface, generally at angles of incidence greater than the critical angle.[3]

Internal Reflection Element (IRE) — The transparent optical element used in internal reflection spectroscopy for establishing the conditions necessary to obtain the internal reflection spectra of materials.[3]

Internal Reflection Attachment (IRA) — The transfer optical system which supports the IRE, directs the energy of the radiant beam into the IRE, and then redirects the energy into the spectrometer or onto the detector. The IRA may be part of an internal reflection spectrometer or it may be placed into the sampling space of a spectrometer.[3]

Attenuated Total Reflection (ATR) — The reflection that occurs when an absorbing coupling mechanism acts in the process of total internal reflection to make the reflectance less than unity.[3] By convention, accessories have been named using this nomenclature and are referred to in the same manner throughout this article.

III. THEORY

In ATR, a type of IRS, the sample is placed in contact against a special crystal, termed an internal reflectance element (IRE). The IRE is composed of a material with a high index of refraction, such as zinc selenide (ZnSe), thallium iodide-thallium bromide (KRS-5), or germanium (Ge). The infrared beam from the spectrometer is focused onto the beveled edge of an

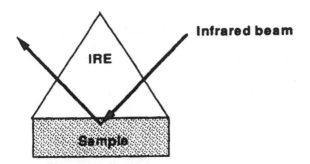

FIGURE 1.   Illustration showing the internal reflection which occurs at the sample-crystal interface.

IRE by a set of mirrors, reflected, generally numerous times, through the crystal, and then directed to the detector by another set of mirrors.[1,2,4-6]

The use of ATR in spectrometry is based upon the fact that although complete internal reflection occurs at the sample-crystal interface, as long as the angle of incidence is less than the critical angle, radiation does in fact penetrate a short distance into the sample (refer to Figure 1). This penetration is termed the evanescent wave, and radiation of selected wavelengths is absorbed by the sample which is in contact with the IRE, at each point of reflectance.[1,7] The resultant absorption spectrum closely resembles that of a transmission spectrum for the same sample. However, the spectrum will depend upon several parameters, including the angle of incidence for the incoming radiation, the sample size (thickness and area), the wavelength of the radiation, the number of reflections, and the refractive indices of the sample and IRE.[1,2,4-7]

The effective "penetration depth" of the evanescent wave can be calculated at any wavelength, λ. To a first approximation, usable data is obtained when this is less than one wavelength. Typical working values are 0.2 to 0.5λ.[4] The depth of penetration (d_p, defined as the distance from the IRE-sample interface where the intensity of the evanescent wave decays to $1/e$ of its original value) can be calculated using the formula[4,8] shown in Equation 1:

$$d_p = \frac{\lambda}{2\pi n_p \left(\sin^2 \theta - n_{sp}^2\right)^{1/2}} \tag{1}$$

where λ = wavelength of the radiation in the IRE, θ = angle of incidence, n_{sp} = ratio of the refractive indices of the sample vs. IRE, and n_p = refractive index of the IRE.

There are several important consequences of Equation 1 that are not immediately obvious. One of the effects of this equation can be illustrated through the examination of a typical organic material, having a refractive index of 1.5, using an IRE having a refractive index of 2.4. This will occur when a sample of vinyl polymer is placed in contact with either of the most commonly used IRE materials, KRS-5 or ZnSe. By using a value of $1.5/2.4 = 0.625$ for the ratio of the refractive indices (n_{sp} in Equation 1), the depth of penetration at 10 μm (1000 cm^{-1}) is calculated to be 2.0 μm, using an incidence angle of 45°, and 1.1 μm, using an incidence angle of 60°. Note that at a wavelength (λ) of 2.5 μm (4000 cm^{-1}), the depth of penetration is only one quarter of these calculated values.

This means that when studying a layered material which has an interface at a depth of 1.5 μm, the light would penetrate only the top layer using a 60° crystal. However, using a 45° crystal would result in a "wavelength dependent" composite spectrum, since the energy at 2.5

APPROXIMATE PENETRATION DEPTH [microns]
[common refractive index range]

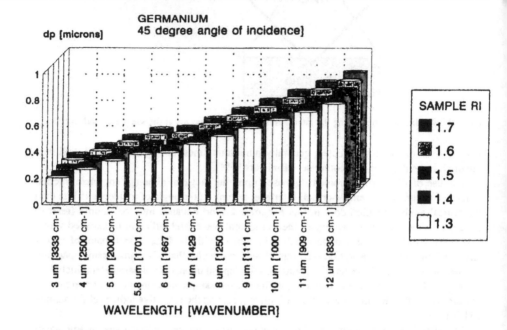

FIGURE 2. Graph illustrating the effects of depth of penetration vs. wavelength for a 45° Ge IRE. The refractive index of the sample is also seen to significantly influence the depth of penetration at each wavelength. (Graph is published with the permission of C. M. Paralusz, Permacel, A Nitto Denko Co., North Brunswick, NJ.)

μm (4000 cm⁻¹) would penetrate only the top layer, but the energy at 10 μm (1000 cm⁻¹) would penetrate some distance into the underlying layer as well. The approximate depths of penetration vs. wavelength for 45° Ge and KRS-5 or ZnSe IREs are shown in Figures 2 and 3, respectively. In each example, the depth of penetration is observed to be dependent upon both wavelength and the refractive index of the sample under examination. Note that the depth of penetration using a ZnSe crystal will vary substantially with the refractive index (RI) of the sample, whereas the effect of sample RI upon penetration depth is less pronounced when using a Ge IRE, possessing a higher RI.

A second consequence of Equation 1 is that a poor choice of crystal can result in band distortions being observed in the spectrum. The denominator includes the expression $(\sin^2\theta - n_{sp}^2)^{1/2}$, which becomes imaginary when n^2_{sp} is greater than $\sin^2\theta$. This is the value of θ called the critical angle (θ_c); at incident angles less than this, a "recognizable" spectrum is not recorded. However, a complicating issue is that the RI of the sample is not a constant value across a spectrum. Rather, the RI varies as a first derivative function at the frequency of an infrared absorbance. As the frequency decreases, the RI decreases sharply as the absorbance maximum is approached, rises to its average value at the absorbance frequency, and then increases abruptly. This means that on the low frequency side of a band, the value of n^2_{sp} may be greater than $\sin^2\theta$. In the above example, this would be a case in which the RI of the sample increased to only 1.7, an increment of only 0.2.

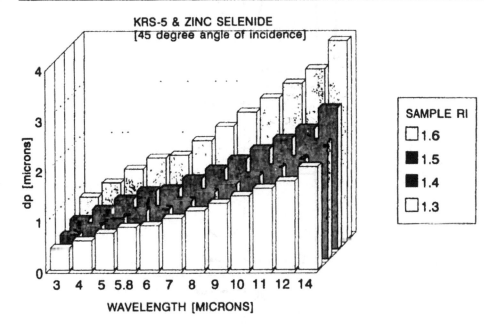

APPROXIMATE PENETRATION DEPTH [microns]
[common refractive index range]

FIGURE 3. Graph illustrating the effects of depth of penetration vs. wavelength for a 45° KRS-5 or ZnSe IRE. The refractive index of the sample is also seen to significantly influence the depth of penetration at each wavelength. (Graph is published with the permission of C. M. Paralusz, Permacel, A Nitto Denko Co., North Brunswick, NJ.)

The spectral results are that "normal" band shapes are observed when the incidence angle of the incoming infrared beam is much higher than the critical angle (Figure 4A). However, when the incidence angle is less than the critical angle, the observed bands have a first derivative band shape (Figure 4C). At angles near the critical angle, the band shape is slightly distorted, being broader, but more importantly, shifted to lower frequency (Figure 4B). The band intensity may also increase due to the resulting deeper depth of penetration. The frequency shift can be more than 10 cm⁻¹, which may cause difficulties with searching against spectral libraries composed of data collected by a transmission mode of analysis.

The problems involved with RI matching in IRS have always existed, of course, but the proliferation of different IRA designs now makes these encounters occur more frequently. The majority of horizontal IRAs use a 45° crystal of ZnSe as a standard IRE. The above discussion shows that severe band distortions will be observed using this crystal in any sample that has an "average" RI greater than about 1.5, since the RI can vary by more than ±10% at an absorbance peak.

An even greater potential problem arises from use of the continuously variable angle IRAs, such as the hemicylindrical ATR or the Seagull™ (Harrick Scientific, Ossining, NY). Analysts using these accessories can examine samples both below and above the critical angle settings, and, therefore, need to become familiar with the theory and consequences in order to avoid recording ambiguous spectral information.

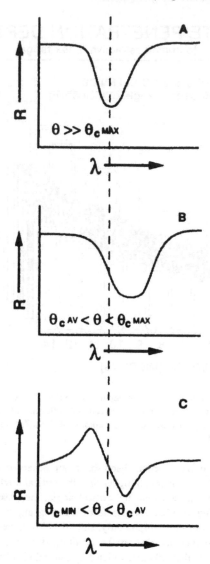

FIGURE 4. Illustration showing the spectral effects of angle of incidence (θ): (A) when the incidence angle is much higher than the critical angle, normal band shapes are observed; (B) when the incidence angle approaches the critical angle, band shape is distorted and shifted to longer wavelength (lower frequency); and (C) when the angle of incidence is less than the critical angle, bands have a first derivative shape.

To demonstrate the effects of changing the angle of incidence over a large range, the spectra of a polymer film were recorded at 60, 45, and 30° (Figure 5A, 5B, and 5C, respectively) using a ZnSe crystal in a hemicylindrical IRA (Bio-Rad, Cambridge, MA). The band intensities in the 60° spectrum are the most reliable since the penetration depth is less than that at 45°, and the angle of incidence is far removed from the critical angle. At a 45° angle of incidence, the

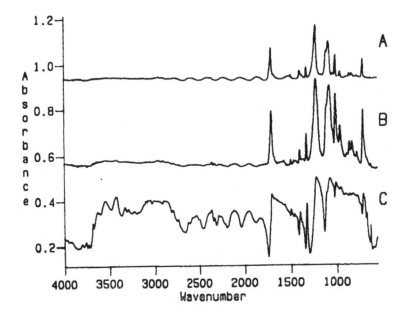

FIGURE 5. Spectra collected for a clear polymer film using the hemicylindrical ATR accessory equipped with a ZnSe IRE at the angle of incidence settings: 60, 45, and 30° (A, B, and C, respectively). The spectra have been offset for clarity. Refer to text for discussion.

penetration depth is, theoretically, 1.8 times greater. Indeed, the recorded spectrum reflects this same factor, depending upon which band is measured. The spectrum recorded at 30° is essentially useless, with all the bands severely distorted.

The novice can avoid many problems by using an accessory that uses a 60° crystal composed of KRS-5 or ZnSe, or a 45° crystal made of Ge. In the latter case, the depth of penetration will be less, but modern FT-IR instruments have a high signal-to-noise ratio (SNR) and can easily record the weaker signal.

The effect of increasing penetration depth with increasing wavelength (as shown in Figures 2 and 3) is also reflected in the relative intensities of the bands in the spectrum of a sample, when compared to its transmission spectrum. The bands at longer wavelengths are seen to be more intense than those at shorter wavelengths, because longer wavelength radiation penetrates further into the sample. This can be seen in Figure 6 where the transmission spectrum (Figure 6C) for a hand lotion sample is compared to that obtained by an IRS method (Figure 6B). Software exists to correct for this wavelength dependence, as seen in the corrected spectrum shown in Figure 6A, so that the IRS spectrum resembles that obtained by transmission. Generally, the only time these band intensity differences pose a problem is when comparing IRS data to data collected by another technique. This occurs most often when performing library searches for identification, as most data bases consist of spectra collected in a transmission mode (cast films, liquid cells, KBr pellets, etc.). If a software routine for "correcting" ATR spectra is not available prior to library searching, better matchings between these two types of spectra can be attained if an algorithm sensitive to peak position rather than intensity (such as a derivative type) is used for performing the comparison during the search routine.

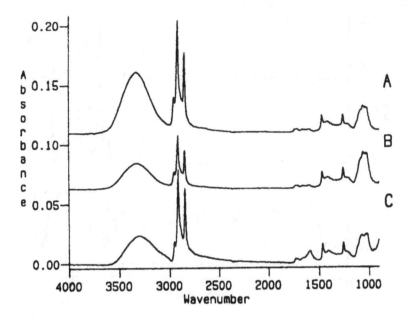

FIGURE 6. Spectra for a sample of hand lotion collected by (A) ATR after correction using the Bio-Rad "atrcorr" program, (B) ATR before correction, and (C) a standard transmission method. The ATR spectra were collected using a 45° ZnSe IRE. Spectra are offset for clarity.

IV. INTERNAL REFLECTION ELEMENTS (IREs)

A. IRE Properties

No universal material exists for IRS. A description of some of the more commonly used IRE materials is given in Table 1. Several properties, however, need to be considered in the selection of a crystal: RI, durability, spectral range coverage, chemical inertness, price, etc. Additionally, once the material has been chosen, the geometry of the IRE (shape, angle, dimensions) must also be defined. Angle and dimension sizes define the number of reflections of the IRE and also affect its fragility. As the intensity of an IRS spectrum is proportional to the number of reflections, IREs allowing multiple reflections are generally used. For instance, an IRE of dimensions 50-mm length by 2-mm thickness has 20 to 25 reflections, while that for a 1-mm thick crystal of similar length has 40 to 45 reflections, both for a 45° angle of incidence.[6,8] The shape of the IRE is generally dictated by the type of IRA to be used. However, since FT-IR spectrometers are available with a very high SNR, there is a tendency towards using IREs with geometries that give a much lower number of reflections, to the extreme that single bounce IREs have been incorporated into a number of accessory designs.

1. KRS-5

KRS-5 is one of the most commonly used materials for an IRE as it is comparatively inexpensive, and it has a spectral window comparable to that exhibited by most transmission methods of analysis (when the crystal is new). However, as the material is rather soft and easily deformed, the spectral range decreases gradually with use and the appearance of scratches on the highly polished faces. This is shown in Figure 7 where the single beam profiles for a new

TABLE 1. Internal Reflection Element Characteristics

Material	Frequency Range (cm⁻¹)	Index of Refraction	Characteristics
Thallium iodide-thallium bromide (KRS-5)	16,500–250	2.4	Relatively soft, deforms easily; warm water, ionizable acids and bases, chlorinated solvents, and amines should not be used
Zinc selenide (Irtran–4)	20,000–650	2.4	Brittle; releases H_2Se, a toxic material if used with acids; water insoluble; electrochemical reactions with metal salts or complexes
Zinc sulfide (Cleartran)	50,000–770	2.2	Reacts with strong oxidizing agents; relatively inert with typical aqueous, normal acids and bases and organic solvents; good thermal and mechanical shock properties; low refractive index causes spectral distortions at 45°
Cadmium telluride (Irtran-6)	10,000–450	2.6	Expensive; relatively inert; reacts with acids
Silicon	9,000–1,550, 400–	3.5	Hard and brittle; useful at high temperatures to 300°C; relatively inert
Germanium	5,000–850	4.0	Hard and brittle; temperature-opaque at 125°C

and relatively well-used IRE are shown on the same scale. The overall throughput has dropped by approximately 50%. Repolishing of the scratched surface is possible, but should be carried out by the supplier. Thallium compounds are toxic, and although the IRE itself is not harmful in normal usage (the material has good chemical resistance to most organic solvents and cold water), the dust from polishing is quite poisonous. However, the ductility of the crystal's surface is often an advantage in establishing good contact with the sample and, thus, the acquisition of good quality spectra.

A 45° angle of incidence is mostly used to analyze organic substances, allowing for a large penetration depth. A 60° IRE of the same material may be used to decrease the penetration depth, yielding a substantially "weaker" spectrum. Alternatively, the penetration depth can be reduced by using the same 45° KRS-5 crystal at a 60° position in a variable angle IRA. The oblique angle of 15° at the entrance face is transformed, according to Snell's Law, into an angle of 6.2° inside the IRE, thus yielding 51.2° at the sample interface.[6] With a continuously variable IRE position, this allows the flexibility of measurement of IRS spectra for the same contact arrangement at differing penetration depths.

2. Silicon and Germanium

Silicon (Si) and germanium (Ge) are good choices for IRE crystals when the samples to be examined are strong scatterers (e.g., black rubbers) or possess high RIs, as these crystals have higher RIs and the depth of penetration afforded by these materials is lower than that for

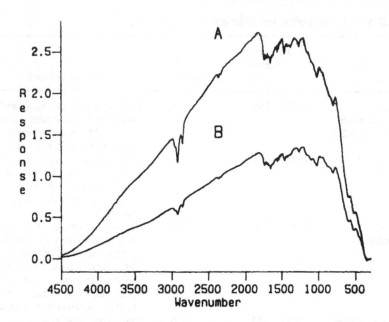

FIGURE 7.   Single beam spectra collected through (A) a new and (B) well-used KRS-5 IRE. The spectra are plotted on the same scale so that the energy profiles and crystal cutoffs can be directly compared.

KRS-5. One drawback with these IREs is their limited spectral range. Germanium has the broadest range, but is quite brittle, whereas silicon is useful for the short wavelength spectral region (and also in the far-infrared region) and is more durable.

3. Zinc Selenide and Cadmium Telluride

For applications where Ge is too fragile and KRS-5 is readily scratched during sample examination, ZnSe or cadmium telluride can be good choices, despite the fact that they are relatively expensive. Their RIs and spectral range coverages are similar to KRS-5, providing good quality, full mid-infrared range spectra. The single beam profiles for a ZnSe (Figure 8A), KRS-5 (Figure 8B), and Ge (Figure 8C) IREs, same angle of incidence (45°) and crystal geometry (10 mm x 5 mm x 1 mm single-pass parallelepiped), are shown for comparison of energy throughputs and expected useful mid-infrared spectral ranges. As crystal length increases, the overall throughput decreases due to loss of energy at each reflection point. Additionally, there is a shift to higher frequency cutoff because a greater length of the crystal is traversed, and the longer wavelengths of light penetrate deeper into the sample and are selectively attenuated.

One significant observation which deals specifically with liquid IRAs is described here. Electrolytic reactions may occur, with the metal body of the accessory serving as an electrode. This is most commonly encountered when examining a solution containing a copper salt (e.g., copper chloride and copper sulfate) or copper complex with a ZnSe IRE. The copper metal tends to plate onto the surface of the crystal, in addition to the IRE undergoing severe pitting. Only a few Angstroms of this coating are sufficient to completely block the evanescent wave. A spectroscopic indication that this reaction is occurring is a steadily decreasing throughput recorded through the IRA over time.

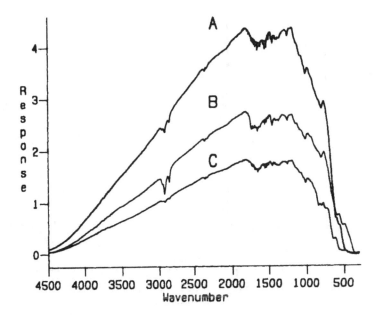

FIGURE 8. Comparison of single beam spectra collected through (A) ZnSe, (B) KRS-5 , and (C) Ge IREs at a 45° angle of incidence (10 mm x 5 mm x 1 mm). The profiles have been plotted on the same scale for ease in comparing energy throughput and useful spectral regions for data collection.

B. IRE Cleaning

Cleaning of an IRE should be performed carefully and with minimal contact with the polished faces of the crystal. A pressure-sensitive tape applied directly onto a solid or powdery sample can be peeled away, removing the majority of the material. A series of solvent rinses (acetone, toluene, and methyl ethyl ketone) with the element held vertically can be very successful in removing organic samples. Swabbing gently with a Rayon swab soaked in a suitable solvent can remove any residual material adhering to the crystal face. Stubborn samples may require a 30-s solvent treatment in an ultrasonic bath. The crystal should be handled only by its edges (i.e., the areas not contacted by the IR beam), and the crystal should be drip-dried. Recommended practices for IRS spectrometry can be found in ASTM Method E 573-90.[4]

C. IRE Contamination

New and "cleaned" IREs should be checked for contamination by collecting a single beam spectrum and looking for significant spectral absorptions, especially in the C–H (3000 to 2700 cm^{-1}) and C=O (1800 to 1600 cm^{-1}) stretching regions. Spectra for two common "grease" contaminants are shown in Figure 9, corresponding to a hand lotion and fingerprint residue. Alternatively, a ratioed spectrum may be collected of the IRE positioned in the accessory by using a single beam spectrum of the empty sample compartment as a background. However, it must be recognized that crystals can possess impurities or have spectral absorbances of their own which may become obvious in this latter case. Typically these will be ratioed out in normal usage (they are "constant" in both sample and background single beam spectra during

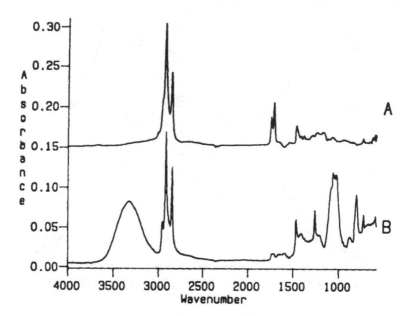

FIGURE 9. Spectra for two common contaminants when collecting data using an IRS technique: (A) fingerprint residue and (B) hand lotion. The top spectrum has been offset for clarity.

data collection, provided that the IRE is not repositioned). However, they do not remain a constant when the IRE is changed (repositioned or replaced), causing significant spectral features in a ratioed spectrum, which would theoretically be a flat baseline provided that the crystals were perfectly matched. This is illustrated in Figure 10A where the spectrum corresponding to two new IREs purchased with the same specifications (size, material, etc.) can be seen. A similar result can also be recorded if the IRE holder is placed backwards (reversed) into the accessory (Figure 10B), even though the crystal has not been replaced. Significant spectral bands appear which may adversely affect interpretation or quantitation by masking the features of the sample. It becomes apparent that reproducibility cannot be overemphasized!

D. Cushioning Pressure Pads

Many of the IRS solid sampling accessories have cushioning pads (polytetrafluoroethylene, neoprene, etc.) on the faces of the plates used to apply pressure to the sample. Therefore, it becomes extremely important, when using such a design, that the whole IRE face is covered by the sample surface so that a spectrum of the pad material is not recorded. A spectrum for such a cushioning pad is shown in Figure 11, which corresponds to a polytetrafluoroethylene material that has absorbed additional contaminants. It is useful to record a reference spectrum of the cushioning pads for each accessory used so that absorptions due to the pad can be immediately recognized in any subsequent analyses. When small samples are examined, one of the methods used to prevent the spectrum of this material from appearing in a ratioed sample spectrum is to cover the pressure plate surfaces during sampling with aluminum foil between the pads and sample. This allows the flexibility of the pad to be retained so that fracture of the crystal becomes less likely, while minimizing the likelihood of recording a spectrum of the pad material. Additionally, the aluminum foil should be cleaned with a solvent before use to

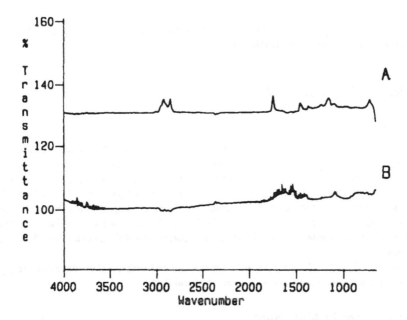

FIGURE 10. Spectra illustrating reproducibility in IRS sampling. The top spectrum (A) corresponds to that obtained when the holder for an IRE is reversed in its placement in an accessory. The spectrum obtained when ratioing the spectral features of two new ZnSe IREs is shown in (B). The top spectrum has been offset for clarity.

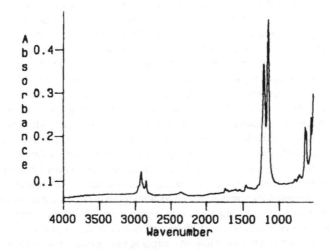

FIGURE 11. The ATR spectrum recorded for a cushioning pad from an IRA pressure plate. This spectrum is easily recognizable as polytetrafluoroethylene.

remove any residual traces of processing oils so that they do not contribute to the sample spectrum or contaminate the sample surface.

V. INSTRUMENTATION

A. Experimental Design

There exists an abundant choice of IRAs for the recording of IRS spectra, and a wide variety of accessories are available to meet nearly every sampling need. In general, accessories can be classified as being used for the examination of (1) liquid and (2) solid and viscous materials, although most accessories may be adapted for multiple uses. The following IRAs were used to examine the samples discussed in this paper: Liquid ATR, Horizontal ATR, Variable Angle macro-ATR, 2-Bounce ATR, Hemicylindrical ATR (Bio-Rad, Cambridge, MA), and Variable Angle micro-ATR/4X Beam Condenser and Prism cell (Harrick Scientific, Ossining, NY).

A Bio-Rad FTS 7 mid-infrared spectrometer, equipped with a DTGS (deuterated triglycine sulfate) detector, was used to collect all the spectral data.

B. Stray Light in Alignment

Stray light (energy not passing through the crystal) should be eliminated by proper alignment of the accessory, for those accessories not possessing fixed alignment, prior to data collection. A common design for an IRA used to examine solid samples is shown in Figure 12. Firstly, when aligning a solid sampling accessory, the pressure plates should remain close to, but not touching, the IRE. This aids in positioning the infrared beam to pass through, rather than by, the IRE (completely missing it) as shown in Figure 13. The plates are placed slightly away from the crystal so that they block the infrared beam from reaching the detector if it does not pass through the IRE and also so that they do not significantly decrease the energy passing through the accessory when attempting to align it. When first attempting to align an accessory to be used with a low throughput IRE, such as Ge, it is often easier to use a KRS-5 or ZnSe IRE to achieve gross alignment and then replace this crystal with the one to be used, prior to performing final alignment. This is particularly useful when using a spectrometer equipped with a red reference laser positioned concentric with the infrared beam, which will illuminate the crystal, thus showing the infrared beampath. Secondly, to assure that the beam is indeed passing through the IRE, the exit end of the crystal should be blocked (refer to Figures 12 and 14) and the infrared energy passing through the system measured (either by data collection or setup/alignment energy measurement). No energy should reach the detector in this situation. Finally, the energy (setup/alignment or data collection) through the system should be measured with the aligned accessory and compared to the energy when a blocking material is placed to the sides (either side and above/below) of the IRE holder (refer to Positions 1, 3, and 4 in Figure 14), making sure that the incoming and outgoing beampaths are not accidently blocked. The energy reading should not change if the accessory is aligned correctly.

C. Liquid Sample IRAs

1. Description

The popularization of the IRS technique for the examination of liquids has eliminated many of the problems encountered with traditional transmission liquid cells — cleaning, filling,

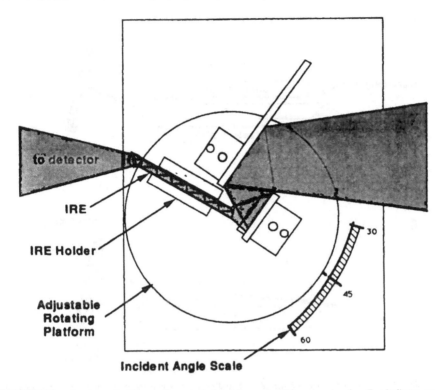

FIGURE 12. A schematic diagram for a typical solid sample IRA not possessing fixed alignment.

short pathlengths, interference fringes, solubility of windows, etc. Liquid IRS is a variant of solid IRS where the IRE crystal is surrounded by a vessel into which the liquid is poured (refer to Figure 15). Additionally, the very reproducible sample contact surface with the IRE yields a constant "effective pathlength" (until a new crystal is used), usually less than 1 to 10 μm (recall that it varies with λ and thus changes by a factor of 10 across the mid-infrared range), which permits the subtraction of the water spectrum in aqueous solutions[9,10] and the quantitative analysis of most liquid systems.[11,13] It should be noted that some liquids, notably latex- and amine-containing solutions, are extremely difficult to remove after examination in an IRA.

2. Aqueous Solutions

Figure 16 shows a typical application for a liquid IRA. An aqueous solution (bottom spectrum) was known to contain organic impurities at a level of 2% by weight. The difference spectrum (top spectrum — greatly expanded) was obtained after spectral subtraction. A search of this difference spectrum vs. the Bio-Rad solvents library revealed one of the materials to be ethyl alcohol. The total analysis time (including 16 s for each spectral data collection) was under 5 min! This experiment is very difficult to perform using a liquid transmission cell with a short pathlength due to the strong infrared absorbance of the water. Examples of solutions are such liquid materials as GPC or LC fractions, solvent extractions, surface washes, and mixtures — inclusive of solvents, beverages, medications, monomer blends, adhesives, oils, and soaps.

3. Experimental Procedure

A typical experiment involving liquid IRS sampling is as follows. First, a single beam spectrum of the clean, dried empty cell, positioned in the accessory, is acquired for ratioing

A

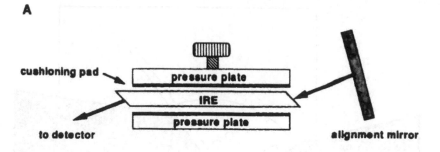

B

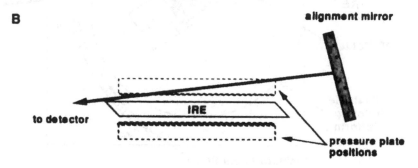

FIGURE 13. Proper alignment of the IRA should be made when the pressure plates are positioned slightly away from the IRE position in the holder. This assures focusing of the infrared beam through the IRE, as shown in (A). It is easy to position the infrared beam so as to completely miss the IRE (or clip the crystal edge), as shown in (B), when the pressure plates are not present (shown as dotted lines) during the alignment procedure.

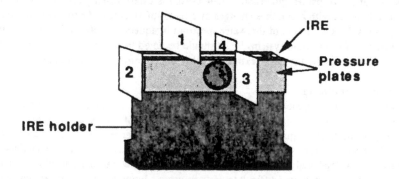

FIGURE 14. After final alignment of the accessory is believed to have been made, the use of blocking material should be made to assure that the beam is truly passing through the IRE. An opaque material should be placed, sequentially, in the indicated positions 1 through 4 while monitoring the energy passing through the accessory at each individual position. Positions 1, 3, and 4 should not affect the infrared signal recorded by the system, while position 2 (exit face) should not allow any infrared energy to reach the detector. The pressure plates should be positioned slightly away from the faces of the IRE.

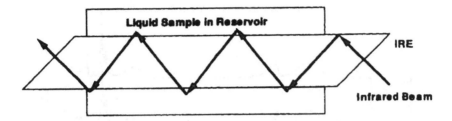

FIGURE 15. Illustration showing the typical beampath followed in an IRE used for the examination of liquids.

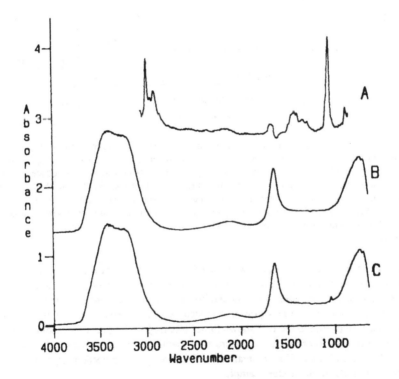

FIGURE 16. Infrared spectra for an aqueous solution obtained using the liquid ATR accessory. (A) difference expanded by a factor of 50, (B) water, and (C) 2% organic impurities. At 8 cm⁻¹ resolution, 16 scans were co-added to yield the data. The spectra are presented on the same ordinate scale, but are offset for clarity.

purposes. This allows the removal of the spectral profiles for both the crystal and the instrument from the ratioed spectra. Next, the first sample ("pure" material) is poured into the vessel and the purge (if appropriate) is allowed to recover. A ratioed spectrum of the liquid material is recorded, using the single beam spectrum collected previously as the background. The cell is emptied and appropriately cleaned prior to the introduction of a new sample ("mixture"). A quick 100% line (i.e., ratioed transmittance spectrum) comparing the status of

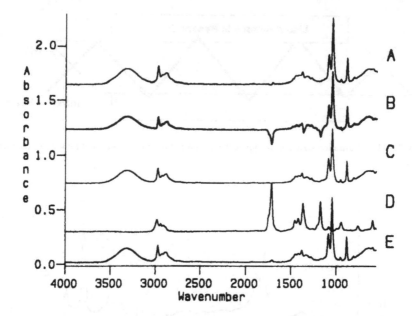

FIGURE 17. Spectra illustrating the choice of an appropriate background spectrum: (A) spectrum corresponding to the difference spectrum obtained from the spectral subtraction of the methyl ethyl ketone absorbance spectrum (D) from the original solution spectrum (E); (B) spectrum collected for the solution, using the single beam spectrum of methyl ethyl ketone as the background (note the negative and derivatized absorption bands); (C) spectrum collected for isopropyl alcohol; (D) spectrum collected for methyl ethyl ketone; and (E) spectrum corresponding to a solution of methyl ethyl ketone in isopropyl alcohol collected using the empty, dry liquid ATR cell as the background. The spectra have been offset for clarity.

the cell vs. the original background can be used to verify its cleanliness and dryness. A new background single beam spectrum is necessary under the following conditions: if the crystal is replaced, following cleaning by disassembly, as this will change the pathlength due to imprecise repositioning; if a new crystal is loaded into the accessory, as it will have its own differing spectral profile; if the IRE shows some "memory" effect; or if the resolution for data collection needs to be changed. A spectrum of the new sample is recorded. This is then followed by spectral subtraction to obtain the difference spectrum needed for identification of unknown materials in the mixture sample.

Spectral subtraction is the preferred method of removal of absorbance features from a common component, and Figure 17 illustrates the reasoning for this. If a single beam spectrum for the common component is used as the background for obtaining a ratioed spectrum of the mixture, the bands are very likely not to ratio completely as the amount of the common species is no longer the same. These differences in band intensities, based upon concentration change, cause a complicated spectrum with band inversions, as seen in Figure 17, making interpretation and identification extremely difficult. Therefore, the ratioed spectrum for each sample should be recorded, using the empty cell as the background, and spectral subtraction invoked to remove the spectral features due to the common species. This creates a much "cleaner" spectrum with interpretable absorption bands and also allows for the observation of band shifts due to interactions of the species in the mixture.

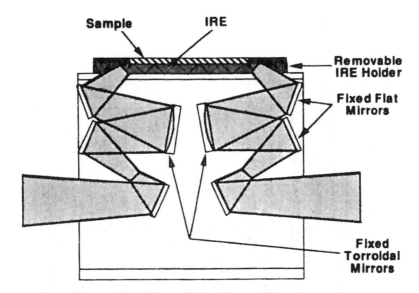

FIGURE 18.   Illustration showing the typical beampath followed in a horizontal IRA.

D. Solid Sample IRAs

1. Description
The appropriate IRS sampling technique can produce good quality spectra for many materials which are difficult to analyze with transmission methods. These materials include powders, pastes, adhesives, coatings, rubbers, fibers, thick films, textiles, papers, grease, and foams. By optimizing the available contact area of the IRE, the physical contact, and the geometry of the IRE (angle of incidence and dimensions), the intensities in the IRS spectrum can be varied although the bands in the high-wavenumber region may remain disproportionately small.

Typically, the spectra of sticky substances or hard, coarse powders are difficult to measure using the transmission technique, as their constituent particles cannot be reduced in size to be comparable to the matrix material for examination by methods such as pressed halide pellets or cast films. In these cases the IRS technique becomes superior, as the semisolids can simply be smeared onto the crystal surface, while the solids are pressed against the face of the crystal for the spectrum to be recorded.

2. Horizontal IRAs
In horizontal IRS, the IRE crystal is beneath the sample with a single exposed face, rather than sandwiched between the sample as in traditional vertical accessories, with the IRE face typically located outside the purge area of the spectrometer (refer to Figure 18). In the horizontal IRA, this allows for the quick application of the sample to the IRE surface, eliminates the loss of optical bench purge due to the out-of-compartment orientation, permits reproducible pathlengths for spectral subtraction and quantitative analysis, and allows for the examination of bulky samples and corrosive materials. However, caution should be used with any horizontal IRA, as the IRE is typically not bonded permanently to the IRE holder but has been made removable for easy cleaning and replacement. This ease in removal allows small

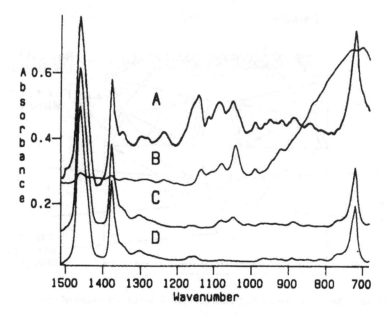

FIGURE 19. Carryover of materials when using a horizontal IRA can be a problem as seen in (A) (ordinate expansion 10X). The bottom three spectra (B through D) correspond to materials examined sequentially by the IRS method. Although good cleaning of the IRE between samples was performed, material became embedded in the crystal-holder interface. A spectral subtraction of the first sample spectrum (D) from its reexamination following the examination of the sample series, yielded the top difference spectrum (A). The spectra have been offset for clarity.

voids to exist on each side of the IRE in which small particulate matter may become caught or into which liquids or greases may flow. Figure 19 shows a series of spectra collected in sequence, ratioed against an accessory background collected prior to the application of the first sample. Thorough cleaning of the crystal top surface was made before the introduction of each sample. The top spectrum corresponds to the difference spectrum obtained from a spectral subtraction of the initial examination of the first material from its reexamination (factor of 10 ordinate expansion). It can easily be seen that spurious bands now appear in this residual spectrum, which should theoretically be flat and featureless. It is apparent that these spectral features correspond to absorption bands present in the second and third samples. Even though the cleaning procedure was thorough, enough material from these samples became embedded along the sides of the crystal to generate anomalous features in the examination of a later sample. Therefore, frequent collections of a background spectrum and thorough cleaning by dismantling should be practiced when using these accessories in order to avoid misinterpretation of spectral features.

3. Short Pathlength Accessories

Alternatives to this design are reflected in the prism cell (refer to Figure 1) and the 2-Bounce ATR accessories. They allow for only one or two reflections to occur at the sample-IRE interface (short effective pathlengths), respectively, opposed to the multi-reflection elements used to increase pathlength and achieve strong absorption band intensities. These two accessories are extremely useful when examining highly absorbing materials. Employing a large crystal means that the whole infrared beam is used, and good absorbances with high signal-to-noise levels are routinely obtained, even though the number of reflections is low.

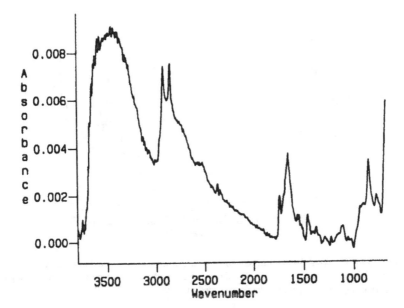

FIGURE 20. Spectrum obtained for a single human hair located perpendicular to the long axis of an IRE at its first reflection point. The spectrum was collected using a 45° Ge IRE in the 4X beam condenser/ micro-ATR accessory, co-adding 64 scans at 8 cm^{-1} resolution. The Bio-Rad "atrcorr" program was applied to the original spectrum to correct for the wavelength dependence of the absorption bands.

4. Sample Types

A multitude of samples fall into the category of solids, as mentioned previously, and their preparations for examination by IRS will vary . A discussion of possible ways of acquiring a spectrum from a variety of sample types follows.

a. Fibers. Fibers can be flattened by use of a hand press to increase their surface area for contact with the IRE crystal. Single fibers are best examined using a micro-ATR/beam condenser accessory (typified by the Harrick micro-ATR/4X beam condenser which uses a 10 x 5 x 1 mm IRE) as the relative crystal coverage by the sample is much higher than in a macro-ATR accessory, which uses a much larger crystal and does not have beam-condensing optics. The spectrum for a single strand of hair (intact, not flattened in a press), positioned perpendicular to the crystal length, is shown in Figure 20, illustrating that excellent signal-to-noise spectra can be obtained from only a single strand of material. If possible, multiple lengths of the fiber should be placed so that they cover the IRE face. If only a macro-ATR accessory is available, the fiber(s) should be placed at the first reflection point in the IRE as the infrared beam is strongest at this location (subsequent bounces reduce the intensity of the beam) resulting in the best quality spectrum being recorded.

b. Large or Irregular Shapes. Irregular shapes, or large pieces of material, can be examined by various means. First, for materials amenable to cutting, a scalpel or sharp knife can be used to create slices of the sample with flat areas to contact the IRE faces. Failing this, a large tooth file or coarse sandpaper can be run over the sample to create a powdered material which can easily be pressed against the IRE. An alternative is to crush the sample with a hammer, load the smaller pieces into a Wig-L-Bug mill (Crescent Dental Manufacturing Co., Chicago, IL), and grind the material into a fine powder for placement against the IRE. The material may be dissolved by suitable solvents and cast onto the IRS crystal. After drying, the

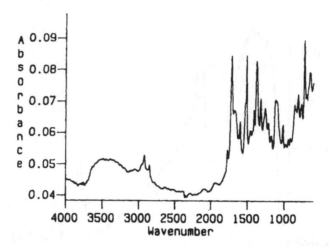

FIGURE 21. A spectrum for the thin coating on a copper wire was collected after flattening the sample between the platens of a hydraulic press. The flattening allowed good contact between the coating and the IRE faces, resulting in this spectrum being recorded in 16 s.

spectrum of the material can be recorded. In this case, since numerous solvents contain impurities or stabilizers (e.g., butylated hydroxytoluene in tetrahydrofuran), a spectrum from a "blank" generated from just casting the solvent should also be recorded. If the sample is pliable, a flat sampling surface can be created using the flat plates of a hydraulic press. This was done for a coated piece of copper wire. Figure 21 shows the spectrum obtained for the coating after a piece of the wire was flattened between the platens of a press so that the surface would make good contact with a Ge IRE.

c. Films. Films are easily examined by covering the IRE surface. However, it is advisable to examine both sides of a film since it may be a laminate. This behavior is illustrated in Figure 22 where the two spectra shown are for the respective sides of a white packaging film. If the analyst does not use caution in examining both sides to verify the homogeneity of the material, or if perchance the same side of the film is not placed against both sides of the IRE during examination (as seen in Figure 22), a valid analysis is not made. This latter case can be extremely confusing, as the combination spectrum from two polymer materials can be extremely difficult to interpret or identify. Multilayer film composition can be verified by examination of a cross-section under a visible microscope. Often laminates may be peeled apart using a scalpel while viewing the "operation" under a microscope so that a spectrum for each layer can be recorded. At times, soaking the sample in an appropriate solvent bath will remove adhesive layers between the film layers, leaving them separated for analysis.

Most spectrometers have infrared beams that are partially polarized (i.e., not equally polarized in both vectors). This may cause a problem when examining films that show directional orientation since the relative band intensities will depend on the film orientation. Therefore, it becomes important to position such a material with the same orientation on each side of the IRE in order that reproducible, homogeneous measurements may be made. In order to exaggerate this effect, spectra were collected for a polymer material examined at 45° in a hemicylindrical IRA fitted with a polarizer, as seen in Figure 23. The only difference between the two spectra is that the setting of the infrared polarizer, present in the accessory design, was

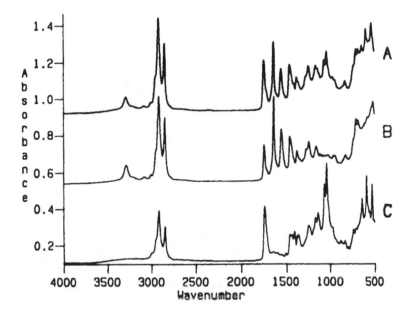

FIGURE 22. Homogeneity cannot be assumed to exist in packaging material. Therefore, it is extremely important that the same side of a sample be placed on both sides of an IRE for examination. The bottom two spectra (B and C) correspond to the differing sides of a food packaging material, while the top spectrum (A) corresponds to a composite obtained when different sides of the sample were placed against the two faces of the IRE during examination. Spectra have been offset for clarity.

changed by 90°. Dramatic differences in band positions, as well as intensities, can be easily observed. This same behavior can also be recorded (albeit not as dramatic) without the use of a polarizer just by simple rotation of the sample.

d. Powders. Powders can be placed directly onto the IRE with application of a pressure plate clamp to achieve good contact. Figure 24 shows the spectra collected for two inorganic powders, titanium dioxide and alumina, when using the micro-ATR/4X beam condenser accessory equipped with a 45° Ge crystal. Alternatively, the sticky side of a piece of adhesive tape can be coated with the powder and, in turn, placed into the accessory with the powder side towards the crystal face. Care must be taken so that the spectral characteristics of the adhesive are not attributed to the material under examination.

e. Surface Treatments. Surface treatments are usually recorded using the measurement conditions designed to achieve shallow penetration depths (e.g., using a Ge crystal). Sometimes the sample is examined at two differing angles, and then spectral subtraction is used to remove the bulk characteristics of the material (deeper penetration) from the surface spectrum (shallower penetration) due mainly to the species on the exterior of the sample. A halide salt abrasion of the surface, followed by the placement of the halide powder against the crystal face, may allow the recording of the spectrum due to surface species. Similarly, a solvent wash of a surface, followed by evaporation of the wash on an IRE, may allow the acquisition of a spectrum.[14] Again, one should record a spectrum for each "blank" (i.e., halide powder, solvent, etc.) to insure that the spectra recorded for the species come from the sample and not from the materials used in sample preparation.

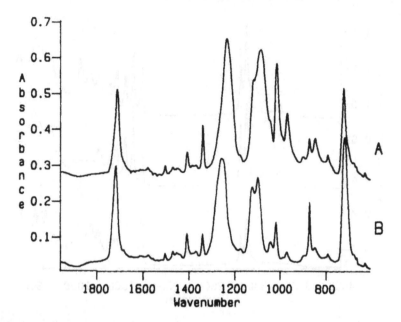

FIGURE 23. Orientation of a sample may result in slightly different spectra being recorded. These two spectra were acquired for the same sample (without repositioning) simply by placing a KRS-5 wire-grid polarizer in the infrared beampath. They correspond to 90° (A) and 0° (B) polarizer settings. Appropriately polarized background spectra were used to compute each ratioed spectrum. The top spectrum has been offset for clarity.

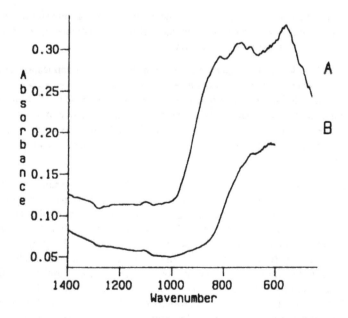

FIGURE 24. Spectra for the inorganic powders (A) alumina and (B) titanium dioxide obtained when using the micro-ATR/4X beam condenser accessory equipped with a 45° Ge IRE. At 4 cm^{-1} resolution, 16 scans were co-added to yield the spectra. The top spectrum has been offset for clarity.

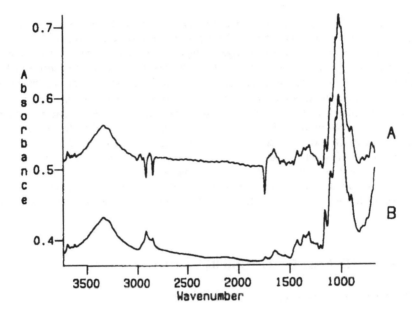

FIGURE 25. Spectra collected for a cellulosic sample using backgrounds acquired with the pressure pads of the IRA positioned against (A) the IRE and away from (B) the IRE.

f. Semisolids. Most semisolids, such as sludge, grease, paste, or wax, can be smeared across an IRE surface. If they are "too solid," a slight heating prior to application on an IRE may make them less viscous. Dissolution or suspension in a solvent, followed by evaporation of the solvent on an IRE, will also allow good quality spectra to be obtained.

Ointments, oils, and gels all fall into the intermediate category (not really solids or liquids) where they may be examined with either liquid or solid sampling accessories, depending upon their viscosities. They simply need to be poured into a vessel or smeared onto an IRE surface with no pressure plate applied, although cleaning of the IRE holder may be challenging after their examination. Some semisolids, notably latex based materials, are almost impossible to remove from a liquid IRA without disassembly and vigorous cleaning methods.

E. Experimental Procedure

A typical experiment involving solid IRS sampling is as follows. First, a single beam spectrum of the clean, dried crystal positioned in the accessory holder is acquired for ratioing purposes. This allows the removal of the crystal and instrument spectral profiles from the final spectrum. The pads on the pressure clamp should not be in contact with or touch the crystal, as a spectrum due to their material will be recorded. Figure 25 shows a comparison of the spectral quality obtained for a packaging material using a background with and without the foil-covered pads touching the IRE faces. Note that the SNR is significantly better in the lower spectrum where the foil covered pads did not attenuate the infrared beam by contacting the IRE during the background collection. Additionally, two spectral anomalies appear in the incorrectly collected spectrum (top). First, the presence of a fringe pattern is superimposed on the baseline, probably due to the measurement of the thin spacing at the IRE-foil interface. Second, the occurence of "negative" bands is attributable to organic residue on the foil covering the pads. Since this species is not present in the sample, the spectral features are not ratioed out in the final spectrum. Next, the IRE holder is removed from the accessory and the sample is placed in contact with the IRE face(s), with total coverage if possible. The pressure clamps are applied and tightened to fingertip tightness to assure good contact at the IRE-sample interface. The holder is returned to the accessory, the purge (if appropriate) allowed to equilibrate, and a ratioed spectrum of the material is recorded, using the single beam spectrum collected previously as the background. The holder is removed, cleaned, and reloaded with a new sample. The crystal needs to be cleaned between samples if there is carryover of material on the IRE surface from one sampling to another. It is highly recommended that a quick 100% line be collected comparing the status of the IRE vs. the original background to verify its cleanliness prior to examination of a new sample. For sensitive work, a new background must be collected if the crystal is replaced or moved during cleaning (change in pathlength, number, and/or position of reflections), if a new crystal is loaded into the accessory (differing spectral profile), or if the resolution for data collection needs to be changed.

VI. RESULTS AND DISCUSSION

A. Surface Examination

The analyst should be aware that only the surface layers of the sample are examined by the IRS technique. Therefore, if the sample is not homogeneous (e.g., emulsions containing differing droplet sizes or sediment aggregates in used motor oil), the spectrum will not be completely representative of the sample as a whole. This also applies to liquids, where micellar behavior can be expected or selective attraction to the crystalline material occurs. For instance, experience has shown preferential adherence and migration of amines, proteins, and latexes to ZnSe IREs. However, this characteristic can be used to an advantage when studying weathering effects on plastic materials or coatings, the migration of species to the surface of a polymer (such as antistats), or in thin coatings on a variety of substrates (e.g., acetates on paper).[15,16] Figure 26 shows the migration of an antistat material to the surface of a packaging film. The first spectrum was acquired following the washing of the film surface with isopropyl alcohol. This spectrum shows bands mainly due to the polymer. The subsequent 2 spectra were collected after 15 and 30 min time delays without removal of the sample from the accessory. The migratory behavior can easily be monitored vs. time since, as the antistat migrates to the surface, it coats the polymer surface, masking the polymer absorbance bands to the point where they finally disappear in the recorded spectrum. This introduces another variable to consider when examining samples by IRS. Since it is a surface technique, the sample should be examined "as is" and then following a wash of the surface with an appropriate organic solvent. In this fashion, migratory species such as antistats, mold release agents, low molecular weight polymers, and plasticizers, etc. will not be erroneously assigned as the bulk compositional material, but their presence will be duly recognized as a part of the whole formulation.

B. Solute Identification

Internal reflection spectrometry can be readily used to identify solutes in volatile solvents since the solvent can be readily evaporated, leaving the solute as a thin layer on the surface of the IRE. A spectrum of an adhesive collected with the prism cell is shown in Figure 27B. This adhesive was extracted with *n*-hexane and the extract evaporated to dryness on the IRE of the prism cell. The spectrum of the resulting extract is also shown in Figure 27A. One of

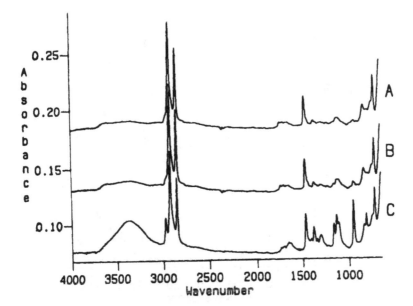

FIGURE 26. Spectra collected for an antistatic packaging material (A) 30 min after washing the surface with isopropyl alcohol, (B) 15 min after the alchohol wash, and (C) immediately after the alcohol wash. The spectra have been offset for clarity.

the components is easily identifiable as 2-ethylhexyl acrylate, when searching the Bio-Rad polymer library for identification.

C. Quick, Repetitive Analysis

One of the more common problems faced by the analyst is the quick qualitative identification of a wide variety of materials. It is in this area that IRS has found great usefulness, as it permits the spectroscopist to obtain infrared spectra on many samples with little or no preparation. Repetitive analysis of liquid samples is made easy by the wipe on/wipe off sampling afforded by the horizontal ATR accessory, provided appropriate measures are taken to avoid sample carryover effects as described above (e.g., epoxy sealing of the crystal into the IRE holder). Figure 28 shows one application, the monitoring of the unsaturate level in a series of edible oils. The infrared band at 3008 cm^{-1} corresponds to the C–H stretching mode of unsaturation. The intensity of this band is seen to increase when progressing from peanut (bottom) to olive (middle) and, finally, to corn (top) oil. This corresponds to an increase in the ratios reported for the unsaturate-to-saturate fatty acids in these oils, 3.7:1, 6:1, and 11:1, respectively.[17] Each spectrum is the result of only 30 s of data collection, allowing for excellent turn-around time for incoming batch analysis.

D. Multilayer Samples

Similarly, the IRS methodology can be exploited through the rapid analysis of a multilayer system, such as a peel-off label assembly. Figure 29 shows the spectra obtained for the four surfaces of such a sample (both sides of the liner and both sides of the label). Each spectrum was recorded in under 30 s using a macro-ATR accessory equipped with a 45° KRS-5 crystal.

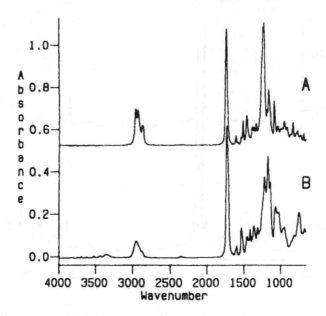

FIGURE 27.   Spectra for an adhesive (B) and its hexane extract (A) examined with the prism cell. At 4 cm⁻¹ resolution, 16 scans were co-added to yield the spectra. The top spectrum has been offset for clarity.

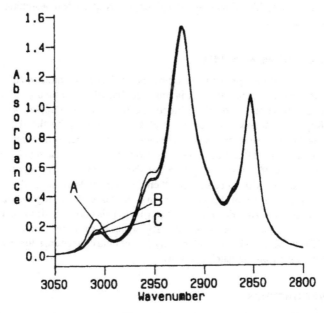

FIGURE 28.   Overlaid spectra for corn, olive, and peanut oils (A, B, and C, respectively) examined with the horizontal ATR accessory. At 4 cm⁻¹ resolution, 16 scans were co-added to yield each of the spectra.

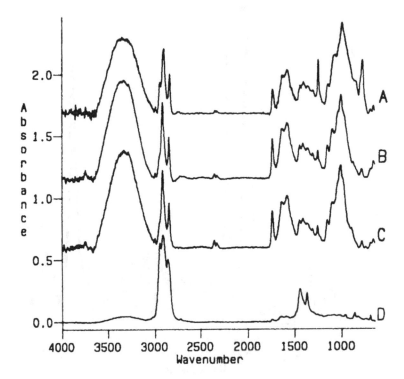

FIGURE 29. The spectra obtained for a peel-off label assembly using the macro-ATR accessory equipped with a 45° KRS-5 IRE. At 4 cm⁻¹ resolution, 16 scans were co-added to yield each of the spectra for the (A) front of the liner, (B) back of the liner, (C) front of the peel-off label, and (D) back of the peel-off label. The spectra have been offset for clarity.

E. Reaction Monitoring

The monitoring of reactions in real time (kinetic experiments) is ideal by IRS as the pathlength and the contact arrangement remains constant during the experiment. Figure 30 presents a series of spectra taken during the cure of an adhesive. The viscous adhesive was simply smeared across the IRE surface of a horizontal ATR accessory and spectra collected at 30 s intervals. Each spectrum is the result of 8 co-added scans at 4 cm⁻¹ resolution. Note the disappearance of the bands at 727 and 693 cm⁻¹ and the growth of the bands at 1448 and 835 cm⁻¹ with time, enabling the analyst to follow the chemical reaction corresponding to the setting of the glue.

F. Black, Carbon-Filled Materials

Black, carbon-filled rubbers have typically been difficult to analyze by infrared spectrometry. However, using a short pathlength Ge crystal permits good quality spectra to be recorded for such totally absorbing materials, as shown in Figure 31. Less than 30 s of data collection were required with the micro-ATR/4X beam condenser accessory to yield this spectrum. Note the broad band at 1640 cm⁻¹ due to the graphitic carbon in the rubber. The reason for using a Ge crystal in the examination of heavily carbon-filled materials is seen in Figure 32. The

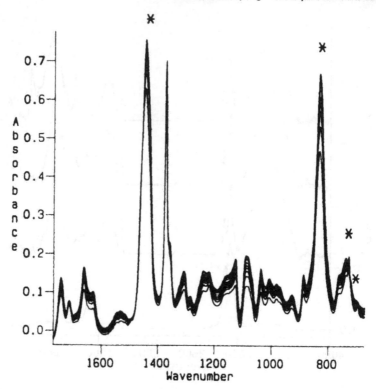

FIGURE 30. Spectra obtained for the cure of a glue monitored with the horizontal ATR accessory at 30-s intervals. Each spectrum is the result of 8 scans co-added at 4 cm^{-1} resolution. Peaks seen to change significantly with time are indicated by an asterisk.

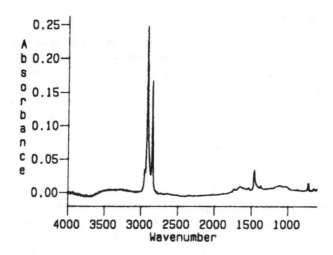

FIGURE 31. Spectrum of a black, carbon-filled rubber examined using the micro-ATR/4X beam condenser equipped with a 45° Ge IRE. At 4 cm^{-1} resolution, 16 scans were co-added to yield the spectrum.

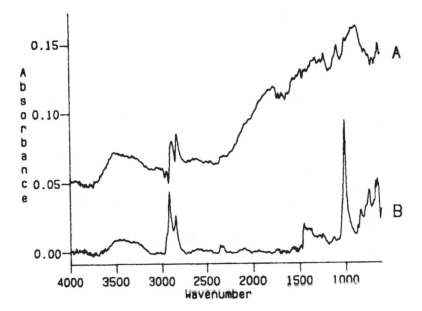

FIGURE 32. Spectra collected for a black, carbon-filled polymer material using (A) a KRS-5 and (B) a Ge IRE. The spectra have been offset for clarity.

spectral features are difficult to discern in the spectrum collected for this black polymer sample when a KRS-5 crystal was used, but appear cleanly in the spectrum collected with a Ge crystal. This is probably due to the RI of the sample approximating that of the KRS-5 IRE, with the depth of penetration giving more scattering and thus an inferior spectrum.

G. Trapped Air Bubbles

One of the most commonly encountered problems when using the liquid IRS cells is that of air bubbles trapped against the IRE element in a reservoir. Figure 33 shows the spectra (directly overlaid on the same scale) for a carbonated and degassed (flat) beverage collected using the liquid IRA. Since not all of the IRE is covered by the liquid in the carbonated example, the band intensities differ from those of the flat drink, as seen in the fingerprint region of the spectra. This can be explained by the pathlength of the sample being shorter due to incomplete crystal coverage. Additionally, the entrapped carbon dioxide (opposed to dissolved CO_2 which gives a single sharp band since it is no longer gaseous) gives rise to a band centered at 2350 cm^{-1} with the familiar P-R contour because it is on the IRE face. Typically, this behavior is not a problem with qualitative analysis, but poses a major obstacle when performing quantitative analysis as both concentration and pathlength then become variables. This situation may also be encountered by accidental introduction of air bubbles into the IRA cell during loading with a syringe or pipette or collection of vapor phase species if the liquid is heated.

H. Inadequate Contact

Inadequate contact of a sample against an IRE face is illustrated in Figure 34 for the examination of a red rubber sample using the macro-ATR accessory equipped with a KRS-5

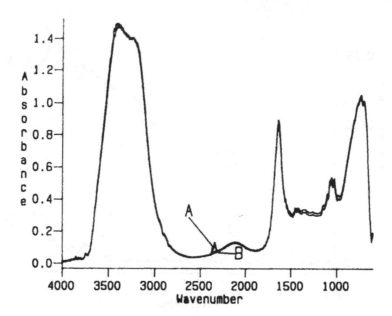

FIGURE 33. Overlaid spectra for (A) a fresh, carbonated and (B) a flat, degassed beverage collected using the liquid IRA. At 8 cm⁻¹ resolution, 16 scans were co-added to yield the spectra.

crystal. The sloping, elevated baseline (with superimposed fringe pattern) coupled with extremely weak spectral absorbances (bottom spectrum) are highly indicative of poor contact between the IRE and sample. Increasing the pressure applied by the pressure plates against the sample (top spectrum) molds the sample, improving the coverage and contact of the sample against the IRE face and removing any vacant air spaces. This may not always be feasible, especially when materials are granular or very rigid in composition.

This poor contact behavior can also occur in many loose-weave materials or open foam packaging. In cases such as these, the sample can be put between the platens of a press and "flattened". This procedure removes many of the air voids and creates a flat film surface to be placed against the IRE faces, as shown in Figure 35.

I. Total Absorption

Totally absorbing spectral regions may be made useful by employing one of several methodologies.

1. Partial IRE Coverage

The analyst's first effort to reduce the absorbance intensities recorded for a sample can simply be to cover less of the IRE face, effectively shortening the sample pathlength. Care must be taken when incomplete crystal coverage is used so that the spectrum of the cushioning pads is not recorded. This method of shortening the pathlength is usually used in those instances where qualitative, rather than quantitative, measurements are to be made, due to difficulties in reproducibility in IRE coverage. Second, the smaller sample should be placed at the first reflection points in the IRE, as the infrared beam is strongest at these locations (subsequent bounces reduce the intensity of the beam) allowing for better SNR in the collected spectrum.

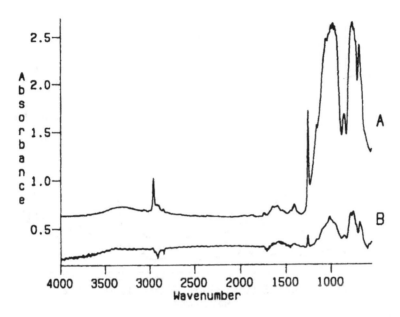

FIGURE 34. Spectra illustrating the degree of contact between sample and IRE. The top spectrum (A) corresponds to good contact of a rubber against the IRE surface, and the bottom spectrum (B) corresponds to poor contact (ordinate expansion X10).

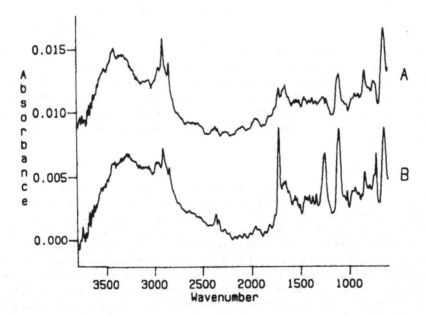

FIGURE 35. Spectra collected for a loose-weave material (A) prior to and (B) after pressing the material between the platens of a hydraulic press. The top spectrum has been offset for clarity.

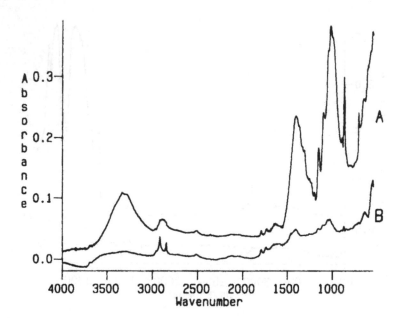

FIGURE 36. Spectra collected for a sample of red construction paper using the macro-ATR accessory equipped with (A) a 45° KRS-5 IRE and (B) a 45° Ge IRE.

2. IRE Selection

The preferred method of analysis for quantitative measurements is to choose a more suitable crystal affording a shorter effective pathlength and, thus, weaker absorbance values. This involves either a change in IRE material to afford shallower penetration depth or a change in crystal length to reduce the number of contact points with the sample, thus reducing the effective sample pathlength in each case. The spectra for a piece of red construction paper, recorded using a KRS-5 and Ge crystal in the macro-ATR accessory with a 45° angle of incidence, are shown in Figure 36A and 36B, respectively. The intensity of the cellulosic and inorganic filler bands in the fingerprint region no longer are totally absorbing (bottoming out) with the shallower penetration depth afforded by the Ge crystal, whereas they are with the use of the KRS-5 IRE.

3. Accessory Choice

Alternatively, a change in accessory may be required. Figure 37 shows the spectra for a sample of liquid paper recorded using the horizontal ATR accessory (top) and the prism cell (bottom), both equipped with a 45° ZnSe crystal. These spectra have been scaled for direct comparison, using the carbonyl band as the normalizing spectral feature. Figure 38A presents the spectrum acquired for a rigid, clear plastic film using a macro-ATR, with Figure 38B showing that acquired with a micro-ATR/4X beam condenser accessory equipped with 45° KRS-5 IREs. In both instances, the shorter pathlengths resulting from the reduction in the number of contact points with the sample present perfectly usable full range mid-infrared spectra for identification and quantitation of components. However, if one were looking for weak bands due to trace additives in a formulation, the longer pathlength spectra would be preferable, offering better signal-to-noise for the observation of weak spectral features at the expense of possessing totally absorbing spectral regions. Also, note that the absorbance bands

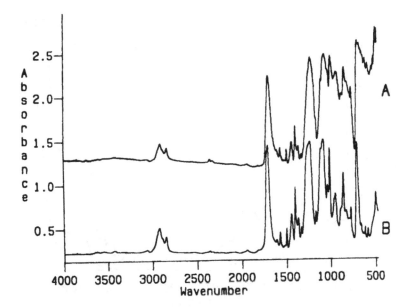

FIGURE 37. Spectra obtained for a sample of liquid paper examined using the horizontal IRA equipped with (A) a 45° ZnSe IRE and (B) the prism cell equipped with a 45° ZnSe IRE. The spectra have been scaled with normalization of the carbonyl band.

below 1000 cm⁻¹ (Figure 37A) show extreme distortions due to their intensities, which approximate total absorption in this region .

4. Angle of Incidence (θ)

A final way of reducing pathlength is to change the angle of incidence when using a variable angle IRA. However, the range in the depth of penetration afforded by this procedure is not as dramatic as that shown by the aforementioned sampling techniques. This effect of angle of incidence can be seen in Figure 39 where a polymer film was examined at 45° and 60° in a hemicylindrical ATR accessory equipped with a ZnSe crystal.

J. Background Selection

One of the most commonly encountered problems is the use of an inappropriate background reference spectrum for obtaining a ratioed spectrum. The proper single-beam background reference spectrum should be of the cleaned, dried IRE in position in the IRS accessory. If spectral contributions due to species in the sample (e.g., major solvent, underlying substrate, etc.) are to be removed, it should be accomplished by spectral subtraction of the individual spectra obtained for the mixture and pure material (species to be removed). A spectrum consisting of a hodge-podge of bands going in both directions will be recorded if the pure material is used as the background, making interpretation an extremely difficult task. This is due to incomplete ratioing of absorption bands, as the pathlength examined for this species differs in each sampling; therefore, the band intensities will not be identical. Additionally, any slight shifts in band positions due to solvent interactions or chemical associations/reactions will be difficult to discern. This is shown in Figure 40 for the examination of an adhesive on a packaging film where only a small amount of adhesive remained on the substrate. The

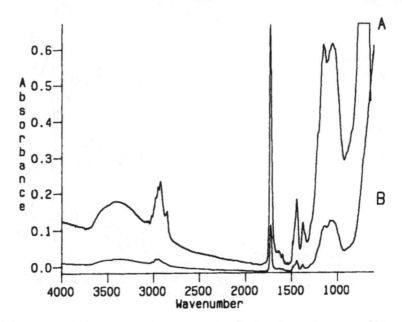

FIGURE 38. Spectra obtained for a rigid clear plastic examined using the macro-ATR accessory equipped with (A) a 45° KRS-5 IRE and (B) the micro-ATR/4X beam condenser equipped with a 45° KRS-5 IRE.

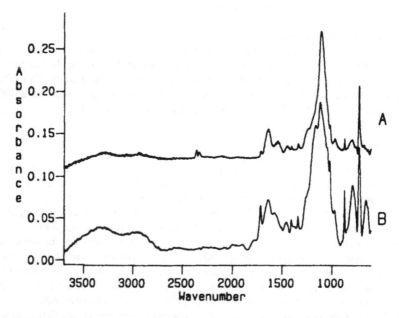

FIGURE 39. Spectra collected for a clear polymer film using the hemicylindrical IRA equipped with a ZnSe IRE at the angle of incidence settings: (A) 60° and (B) 45°. The spectra have been offset for clarity.

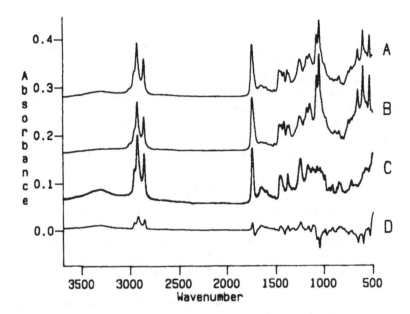

FIGURE 40. Spectra illustrating the correct IRS sampling method for examination of solid materials: (A) spectrum corresponding to a small anount of adhesive on a polymer substrate ratioed against the clean IRE; (B) spectrum of the substrate; (C) difference of spectrum obtained from the spectral subtraction of the substrate (B) from that of the adhesive on substrate (A); and (D) spectrum of the sample of adhesive on substrate ratioed against a single beam spectum of the substrate. The spectra have been offset for clarity.

difference spectrum (top) obtained from applying spectral subtraction is much superior to that obtained by inappropriate ratioing of the spectrum (bottom), where the single beam spectrum of the film was used for the background.

VII. CONCLUSIONS

The results obtained by using the methods described in this chapter demonstrate that the potential applications for attenuated total reflectance spectrometry are enormous. Many factors have been identified that influence the infrared spectrum acquired by IRS, but each factor can be taken into account, and a good quality spectrum obtained for most types of samples encountered by the spectroscopist. The versatility of the technique makes it useful for the identification and quantitation of many condensed phase materials. It is apparent the best IRS results will be obtained provided the analyst has access to several IRAs possessing different geometries, as well as IREs having different RIs, so that the conditions used for each sample examination can be optimized.

ACKNOWLEDGMENTS

The authors would like to thank Concetta M. Paralusz (Permacel, A Nitto Denko Co., North Brunswick, NJ) for her constructive review of the contents of this chapter and for contributing the graphic artwork shown in Figures 2 and 3.

REFERENCES

1. Harrick, N. J., *Internal Reflectance Spectroscopy*, John Wiley & Sons, New York, 1967.
2. Mirabella, F. M. and Harrick, N. J., *Internal Reflection Spectroscopy: Review and Supplement*, Harrick Scientific Corporation, Ossining, NY, 1985.
3. *Terminology Relating to Molecular Spectroscopy*, Standard Practice E131-90a, American Society for Testing and Materials, Philadelphia, PA, 1990.
4. *Standard Practices for Internal Reflection Spectroscopy*, Standard Practice E573-90, American Society for Testing and Materials, Philadelphia, PA, 1990.
5. *Optical Spectroscopy Sampling Techniques Manual*, Harrick Scientific Corporation, Ossining, NY, 1987.
6. *Laboratory Methods in Vibrational Spectroscopy*, Willis, H. A., Van Der Maas, J. H., and Miller, R. G. J., Eds., John Wiley & Sons, New York, 1987.
7. Smith, A. L., *Applied Infrared Spectroscopy: Fundamentals, Techniques, and Analytical Problem-Solving*, John Wiley & Sons, New York, 1979.
8. Westra, J. G. and Woerkom, P. C. M., Reflection Spectroscopy, in *Laboratory Methods in Infrared Spectroscopy*, Miller, R. G. J. and Stace, B. C., Eds., Heyden & Son, New York, 1972, chap 10.
9. Scheuing, D. R., *Bio-Rad FTS/IR Note #52, FT-IR Technique for Monitoring the Interfacial Interactions of a Surfactant Solution and a Solid Hydrocarbon*, November 1987.
10. *Chemical, Biological and Industrial Applications of Infrared Spectroscopy*, Durig, J. R., Ed., John Wiley & Sons, New York, 1985.
11. Compton, S. V. and Compton, D. A. C., *Bio-Rad FTS/IR Note #49, Quantitative Analyses of Mixtures by FT-IR Spectroscopy, I. Liquid and Solid Phase Samples by the P-Matrix Method*, June 1987.
12. Cahn, F. and Compton, S., Mulitvariate Calibration of Infrared Spectra for Quantitative Analysis Using Designed Experiments, *Appl. Spectrosc.*, 42(5), 865, 1988.
13. Compton, S. V., *Bio-Rad FTS/IR Note #69, Quantitative Analyses of Mixtures by FT-IR Spectroscopy, II. The Specfit Method*, January 1989.
14. Griffiths, P. R. and de Haseth, J. A., *Fourier Transform Infrared Spectrometry*, John Wiley & Sons, New York, 1986.
15. *Fourier Transform Infrared Spectroscopy, Applications to Chemical Systems*, Vol. 2, Ferraro, J. R. and Basile, L. J., Eds., Academic Press, New York, 1979.
16. *Advances in Applied Fourier Transform Infrared Spectroscopy*, Mackenzie, M. W., Ed., John Wiley & Sons, New York, 1988.
17. Robinson, C. H., *Normal and Therapeutic Nutrition*, Macmillan, New York, 1972, 664.

Chapter 4

DIFFUSE REFLECTANCE INFRARED SPECTROSCOPY: SAMPLING TECHNIQUES FOR QUALITATIVE/ QUANTITATIVE ANALYSIS OF SOLIDS

Scott R. Culler

CONTENTS

I. Introduction 94
II. Theory of DRIFTS 94
III. Sampling Basics of DRIFTS 95
   A. Accessory Alignment 95
   B. Background Spectra 96
IV. Qualitative Applications of DRIFTS to Different Sample Types 97
   A. Particulates 97
   B. Liquids 98
   C. Fibers 98
   D. Films or Solids 98
      1. Nondestructive Analysis 98
      2. New Si-Carb Sampling Technique for Non-Powdered Samples 98
V. Quantitative Applications of DRIFTS 99
   A. Requirements For Quantitative DRIFTS Analysis Using the Example of
      Silane-Treated Fillers 99
      1. Linear Calibration Plot 99
      2. Reproducibility 101
         a. One Instrument 101
         b. Multiple Instruments 102
      3. Internal Reference Material 103
   B. The Example of the Degree of Cure of Composite Resins 103
      1. Grinding as the Sample Preparation Technique 103
      2. Si-Carb as the Sample Preparation Technique 104
VI. Conclusions 104
References 104

0-8493-4203-1/93/$0.00+$.50
© 1993 by CRC Press Inc.

I. INTRODUCTION

Recently, the Diffuse Reflectance Infrared Fourier Transform Spectroscopy (DRIFTS) sampling technique has gained popularity for the study of powders, solids, and species adsorbed on solids. The proper application of any technique requires an understanding of the theoretical limitations of that technique which are discussed. This chapter also discusses some of the sample techniques that can be utilized to obtain high quality infrared spectra using DRIFTS. Quantitative results of a powdered sample can be obtained if the proper sample preparations are made. These powdered sample preparation techniques are discussed. Also, a new sampling method has been introduced which utilizes silicon carbide abrasive paper to collect spectra of nonpowdered samples which greatly increases the usefulness of the DRIFTS technique. This method will be explained. Variables which effect the reproducibility of collecting spectra are also discussed. Examples are provided to demonstrate the qualitative and quantitative nature of the technique.

II. THEORY OF DRIFTS

The theory related to DRIFTS has been around for many years as reported by Kubelka and Munk.[1,2] The theory is based on incident radiation on a powder sample being scattered in all directions. The scattered radiation is collected with the proper optical setup and directed to the detector. Radiation absorbed by the sample yields the spectrum, as with most other infrared techniques. A background spectrum of a non-absorbing powder — potassium bromide (KBr) or potassium chloride (KCl) — is required. Many commercial DRIFTS cells are available for collecting infrared spectra. The Kubelka-Munk (K-M) theory defines the reflectance spectrum as the ratio of the sample concentration to the scattering intensity of the sample. The K-M equation is described as Equation 1:

$$f(R_\infty) = \frac{(1-R_\infty)^2}{2R} = \frac{K}{s} \tag{1}$$

where R is the absolute reflectance of the layer, s is the scattering coefficient, and K is the molar absorption coefficient. This relationship is the Beer-Lambert law for DRIFTS. Fuller and Griffiths[3] discussed the use of the K-M theory for the quantitative analysis of certain samples. The theory predicts a linear relationship between the molar absorption coefficient and the maximum value of $f(R_\infty)$ for each peak if the scattering coefficient, s, is constant. The scattering coefficient is dependant on the particle size of the sample and must be constant to obtain quantitative results. Therefore, the K-M relationship can be written as Equation 2:

$$f(R_\infty) = \frac{(1-R_\infty)^2}{2R} = \frac{c}{k'} \tag{2}$$

where c is the concentration of the sample and k' is related to the particle size and molar absorptivity of the sample by k' = s/2.303e. This relationship becomes nonlinear at high concentrations, thus limiting the quantitative range which can be studied.

The major drawbacks with the DRIFTS technique arise from the specular component of the reflected radiation. The specular component does not penetrate the sample. This becomes a particular problem in spectral regions where there is a highly absorbing species and the surface

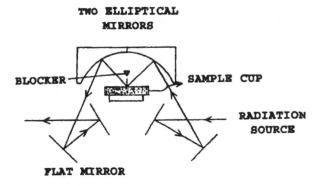

Figure 1. Schematic diagram of the "Collector" DRIFT cell showing four flat mirrors, two elliptical mirrors, and the blocker to help eliminate specular reflected radiation with the path of radiation indicated.

effects dominate the bulk effects leading to Reststrahlen bands. The Reststrahlen phenomenon can lead to distortion or inversion of the absorption bands. Also, if the sample has a very high absorptivity and little radiation is relected, the near absence of bands, when a very large band would normally be expected, may occur. When these conditions occur, quantitative analysis is very difficult and often impossible. This problem can be helped by diluting the sample with potassium bromide or potassium chloride powder and grinding to smaller particle sizes.[4]

The DRIFTS technique is powerful for the analysis of powders because of the internal scattering of radiation that occurs. The effective pathlength of the infrared radiation is increased many fold by this scattering. Therefore, it is possible to detect very low concentrations of species on powder samples compared to the standard one-pass transmission method.[4]

III. SAMPLING BASICS OF DRIFTS

Many DRIFTS accessories are commercially available, including ones for heating samples to very high temperatures for studying adsorbed species on catalysis and degradation reactions. The DRIFT accessories used in this paper were purchased from Spectra-Tech, Inc. (Collector model). A schematic diagram of the cell is shown in Figure 1. The accessories are designed to focus the infrared radiation on the sample and to collect the scattered radiation from the sample and focus it on the detector. For best results, an accessory that has an adjustable height sampling stage should be used because it allows every sample to be analyzed with a maximum energy throughput and minimal sample preparation.

A. ACCESSORY ALIGNMENT

Alignment of the DRIFTS accessory is very critical for the proper analysis of samples because DRIFTS is an energy loss technique. The additional mirrors of the DRIFTS accessory that focus the infrared radiation onto the sample and onto the detector all contribute to the loss of energy. Typically, only 10 to 15% of the energy throughput is available for DRIFTS analysis, compared to the transmission mode. If the Fourier Transform-Infrared (FT-IR) has multiple aperture settings, the instrument should be set at its maximum aperture so that as much of the infrared energy available can be used. The accessory is adjusted level to the optics bench by using a circular level. Next, adjust the height of the accessory to make sure the incident beam hits the center of the first mirror. The rest of the steps are aimed at optimizing

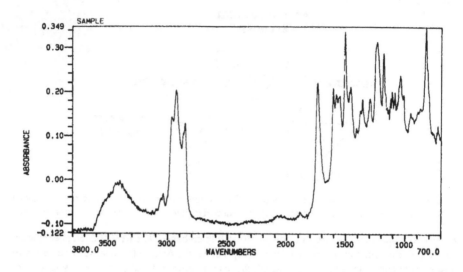

FIGURE 2. The effect of KBr particle size on the energy throughput to MCT detector.

the energy throughput to the detector, as monitored by the peak counts or intensity of the interferogram center burst. Place the alignment mirror in the sample cup and adjust its height. Using the alignment tools provided with the accessory, adjust the rotation and in/out of plane location of each mirror position until no further increases in energy throughput are obtained. Because an alkali halide powder is used to generate the background spectrum, it is a good idea to make the final adjustments to the alignment of the accessory to maximize the energy throughput from the alkali halide powder. This powder should be very dry and under 40 μm in particle size. Avoid touching the mirrors and be very careful not to scratch the mirrors when cleaning off powder. For cleaning, use a camel-hair brush, camera lens air-bulb cleaner, or gentle flow of dry, oil-free air. The cell should be realigned every time the alignment of the instrument is changed or the detector position is moved. This is very critical for obtaining consistent quantitative results.

B. BACKGROUND SPECTRA

As with any infrared sampling technique, a background spectrum is required for DRIFTS. The background is typically a nonabsorbing, highly scattering material (e.g., an alkali halide powder). Figure 2 shows the effect of particle size on the energy throughput for KBr powder. The finer the particles, the higher the energy throughput. However, it is very difficult to grind KBr below 44 μm in under 5 min of grinding with a mortar and pestle. Between 74 and 44 μm is a practical range of particle sizes to use for collecting DRIFT background spectra, as little additional throughput is gained below this particle size range. Water absorbs infrared radiation; therefore, the powder used for the background should be very dry. A new KBr background must be collected when the alignment of the instrument has been altered, or the alignment of the DRIFT accessory has changed. The background spectrum can be stored on disk and recalled for use at any time. Maximum precision would require that the finely ground

KBr powder be stored in a desiccator, and a new background spectrum collected each day samples are run. For powdered samples, load the sample into the sample cup using a spatula. Level the sample by tapping the cup on the counter top a few times. Avoid packing the sample into the cup as this will reduce the surface area exposed to the infrared beam and can reduce the intensity of the scattered radiation.

IV. QUALITATIVE APPLICATIONS OF DRIFTS TO DIFFERENT SAMPLE TYPES

DRIFTS has been used extensively for the characterization of particulate solids and liquids that can be adsorbed or deposited onto solids.[3, 4] However, any sample which scatters infrared radiation can be studied using the DRIFT cells, although other sampling techniques for the IR may be more appropriate or preferable depending upon the type of sample.[5] Alternative sampling methods for the analysis of particulate solids include grinding the sample in KBr powder and pressing out pellets, or dispersing it in Nujol for transmission analysis.[6, 7] Both of these techniques are discussed in Chapter 2. These approaches require a lot of time in sample preparation, and reproducible spectra can be difficult to obtain. The photoacoustic method can also be used.[8] This technique is discussed in Chapter 5.

DRIFTS is gaining popularity because it requires very little sample preparation for powdered samples and because spectra with high signal-to-noise ratio (SNR) can be collected in minimal time. One natural application for DRIFTS is particulate minerals and fillers because the nature of the surfaces can easily be determined. The nature of free hydroxyl and hydrogen bonded species are especially easy to detect, as has been reported using transmission IR sampling techniques.[9] Another useful application for DRIFTS is the study of coupling agent interactions with fillers/fibers used in the manufacture of high strength-reinforced composite materials.[10-12]

A. PARTICULATES

A majority of the time we are interested in knowing the functional groups present or absent in a sample, or whether a major contaminate is present. Usually a qualitative analysis will answer these questions. For qualitative analysis of solid/particulate samples,[3,4] the following options may be applied: (1) run the sample as received without grinding or diluting it in an alkali halide powder; (2) grind the sample and run it neat; or (3) grind the sample and dilute it 1 to 5% in an alkali halide powder. These options are listed in order of minimal sample preparation to maximum sample preparation. If the sample can be analyzed in a nondestructive mode, then other tests can be performed on the same sample. The number of samples not expected to yield acceptable DRIFTS spectra and produce quality spectra with virtually no sample preparation is surprising. It is not necessary to generate the K-M reflectance spectrum for such qualitative analysis. Utilization of the ratio spectrum, which is in transmittance units, saves time and yields higher SNR. Any functional group information is readily available using this approach. A major limitation with the DRIFTS technique arises from functional groups that have high molar absorptivity coefficients. These species tend to absorb more radiation than they reflect and usually all the detailed spectral information is either lost or greatly distorted in these regions. This is particularly true of silica filler systems which very strongly absorb infrared radiation from 1300 to 800 cm^{-1} when studied by DRIFTS. However, this does not mean that infrared information is not obtainable. The region from 4800 to 1400 cm^{-1} will yield very useful information about these systems. It is possible to keep the 1400 to 800 cm^{-1} region on scale by diluting the sample to a very low concentration in an alkali halide powder.

B. LIQUIDS

Trace quantities of liquids can be studied using the DRIFTS technique.[3] The samples need to be deposited on a particle which will produce high scattering, such as KBr or KCl. This approach is particularly useful when trying to identify different fractions from a liquid chromatography run. The excess solvent can be removed by evaporation, and the trace amounts of sample can be analyzed with the DRIFTS technique.[3] If large amounts of liquid are available, the standard transmission collection mode would be advised.

C. FIBERS

Fibers can be characterized by the DRIFTS technique. However, the orientation of the fibers will affect the scattering intensity from the sample and adversely affect the spectra of strongly absorbing functional groups. A method was developed to help eliminate the anisotropic scattering of radiation from oriented fibers.[13] This method involves the use of KBr overlayers on top of the sample fibers to provide more uniform scattering. The use of a KBr overlayer successfully reduces the orientation effect of glass fibers, allowing the spectral contributions from the glass to be subtracted so the nature of the adsorbed species on the surface could be studied.

D. FILMS OR SOLIDS

1. Nondestructive Analysis

Bulk solids can be analyzed if they reflect enough energy. The results generated would not be true diffuse reflectance in its purest form. This technique is worth trying, especially if an Attenuated Total Reflectance (ATR) accessory is unavailable or if the material is very difficult to grind. The KBr overlayer method previously outlined can be useful for obtaining spectra from films and solids that are not naturally diffuse-scattering samples.[13]

2. New Si-Carb Sampling Technique for Non-Powdered Samples

Recently, a new sampling technique has been reported that greatly increases the utility of the DRIFTS technique.[14] This technique is called the Si-Carb sampling method. It involves the use of 340- and 400-grit Silicon Carbide abrasive paper. The sample to be analyzed is abraded onto the surface of the silicon carbide paper. A spectrum of the abrasive paper is used as the background. The sample which has been abraded onto the silicon carbide paper is now able to scatter the infrared radiation in all directions so the basic requirements of the K-M theory are fulfilled. The spectral contributions of the resin adhesive and binders in the abrasive paper (background spectrum) are cancelled out when the sample is scanned. The resulting spectrum is of the material that has been abraded onto the paper. This technique is the most exciting development in recent years in the area of DRIFTS. Virtually any solid sample that adsorbs infrared radiation can conveniently be studied using this approach. Even coatings such as

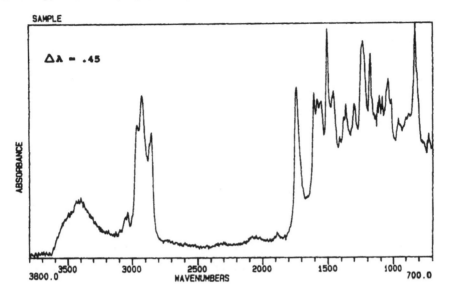

SAMPLE

Δλ = .45

ABSORBANCE

3800.0 3500   3000   2500   2000   1500   1000 700.0
                              WAVENUMBERS

FIGURE 3.   DRIFT spectra of black paint on jar lid collected using the Si-Carb sampling technique.

paints can easily be studied. The abrasive paper can be taken to the sample, and with a few strokes the sample is ready to be analyzed.

Figure 3 shows a spectrum of black paint on a jar lid collected using the Si-Carb technique. The spectral quality allows easy identification of the major functional groups that are present. Additionally, this method can be used to provide depth profiling of solid samples to determine differences in top/middle/bottom layers of a bulk solid material. The spectrum of the middle layer can be generated by abrading the sample on abrasive paper until the desired layer is reached, or the sample can be cut with a scalpel to expose the middle of the sample.

V. QUANTITATIVE APPLICATIONS OF DRIFTS

Another valuable use of DRIFTS is to compare similar samples to determine their exact molecular differences. For quantitative analysis, it is necessary to satisfy the basic requirement of the K-M theory: keep the scattering from the samples constant. Therefore, it requires more effort to obtain quantitative results, but these efforts are usually rewarded because the information obtained in quantitative analysis is so valuable in product development, quality control, and basic research.

A. Requirements for Quantitative DRIFTS Analysis Using the Example of Silane-Treated Fillers

1. Linear Calibration Plot
The major application of DRIFTS has been the analysis of powders and the interaction of species on fillers. Many studies have been carried out using the DRIFTS technique to study the interaction of silane coupling agents with fillers used to manufacture high-strength reinforced composite materials.[10-12] The K-M theory suggests that a linear relationship should exist

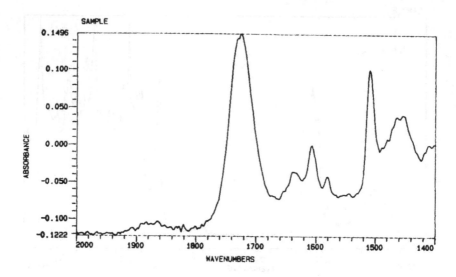

FIGURE 4. Calibration plot for DRIFT quantification method of C–H stretching area vs. treating solution concentration for A-174 silane coupling agent on an amorphous silica filler.

TABLE 1. Precision Statement for DRIFTS Quantification Method for the C–H Stretching Area of A-174 Silane Coupling Agent Normalized to the Si–O–Si Absorbance of the Silica Filler

Sample	Operator 1	Operator 2	Operator 3
1	3.86	3.95	4.04
2	3.86	3.95	4.02
3	3.96	3.87	4.03
4	4.14	3.95	4.02
5	3.93	3.96	4.03
6	3.87	4.03	4.07
7	3.97	3.89	4.02
8	3.94	3.89	4.09
9	4.27	3.96	4.01
10	3.89	3.96	4.07
Mean (SD)	3.91 (0.045)	3.94 (0.046)	4.04 (0.027)

Note: Overall precision N = 30; Mean (SD) = 3.96 (.068).

TABLE 2. Nested Design Results of DRIFTS Quantification Method for C–H Stretching Area and SiOH Stretching Area Normalized to the Si–O–Si Absorbance of the Silica Filler.

C–H Stretching Area of Silane/Si–O–Si Area of Filler

		Mean (SD)	% Variance
Location 1	3.98 (.07)	Machine	61.04%
Location 2	3.85 (.06)	Operator	3.44%
Location 3	3.76 (.04)	Sample	35.31%
All locations	3.87 (.12)		

SiOH Stretching Area of Silane/Si–O–Si Area of Filler

		Mean (SD)	% Variance
Location 1	12.46 (.36)	Machine	32.22%
Location 2	12.27 (.40)	Operator	6.56%
Location 3	11.91 (.34)	Sample	61.21%
All locations	12.22 (.43)		

Note: The contribution of variance due to machine, operator, and sample are included.

between the concentration of the silane coupling agent on the filler and the intensity of the reflectance spectrum for each functional group that absorbs infrared radiation. A simple experiment can be performed to test this basic premise. Figure 4 shows the results of the integrated area of the C–H stretching region vs. the concentration of 3-methacryloxypropyltrimethoxy silane coupling agent (A-174) deposited on an amorphous silica filler. Two hundred scans were collected on a Nicolet IR-44 system spectrophotometer purged with nitrogen and equipped with a wide band MCT detector. (All results discussed in this paper used these instrumental conditions.) Note that the spectra are reported in absorbance units and not Kubela-Munk units. I have found that this saves time and does not affect the quantitative results.[15] The filler in this system can be used as an internal reference as the absorbance band at 1865 cm[-1] is due only to the Si–O–Si of the amorphous silica. Therefore, the sample weights added to the sample cup do not have to be precisely controlled. The C–H stretching region is directly related to the amount of silane present on the filler. From the calibration plot in Figure 4 it is seen that a linear relationship was obtained. This basic test should be performed to validate the quantitative method developed. It is also necessary to determine that the range over which the quantification method is going to be used is linear (i.e., it must exceed the range of interest). The scattering requirements are satisfied by grinding the samples into a fine powder and screening to a constant size range (less than 74 μm in this example). Both the theory and experience show that when the particle size is not held constant, major changes in the DRIFTS spectra occur, and reproducible results are not obtained.

2. Reproducibility

a. One Instrument A basic requirement for quantitative analysis is that the method must yield reproducible results. A convenient way to determine reproducibility is to determine the precision of the test. Table 1 shows the results for three different operators using the same

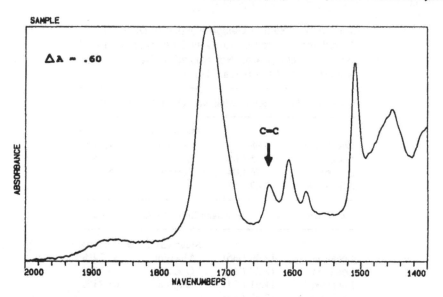

FIGURE 5. DRIFT spectrum of top surface of 2-mm thick light cured composite material after 20-s cure time using the Si-Carb sampling technique.

FT-IR. Each operator collected one spectrum each day on ten different days. All 30 samples were from the same batch of silane-treated filler. All spectra were collected within a 3-week period of time. The same background KBr spectrum was used for every sample that was collected and was stored on disk for easy retrieval. The standard deviation was found to be about 1.5%, which is excellent for any FT-IR sampling technique.

b. Multiple Instruments In an industrial setting, the laboratory and production facilities can often be hundreds of miles apart. It is then necessary to be able to obtain quantitative results on more than one instrument and DRIFT accessory. A three-factor, fully nested design[16] for the silane-treated silica filler was run to determine the reproducibility of the DRIFTS technique, and the factor or factors that contributed to the variability in the results obtained. The variables studied were operator, sample, and instrument (FT-IR). Two instruments were Nicolet IR-44s and one was a Nicolet IR-32 FT-IR. Before the design was started, each DRIFT accessory was realigned using the same ground KBr powdered sample, a background spectrum was collected using this KBr powder on each instrument. It was stored on disk and recalled for all samples that were collected. All three FT-IRs were purged with nitrogen and were equipped with mercury cadmium telluride (MCT) detectors. Two hundred scans were collected of each sample and background spectrum. Each operator collected one sample spectrum on each of ten different days. Each sample was purged for 5 min after optimum alignment for energy throughput before the spectrum was collected. A summary of the results are shown in Table 2 along with the components of variance. Note that the magnitude of error is in the 3% range for all the values calculated. For the SiOH area, the main component of variance is due to the sample. The main component of variance was due to the FT-IR for the C–H stretching region. For any test it is desirable for the sample to be the biggest component of variance so minor differences can be detected. These results are over nine different operators, four of which had never previously collected DRIFT spectra. Examination of the background spectra from the three FT-IRs used indicated there were differences in the intensities of a small band

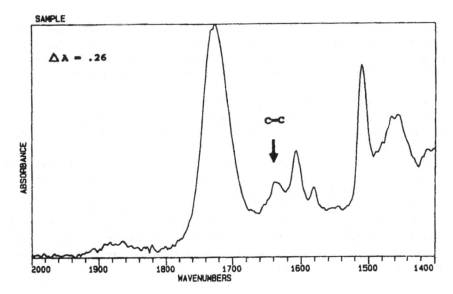

FIGURE 6.   DRIFT spectrum of bottom surface of 2-mm thick light cured composite material after 20-s cure time using the Si-Carb sampling technique.

in the C–H stretching region which is attributed to icing over of the MCT detectors. The results shown in Table 2 indicate just how reproducible the DRIFTS technique can be over nine operators, three FT-IRs, and three DRIFT accessories over a 3-week period.

3. Internal Reference Material

Another requirement for a good quantitative test is an Internal Reference Material (IRM) which can be run to control chart the test method and indicate when the test is out of control and a correction needs to be made. The IRM material cannot change over time. The silane-treated silica batch studied in the nested design was chosen as an IRM for the DRIFTS technique. This material has been used in our laboratory to determine if the instruments are in control over time. When the FT-IR alignment is changed, or the DRIFT accessory alignment is changed, an IRM spectrum can be collected to determine if further adjustments must be made to continue to obtain valid results.

B. The Example of the Degree of Cure of Composite Resins

1. Grinding as the Sample Preparation Technique

Another example of the quantitative nature of the DRIFTS technique is to determine the degree of polymerization of a composite resin matrix after curing. The resin matrix used in this example is a difunctional methacrylate resin blend of an isomer of Bisphenol-A-diglycidylmonomethacrylate (Bis-GMA) and triethylene glycol dimethacrylate (TEGDMA) filled with an amorphous silica filler. The cured composite sample is ground into a powder and screened to below 74 μm in size. The bands of interest are the vinyl absorbance at 1635 cm^{-1} due to both resins and the phenyl absorbance at 1510 cm^{-1} due to the Bis-GMA molecule (same bands as shown in Figures 5 and 6). The uncured composite spectrum can be collected in transmission by pressing the uncured paste between two salt plates. The polymerization reaction does not affect the absorbance due to the ring mode at 1510 cm⁻¹. This allows the ring mode absorbance to be used as an internal reference band for the system. Integration of the area of these two bands allows the relative degree of polymerization to be calculated for each sample when compared to the values obtained for the uncured composite paste.

2. Si-Carb as the Sample Preparation Technique

The Si-Carb technique can also be used to generate this information. It is well-known that light-cured composites do not cure to a uniform degree at the top and bottom of thick samples immediately after exposure to light. Figures 5 and 6 visually show this difference in the C=C absorbance band at 1635 cm⁻¹ when the sample is irradiated with a visible light source from the top side only. It is much easier to abrade the composite onto silicon carbide paper than to grind the cured composite into small particles by hand or even with a wig-l-bug (which can generate heat that increases the cure). The area of the 1510 cm⁻¹ and 1635 cm⁻¹ bands can be integrated, and the normalized results reveal that the bottom of the sample (which receives the least amount of light because the filler in the composite scatters and absorbs light) has 50% more unreacted C=C groups than the top of the sample (which receives all the visible light) after a 20-s exposure to visible light.

VI. CONCLUSIONS

Both qualitative and quantitative research applications of the DRIFTS technique have been discussed. Almost any type of sample can be analyzed qualitatively using one or more of the approaches discussed in this paper. The Si-Carb sample preparation method greatly extends the utility of DRIFTS for qualitative analysis of almost any solid. DRIFTS can become an invaluable research tool for quantitatively analyzing similar samples to determine their molecular differences. For quantitative analysis, the scattering must be held constant. This can be accomplished by carefully screening samples, establishing an IRM for the system, and keeping a control chart. By using one or more of the various sample preparation techniques available and following the steps necessary to obtain reliable data, DRIFTS is an easy way to answer questions and solve problems in the product development, quality control, or basic research laboratory.

REFERENCES

1. Kubelka, P. and Munk, F., Z. Tech. Phys., 12, 593, 1931.
2. Kubelka, P. and Munk, F., J. Opt. Soc. Am., 38, 448, 1948.
3. Fuller, M.P. and Griffiths, P.R., Anal. Chem., 50, 1906, 1978.
4. Fuller, M.P. and Griffiths, P.R., Appl. Spectrosc., 43, 533, 1980.
5. Culler, S.R., Ishida, H., and Koenig, J.L., The use of infrared methods to study polymer interfaces, in Annual Review of Material Science, Huggins, R.A., Ed., Annual Reviews, Palo Alto, 1983, 363.
6. George, W.O. and McIntyre, P.S., Infrared Spectroscopy Analytical Chemistry by Open Learning, Mowthorpe, D.J., Ed., John Wiley & Sons, New York, 1987, 116.
7. The Coblentz Society Desk Book of Infrared Spectra, Craver, C.D., Ed., U.S., 1982, 13.
8. Rosencwaig, A., Photoacoustics and Photoacoustic Spectroscopy, Chem. Anal., Vol. 57, Wiley, New York, 1980.
9. Iler, R.K., The Chemistry of Silica, John Wiley & Sons, New York, 1979, 634.

10. Ishida, H., Sturctural gradient in the silane coupling agent layers and its influence on the mechanical and physical properties of composites, in *Polymer Science and Technology*, Vol. 27, Ishida, H. and Kumar, G., Eds., Plenum Press, New York, 25.

11. Graf, R.T., Koenig, J.L., and Ishida, H., *Anal. Chem.*, 56, 773, 1984.

12. Vagberg, L., De Potocki, P., and Stenius, P., *Appl. Spectrosc.*, 43, 1240, 1989.

13. McKenzie, M.T., Culler, S.R., and Koenig, J.L., *Appl. Spectrosc.*, 38, 786, 1984.

14. Spragg, R.A., *Appl. Spectrosc.*, 38, 604, 1984.

15. Porro, T.J. and Pattacini, S.C., *Appl. Spectrosc.*, 44, 1170, 1990.

16. Kotnour, K.D., *Process Capability*, 1986, Internal 3M publication.

Chapter 5

A PRACTICAL GUIDE TO FT-IR PHOTOACOUSTIC SPECTROSCOPY

J. F. McClelland, R. W. Jones, S. Luo, and L. M. Seaverson

CONTENTS

I.   Introduction — The Opacity Problem and its Solutions 108
II. Photoacoutstic Signal Generation 108
     A. Signal Generation Model 108
     B. Variation of Sampling Depth 110
III. Instrumentation 112
     A. Photoacoustic Detector 112
     B. FT-IR Spectrometer 113
     C. System Checks 114
IV. Sample Handling 115
     A. General Considerations 115
     B. Sample Loading 115
V.   FT-IR/PAS Methods for Specific Analyses 118
     A. Qualitative Analysis of Macrosamples 118
        1. Polymer Identification by Computer Search 118
        2. Adhesive Spectrum by Spectral Subtraction 120
     B. Qualitative Analysis of Microsamples 121
        1. Single Textile Fibers 121
        2. Single Coal Particles 125
     C. Quantitative Analyses 126
        1. Vinyl Acetate Concentration in Polyethylene Copolymer Pellets 129
        2. Ash Concentration in Coal 130
     D. Variation of Sampling Depth 132
        1. Homogeneous Polycarbonate 132
        2. Mylar-Coated Polycarbonate 133
        3. Cure of an Acrylic Coating 133
        4. Plastic-Coated Paper 134
        5. Chemically Treated Surfaces of Polystyrene 135
     E. Carbon-Filled Materials 137
        1. Composite Material Prepreg 138
        2. Automobile Tire 138
     F. Polymer Films 139
        1. Elimination of Interference Fringes by FT-IR/PAS 139
        2. Polarized Beam Measurements on Oriented Films 139
        3. Polymer Coatings on Metal Containers 140
     G. Semisolids, Gels, and Liquids 141
VI. Conclusion 143
Acknowledgments 143
References 143

0-8493-4203-1/93/$0.00+$.50
© 1993 by CRC Press Inc.

I. INTRODUCTION — THE OPACITY PROBLEM AND ITS SOLUTIONS

The main sample handling problem in Fourier Transform-Infrared (FT-IR) analysis of solid and semisolid materials is that nearly all materials are too opaque in their normal forms for direct transmission analysis in the mid-infrared spectral region. Traditionally, the opacity problem has been remedied by reducing the optical density of samples to a suitable level by various methods of sample preparation.[1-3] This approach, however, leaves much to be desired due to the time and labor involved, the risk of sample alteration and preparation error, and the destructive nature of the process. Consequently, various other approaches have been tried to avoid or to minimize sample preparation (for additional details see Chapter 2 and Chapter 9.)

One such approach is to simply avoid the opacity of the mid-infrared spectral region by using overtone and combination absorbance bands for analysis in the near-infrared spectral region where absorption coefficients are significantly lower and samples are less opaque.[4] Near-infrared spectra combined with factor analysis software are successful in many analytical applications. Unfortunately, less detailed information is extractable from near-infrared spectra due to the broad, overlapping character of near-infrared absorbance bands. Consequently, the mid-infrared, especially the fingerprint region, remains as the most information-rich spectral region for analytical purposes.

Another approach is to select one of the alternative mid-infrared sampling techniques to transmission such as specular reflectance, diffuse reflectance (DRIFTS), or attenuated total internal reflectance (ATR).[1-3] All of these are, however, limited in their applicability. Specular reflectance requires mirror-like surfaces. DRIFTS often involves particle size reduction and dilution in KBr (for additional details, see Chapter 4). ATR presents problems related to the reflection element in terms of clean-up, crystal damage, and reproducible optical contact.

The most broadly applicable mid-infrared solution to the opacity problem is photoacoustic spectroscopy (FT-IR/PAS).[5] Its signal generation process automatically and reproducibly isolates a layer extending beneath the sample surface which has suitable optical density for analysis without physically altering the sample. FT-IR/PAS directly measures the absorbance spectrum of the layer without having to infer the desired absorbance spectrum based on a reflection or transmission measurement. This chapter will discuss FT-IR/PAS signal generation, instrumentation, sample handling, and results obtained with various classes of samples.

II. PHOTOACOUSTIC SIGNAL GENERATION

The photoacoustic signal is generated when infrared radiation absorbed by the sample converts into heat within the sample. This heat diffuses to the sample surface and into an adjacent gas atmosphere. The thermal expansion of this gas produces the photoacoustic signal.

A. Signal Generation Model

The photoacoustic signal generation sequence is shown schematically in Figures 1 and 2. Figure 1 shows the infrared beam intensity incident on the sample with intensity I_o. The beam intensity is modulated at frequencies, $f = v\nu$, by the FT-IR interferometer when the mirror moves with optical path difference velocity v resulting in a unique modulation frequency corresponding to each wavenumber ν. Alternatively, the modulation frequency can be set independent of v and ν if the FT-IR interferometer has step-scan capability.[6-8]

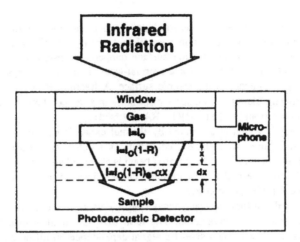

FIGURE 1. Schematic of photoacoustic signal generation showing the infrared beam intensity changes upon reflection and absorption by the sample.

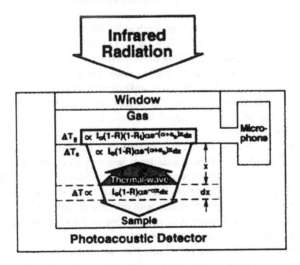

FIGURE 2. Schematic of one-dimensional photoacoustic signal generation showing the temperature changes that occur in the sample and adjacent gas.

After the infrared beam passes through the detector window and a transparent gas (typically helium), a fraction, R, is reflected at the sample surface. The infrared beam intensity is given by $I_o(1 - R)$ at a depth $x = 0$ inside the sample and decays to a value I_o $(1 - R)e^{-\alpha x}$ at depth x due to absorption of infrared radiation in the sample which has an absorption coefficient α.

Each layer dx of the sample that absorbs infrared radiation experiences an oscillatory heating at frequencies f with temperature change amplitudes, ΔT, proportional to $I_o(1 - R)$ $\alpha e^{-\alpha x}dx$ as shown in Figure 2. Each sample layer that oscillates in temperature is a source of thermal-waves.[9] In the one-dimensional energy flow schematic of Figure 2, thermal-waves propagate from the sample bulk to the irradiated surface and into the adjacent gas. During propagation, thermal-waves decay with a coefficient $a_s = (\pi f / D)^{1/2}$ where D is the sample thermal diffusivity.[10] Consequently, the surface temperature oscillation amplitude ΔT_s is proportional to $I_o(1 - R)\alpha e^{-(\alpha + s)x}dx$ for a thermal-wave generated at depth x just before it crosses into the gas adjacent to the sample surface. A fraction, R_i, of the thermal-wave is reflected back into the sample at the surface resulting in a temperature oscillation amplitude in the gas ΔT_g proportional to $I_o(1 - R)(1 - R_i)\alpha e^{-(\alpha + s)x}dx$.

The photoacoustic signal is generated by thermal expansion of the gas caused by heat associated with the sum of all of the ΔT_g contributions. Contributions come from each of the sample layers in which energy of the infrared beam is absorbed and which is close enough to the surface so that the thermal-wave amplitude has not decayed to a vanishing contribution after crossing the sample-gas interface.

Both the infrared and thermal-wave decay coefficients, α and a_s, respectively, play a key role in photoacoustic signal generation. The term $\alpha e^{-(\alpha + s)x}$ in the temperature oscillation expression leads to a linear photoacoustic signal dependence on infrared absorption when $\alpha \ll a_s$. In this situation a layer of sample extending a distance $L = 1/a_s$ beneath the surface contributes 63% (=$e^{-s L}$) of the signal with the other 37% coming from deeper layers of the sample. The thermal-wave decay length L is referred to as the sampling depth of a FT-IR/PAS measurement.

The possibility of varying the sampling depth of FT-IR/PAS measurements via the mirror velocity is apparent when L is written as $L = 1/a_s = (D / \pi v v)^{1/2}$ where the substitution $f = vv$ has been made. Note also that the sampling depth increases with decreasing wavenumber by more than a factor of 3 across the spectrum of a sample going from 4000 cm^{-1} to 400 cm^{-1}. A constant sampling depth, $L = (D / \pi f)^{1/2}$, vs. wavenumber is obtained with FT-IR systems operating in the step-scan mode that allows a constant modulation frequency to be selected and detected independent of the mirror-imposed modulation.

When $\alpha \ll a_s$, the photoacoustic signal increases linearly with α because as α increases, more infrared absorption occurs within the layer thickness, L, which is efficient in transporting heat into the gas. The photoacoustic signal loses linearity (the onset of saturation) as the infrared absorption continues to increase and as a high fraction of the absorption occurs within the layer. At very high infrared absorption, heating occurs very close to the sample surface and the photoacoustic signal no longer increases with infrared absorption (full saturation). After full saturation, some spectral features may still be observed due to spectral variations in the (1-R) term of the photoacoustic signal expression.

FT-IR/PAS spectra usually contain bands with infrared absorption coefficients in the ranges both before and after the onset of saturation, but spectral features due to reflectivity effects associated with full saturation are a very minor presence and often are not discernable. The onset of saturation can be moved to higher absorption by increasing modulation frequency and thereby decreasing the sampling depth. In most quantitative analyses, however, signal saturation does not present a problem due to the presence of weaker bands that are linear with concentration and the tolerance of commercial FT-IR factor analysis software to nonlinearities in spectra.

B. Variation of Sampling Depth

In the analysis of samples with concentration gradients or layers, it is particularly desirable to be able to vary the sampling depth. The most straightforward situation is one where: (1) the thermal diffusivity is homogeneous, (2) the weak absorption bands (such that

TABLE 1. Sampling Depths for Typical Thermal Diffusivities and
Modulation Frequencies

D(cm²/s)	0.1 Hz	1 Hz	10 Hz	100 Hz	1000 Hz	10000 Hz
0.1	6000	2000	600	200	60	20
0.01	2000	600	200	60	20	6
0.001	600	200	60	20	6	2
0.001	200	60	20	6	2	0.6

Note: Sampling depths in μm for typical thermal diffusivities from 0.1 to 0.0001 cm²/s and modulation frequencies from 10^{-1} to 10^4 Hz. D is given by K/ρC where K, ρ, and C are the thermal conductivity, density, and specific heat of the material, respectively.

α values at absorbance band peaks are $\ll a_s$) can be used both for monitoring the depth-varying concentration, and (3) an internal standard absorbance band is available which is associated with a species that does not vary with depth. Furthermore, the sample's thermal diffusivity D and the modulation frequency f must allow the sampling depth $L = (D / \pi f)^{1/2}$ to be adjusted over the range of interest by setting f via the mirror velocity or the step-scan modulator.[8, 11]

Table 1 gives sampling depths for typical values of thermal diffusivities and modulation frequencies. The thermal-wave decay coefficient, which should be larger than the analysis peak absorption coefficient value, is just the reciprocal of the sampling depth, as discussed in the last section. FT-IR/PAS sampling depths for polymers are typically in the micrometer to hundreds of micrometers range. This range nicely compliments ATR measurements which typically have sampling depths ranging from fractions of micrometers to micrometers.[12]

Nonhomogeneous samples are more difficult to analyze when the thermal diffusivity varies significantly with depth and when the absorption coefficient is not small relative to the thermal-wave decay coefficient. Variations in the thermal diffusivity between thin layers of a material result in variations in sampling depth within the sample and in thermal-wave reflection effects[9] at layer interfaces, both of which make interpretation of results more complicated. Analyses based on stronger absorption bands also can lead to ambiguities since the sampling depth is now determined by the decay of both the infrared beam and thermal-waves within the sample. Furthermore, when the sampling depth is reduced by increasing the modulation frequency, reduction occurs in the saturation-induced truncation of strong bands. This phenomenon can be incorrectly interpreted as an increase in species concentration near the surface (see Section V.D.).

Very useful practical information can be obtained from varying the sampling depth of FT-IR/PAS measurements despite the above cautionary considerations and the frequent lack of known infrared and thermal-wave absorption coefficients. Such coefficients, when available, are used to estimate sampling depth as the reciprocal of whichever absorption coefficient (infrared or thermal-wave) is larger. In practice, it is best to use spectral regions where absorption coefficients are known to be low by comparison with other spectral bands and to estimate sampling depth from the expression $L = (D/\pi f)^{1/2}$. The thermal diffusivity is given by $D = k/\rho C$ where k, ρ, and C are the thermal conductivity, density, and specific heat, respectively. These properties can usually be found in the literature for general classes of materials, if not for a specific sample.

FIGURE 3. MTEC Model 200 photoacoustic detector mounted on a Nicolet sample compartment baseplate.

III. INSTRUMENTATION

A. Photoacoustic Detector

A FT-IR photoacoustic detector is a sampling accessory which mounts directly in the sample compartment of the FT-IR optical bench. Figure 3 shows a photoacoustic detector positioned on the baseplate of an FT-IR spectrometer.

The spectral range of a photoacoustic detector depends only on the transparency of the sample chamber window. With suitable windows a detector can operate from the ultraviolet (UV) to the far-infrared. The most common window material is KBr (UV through mid-infrared), but quartz (UV through near-infrared), CsI (UV - 200 cm⁻¹), ZnSe (near-infrared - 560 cm⁻¹), and polyethylene (far-infrared) are also frequently used.

Typical sample holder dimensions are 10 mm in diameter by 9 mm deep. Most analyses, however, require a much smaller sample volume. Sample holder inserts are used to displace unused gas volume and boost the signal level which is proportional to the reciprocal of the gas volume.

Usually, only the central region of the sample chamber is irradiated by the infrared beam because the photoacoustic detector mirror typically causes a factor of two diameter reduction in the FT-IR beam focal-spot size between the normal focus in the FT-IR sample compartment and the focus in the photoacoustic detector sample holder. For instance, if the FT-IR normally has a 10-mm focal-spot diameter, this size is reduced to a 5-mm diameter inside the detector sample holder at the focal plane position which is approximately 1.0 to 1.5 mm below the detector window. Consequently, the sample volume that is actually analyzed is approximated by the focal area times the sampling depth, L, which depends on the modulation frequency.

When helium is used in the sample chamber, the modulation frequency range of a photoacoustic detector should extend from below 1 Hz to nearly 10 kHz. The use of air results in a high frequency cut-off at considerably lower frequency and limits capability to probe close to the sample surface. Measurements at higher frequencies in air are difficult due to a substantial loss in sensitivity after the detector goes through its first Helmholtz resonance.[13] This type of resonance is due to gas oscillations in the tube between the sample and microphone chambers. Acoustic resonances in the sample chamber itself do not occur because the chamber dimensions are too small when a sample is in place.

Helium purging of the sample and microphone volumes enhances sensitivity by a factor of 2 to 3, allows higher frequency operation, and removes moisture and carbon dioxide (CO_2). Moisture and CO_2 cause spectral interference as well as photoacoustic signal generation interference. The latter interference is caused by a phase difference between photoacoustic signals generated by absorption in gases or vapors vs. solids. The phase shift of gas or vapor signals leads to increased noise in spectra.

Many samples such as coal evolve water vapor after being sealed in the sample chamber. In such cases, a cup of desiccant is placed in the sample holder beneath the sample cup. Magnesium perchlorate is an excellent desiccant for this purpose and typically can be used for a day of operation without renewal.

It is important to note that moisture and CO_2 bands in spectra can be due to the presence of vapor and gas in both the FT-IR optical path and the photoacoustic detector. The source location can be identified by recognizing that contamination in the FT-IR causes negatively pointing (transmission-like) moisture and CO_2 bands in FT-IR/PAS spectra, whereas bands are positively pointing (absorbance-like) if contamination is in the detector itself. These observations should be used as a guide in purge and desiccant operations.

A final instrumental consideration is provision for normalizing spectra to account for spectral variations in the FT-IR source and optics, and for any sensitivity changes that may occur from day-to-day due to changes in source intensity or optical alignment. Normalization is performed by computing the ratio of the sample spectrum to a carbon black spectrum. The latter spectrum is best obtained with a reference standard consisting of an absorber element with a stable carbon black coating that is permanently mounted and protected in a dedicated sample holder. Loose carbon black is not a good standard for general use because its signal intensity varies as the powder settles, and it is easily spilled or blown out of the sample cup. In some instances, a glassy carbon, carbon-filled rubber (at least 60%), or graphite standard is desirable (see Section V. E.).

Many FT-IR data systems erroneously label the normalized PAS spectrum as a "transmittance" spectrum rather than an absorbance spectrum. The FT-IR data system should be commanded to change the label to absorbance prior to processing the data because many computer processing routines will either not operate or will produce incorrect results if a spectral file carries the wrong ordinate axis designation.

B. FT-IR Spectrometer[14]

The signal-to-noise ratio (SNR) of FT-IR/PAS measurements depends on the performance level of the FT-IR spectrometer as well as the detector sensitivity and noise level and signal generation efficiency of the sample. Low FT-IR mirror velocities produce low modulation frequencies and more efficient signal generation due to the slow thermal response of samples. Low modulation frequencies yield higher signal-to-noise spectra for a given acquisition time provided that the mirror velocity is held stable at low values by the FT-IR mirror servo-control system. A high infrared beam intensity is also beneficial and requires a large source aperture, high source intensity, and low f number optics. All commercial FT-IR systems provide combinations of these beneficial features to an extent that good FT-IR/PAS measurements can be performed, assuming that the FT-IR is in good operating condition.

Typical default operating parameters for FT-IR/PAS measurements with a fast-scan interferometer are given in Table 2.

Mirror velocity is given in Table 2 as an optical path difference velocity, v, which allows the modulation frequency, f, at a given wavenumber, v, to be calculated from the formula, $f = vv$. As discussed in the last section, the mirror velocity can be increased to decrease the

TABLE 2. Commonly Used FT-IR Operating Parameters for a Fast-Scan
Interferometer.

Mirror velocity = lowest available stable velocity (0.05, 0.1, 0.25 cm/s are typical)
Resolution = 8 or 16 cm^{-1}
Source aperture = maximum
Spectral range = 400–4000 cm^{-1}
Number of scans = 32–256

---

sampling depth and vice versa. Step-scan systems allow a desired modulation frequency to be
selected that is constant at all wavenumbers, thus providing a constant sampling depth for
values of absorbance below the onset of signal saturation.[8, 11] Resolution can also be adjusted
for specific needs but should not be set higher than necessary to resolve the structure of interest
because as the resolution is increased, noise will also increase for a set number of scans. The
spectral range is dictated by the spectrometer source and beam splitter, provided that the
photoacoustic detector window has suitable transparency. In the near-infrared, visible, and
ultraviolet spectral regions, very low mirror velocity and step-scan interferometers are particularly useful in order to keep modulation frequencies from becoming too high at the high
wavenumbers of these spectral regions. The SNR increases proportional to the square root of
the number of scans, therefore the number of scans can be adjusted to provide the SNR
necessary for a given application.

C. System Checks

At initial set-up the optical alignment of the photoacoustic detector to the FT-IR infrared
beam should be checked and adjusted if necessary. A liquid crystal infrared imager is useful
for this purpose. The imager is placed in the sample holder and provides a clear direct image
of the infrared beam's focal spot location. The detector should be aligned in the FT-IR sample
compartment by centering the focal spot in the sample holder. Kinematic detector mount
positioners should be secured in place to register the detector alignment position so that future
reinstallation does not require alignment. Note that if the FT-IR spectrometer in use has
provision for automatic alignment of the interferometer, this alignment routine should be done
prior to placing the photoacoustic detector in the FT-IR. The spectrometer's own detector
should be used because it has the correct position and active area size for the alignment
procedure.

An FT-IR/PAS SNR performance test should yield a peak-to-peak noise of less than 0.2%
at 2000 cm^{-1} for the conditions given in Table 3.

Some FT-IR systems may not allow exact duplication of these parameters and may not
perform well enough to fully meet this noise specification. Others will exceed this specification. The adjustments that will most likely improve FT-IR/PAS system performance (when
necessary) are alignment of the FT-IR beam splitter and tune-up of the FT-IR's mirror velocity
stabilizing servo-loop at low velocities.

The reproducibility of helium purging can be tested by reloading and purging the carbon
black reference several times and comparing the reproducibility of spectra. Purge times
of 10 s before and after sealing the sample cup at a flow rate of 10 cc/s of helium are
adequate for most applications where the sample has a low moisture level. The reproducibility of sample loading can also be checked by reloading and again comparing the reproducibility of spectra.

TABLE 3. Test Parameters for Measuring the FT-IR/PAS System Noise

Mirror velocity (OPD) = 0.1 cm/s
Resolution = 8 cm^{-1}
Source aperture = maximum
Spectral Range = 400–4000 cm^{-1}
Number of scans = 8
Apodization = medium Beer-Norton
Sample = carbon black reference
Detector gas atmosphere = helium

FIGURE 4. Sample holder with a polymer chunk sample ready to be inserted into the photoacoustic detector.

IV. SAMPLE HANDLING

A. General Considerations

The only sample preparation necessary with FT-IR/PAS measurements is to have a sample size smaller than 10 mm in diameter in order to fit into the sample holder. The thickness should not exceed 6 mm. Any sample geometry can be directly analyzed including disks, slabs, chips, chunks, pellets, powders, fibers, films, and sheets. Semisolids and liquids can also be analyzed. Larger samples are cut or abraded to suitable dimensions, typically using a razor blade, scissors, hacksaw, cork borer (#6), or a file or abrasive.

Figure 4 shows a typical sample placed in the sample holder and ready to be inserted in the photoacoustic detector for analysis.

B. Sample Loading

A sample holder is shown in Figure 5 with large and small stainless steel sample cups and brass spacers. The latter are used to minimize the gas volume and thereby increase the photoacoustic signal. In Figure 6, a small cup is shown placed in the sample holder. A funnel for loading powdered samples and other components for handling samples are shown in the figure also. The sample should be approximately 1 mm below the rim of the sample holder in

FIGURE 5. Sample holder, with brass inserts, and small and large stainless steel cups.

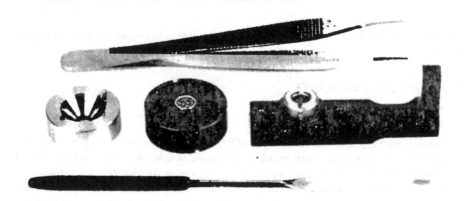

FIGURE 6. Sample handling tools including sample holder loaded with small stainless steel cup, fixture and funnel for loading cups, and cup-handling tweezers and spatula.

order to allow a layer of gas between the sample and the detector window for photoacoustic signal generation in the gas. If very high or low modulation frequencies are used, an approximate distance that the sample should be below the cup rim can be calculated from the formula $(D/\pi f)^{1/2}$ where $D = 1.51$ cm²/s for helium and $D = 0.187$ cm²/s for air. The value used for f, when a rapid scan FT-IR spectrometer is used, should be calculated from $f = vv$ where v is the lowest wavenumber value in the spectrum.

Many samples have moisture contamination which will cause the presence of water vapor in the detector's sample chamber and the appearance of water vapor absorbance bands in spectra. A desiccant for water vapor reduction can be put into a large stainless steel cup (as

FIGURE 7.   Background: large stainless steel cup loaded with desiccant resting on a cup fixture. Foreground: desiccant cup has been placed underneath a slotted brass spacer insert in the sample holder. Sample cups are placed on the spacer. (See Figure 8.)

FIGURE 8.   Polymer pellets in sample cup placed above desiccant cup.

shown in Figure 7) which is placed in the sample holder underneath the sample cup. Figure 8 shows a large cup with polymer pellets which is placed above the desiccant cup. As mentioned in Section III. A., magnesium perchlorate is a very effective desiccant. Water vapor can be eliminated with the combination of purging the detector with dry helium, allowing

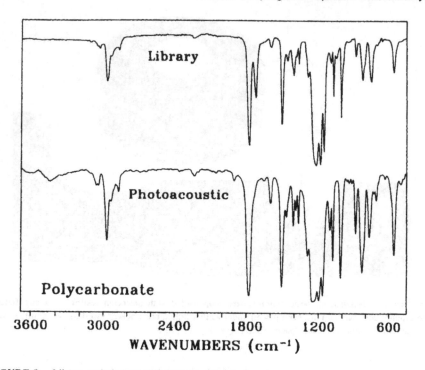

FIGURE 9.   Library and photoacoustic spectra of polycarbonate.

adequate time for the desiccant to work, and minimizing the amount of sample in the cup. Use only enough material to cover the bottom of a small cup if water vapor is a problem. When not in use, desiccant should never be left in cups open to room atmosphere because moisture from the room will collect in the desiccant and form a corrosive liquid.

V. FT-IR/PAS METHODS FOR SPECIFIC ANALYSES

This section provides detailed information on how to measure spectra of important classes of samples and how to process and interpret photoacoustic spectra. All of the spectra were measured at a resolution of 8 cm⁻¹, maximum source aperture, and with a helium gas atmosphere in the detector. The spectra have been normalized by computing a ratio of the sample spectrum to the spectrum of a carbon black standard unless otherwise stated.

A. Qualitative Analysis of Macrosamples

1. Polymer Identification by Computer Search

FT-IR/PAS allows spectra of polymers in powder, pellet, sheet, and chunk form to be directly measured and searched against standard commercial spectral libraries. Figures 4 and 8 show a chunk and pellets, respectively, that have been placed in the sample holder for analysis. Figures 9 through 12 show commercial library spectra[15] and photoacoustic spectra of four common polymers. The photoacoustic spectra have been converted to transmittance to be

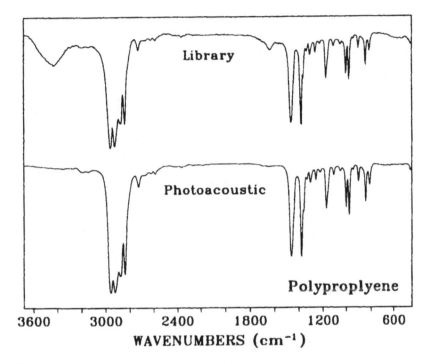

FIGURE 10. Library and photoacoustic spectra of polypropylene.

compatible with the Perkin-Elmer commercial SEARCH™ program. Because different polymer specimens were used when the library and photoacoustic spectra were measured, there are
some variations between them. The degree of band saturation at low values of transmission
vary from spectrum to spectrum but is generally slightly more pronounced in the photoacoustic
spectra since the library samples were in the form of thin films.

The photoacoustic spectra were measured at FT-IR OPD mirror velocities from 0.1 to 0.5
cm/s. A wider velocity range should also be suitable. The spectra were searched against the
Perkin-Elmer SEARCH™ library using the following procedure:

1. After the ratio of sample spectrum to carbon black spectrum is calculated, most FT-IR
   data systems erroneously label the ordinate of the resulting spectrum as "transmittance".
   Scale this spectrum so that its maximum and minimum values are around 90 and 5%
   transmission, respectively.
2. Command the computer to relabel the ordinate in absorbance units.
3. Convert the spectrum to transmission via the FT-IR software.
4. If necessary, rescale the spectrum so that the maximum and minimum are around 90 and
   5% transmission, respectively.
5. Set the search threshold value at 5%.
6. Activate the search which should be based on peak positions only.

The success of this procedure was evident by higher search rankings that the Perkin-Elmer
SEARCH™ software assigned to searches of the photoacoustic spectra than to searches of the
library spectra shown in Figures 9 through 12.

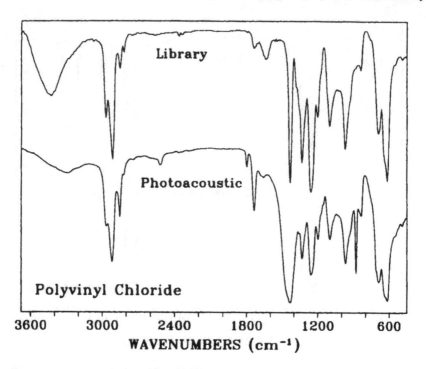

FIGURE 11.   Library and photoacoustic spectra of polyvinyl chloride.

Procedures similar to the one above should be used with other FT-IR systems. A suite of known polymer samples should be used to provide a basis for any refinements that might be necessary for a particular FT-IR software package.

2. Adhesive Spectrum by Spectral Subtraction

Spectral subtraction is a very useful procedure in qualitative analyses in order to separate one component from a background spectrum. If the sample is a two-layer system (such as a coating on a substrate), spectra can be measured separately of the coated and bare sides followed by a spectral subtraction to isolate the coating spectrum. The FT-IR mirror velocity should be set high enough to perform shallow sampling in order to increase the intensity of coating bands in the spectra. Figure 13 shows spectra of an adhesive-coated polyethylene film material, the bare polyethylene, and the adhesive spectra obtained by spectral subtraction. This spectral subtraction approach should be used when it is not possible to isolate the coating sufficiently by increasing mirror velocity alone.

The upper adhesive spectrum of Figure 13 was obtained by a straight spectral subtraction of the two top spectra without scaling one relative to the other. This results in small negative pointing features denoted by stars in the upper adhesive spectrum due to a weakening of the polyethylene substrate photoacoustic signal by the adhesive coating. The weakening is a consequence of the decay of thermal-waves as they cross the coating before reaching the gas. This reduces the amplitude of the polyethylene bands in the top spectrum of the coated-side relative to the bare-side spectrum.

The lower adhesive spectrum of Figure 13 was obtained by interactive spectral subtraction with a small scale change between the two top spectra to correct for the quenching phenomenon prior to subtraction. The correction to the starred features of the lower adhesive spectrum is not exact, probably due to a phase shift associated with the transit time of the

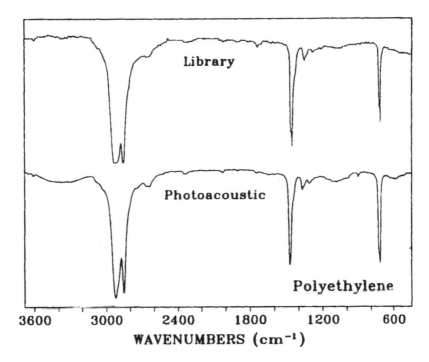

FIGURE 12. Library and photoacoustic spectra of polyethylene.

thermal-wave crossing the coating. These small, remaining artifacts have a derivative-like character associated with slight band shifts. Such small spectral distortions, however, do not prevent a spectral search from predicting the primary adhesive component to be vinyl acetate.

Interactive subtractions may also be necessary when nonlayered samples of varying composition are analyzed if their thermal properties vary significantly with composition. When interactive FT-IR/PAS spectral subtraction is used, it is necessary with many FT-IR data systems to change the data-file ordinate label from transmittance to absorbance units prior to performing the interactive subtraction.

B. Qualitative Analysis of Microsamples

1. Single Textile Fibers

FT-IR/PAS offers some unique advantages for microsample analysis over FT-IR microscopy (for additional details see Chapter 6):

1. No pressing of samples is required to reduce optical density. This avoids the destruction of evidence in forensic analyses and questions of reproducibility and sample alteration posed by the pressing operation. The nondestructive character of FT-IR/PAS allows analyses confirmation by other techniques such as optical or electron microscopy.
2. The delicate optical alignment and aperturing of FT-IR microscopy is avoided because FT-IR/PAS microsampling can be done with a beam focal-spot size that is much larger than the sample since only the portion of the beam that impinges on the sample generates a photoacoustic signal.

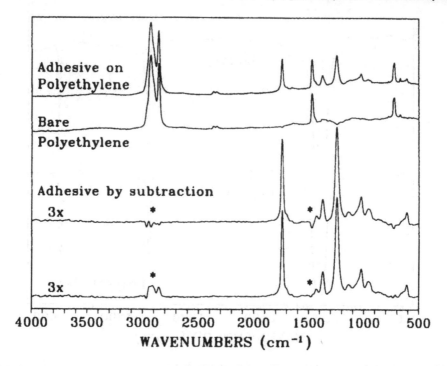

FIGURE 13. Spectra of an adhesive-coated polyethylene material, of bare polyethylene, and of the adhesive obtained by spectral subtraction. The upper subtraction spectrum was obtained by a direct subtraction and the lower by interactive subtraction to remove the negative features at the starred locations.

3. The spectral range of measurements is not limited by the cut-off of narrow band mercury cadmium telluride (MCT) detectors commonly used in FT-IR microscopy. The extended spectral range of FT-IR/PAS microsample spectra allows better spectral differentiation between difficult samples such as nylons.

4. FT-IR/PAS spectra of fibers have more easily discernable detail because normally weak but definitive bands are more prominent due to truncation of strong bands by photoacoustic signal saturation and to enhanced structure at low wavenumbers from more efficient signal generation at low frequencies.

A key consideration in FT-IR/PAS microsample analyses is mounting the sample so that it is surrounded by helium gas for efficient signal generation. Figure 14 shows devices that are used to mount single fibers and particles for FT-IR/PAS analyses. Tweezers are used to mount fibers in spring-loaded split rings before placement in the sample cup. In front of the sample holder is a fixture which holds the ring with either a white or black centering pin to provide a contrasting background during mounting of fibers. In the procedure, fibers are either attached by pressure-sensitive adhesive to the arms of the device shown in Figure 14 in front of the ring fixture, or fibers can be manipulated with tweezers during mounting. This operation can be done using an illuminated bench magnifier with a contrasting background. A microscope is not required. In Figure 15, a black thread depicts a fiber being mounted and placed in the sample holder over a covered desiccant cup. The desiccant cup is shown in Figure 14 on the left side of the ring fixture. A magnesium perchlorate desiccant and a dry helium gas atmosphere are required in microsampling because otherwise the water vapor

FIGURE 14. Components of the MTEC microsampling system for particles and fibers as described in the text.

signal, which is out-of-phase with the sample signal, can dominate the sample signal and seriously degrade the spectrum.

Figure 16 shows an alternative fiber mounting method that uses rings with an adhesive to attach fibers. In this approach, the fiber is first stretched between two pieces of double-stick tape that are attached to a light or dark release paper to give a contrasting background for the fiber. A ring with an outer ring of double-stick tape is centered over the fiber and pressed against it. The ring sticks to the fiber but not to the release paper. The fiber is then cut at the outer diameter of the ring, and the ring is placed in the sample holder with a second ring on top of it to mask the adhesive from the infrared beam. This second ring thereby prevents generation of a photoacoustic signal from the adhesive.

The brass rings used for this mounting technique are 1.5 mm thick and have outer and inner diameters of 10.6 and 5.0 mm, respectively. The adhesive used is 3M double-stick tape which is cut into rings by first applying it to a release paper of the type commonly used to cover adhesive surfaces of labels. Cork borers (sizes 4 and 6) are used to cut the adhesive rings which are then transferred to the brass rings. The inner diameter of the adhesive ring is larger than that of the brass ring. This helps to prevent the infrared beam from impinging on the adhesive and generating an adhesive spectrum. Adhesive-coated rings can be reused a number of times if care is taken to remove the previously mounted fiber.

The best results on single fibers are achieved with low FT-IR mirror velocities which provide a relatively stronger fiber vs. vapor signal. Fibers as small in diameter as 10 μm are run routinely. Smaller diameter fibers should also be practical for FT-IR/PAS analysis but are not commonly presented for analysis.

FIGURE 15.   Black thread depicting a fiber being mounted and placed in the sample holder over a covered desiccant cup.

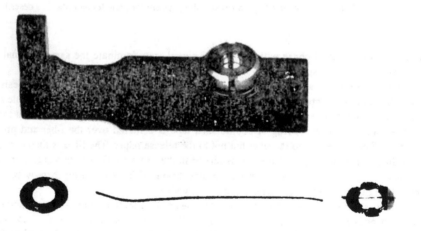

FIGURE 16.   Fiber mounting using a ring with an adhesive to secure the fiber.

FT-IR/PAS spectra of the common types of nylon fibers appear in Figure 17. These fibers are specimens from a fiber library assembled by the U.S. Federal Bureau of Investigation (FBI) forensic science program.[16] The strong absorbance bands of the FT-IR/PAS spectra are truncated due to photoacoustic signal saturation, but this does not detract from their value for qualitative analyses. Spectra of nylon fibers which are not in the FBI collection have been identified successfully using a FT-IR/PAS spectral library of the FBI fibers for computer searching. The fibers used to produce the two top spectra of Figure 18 are not in the FBI

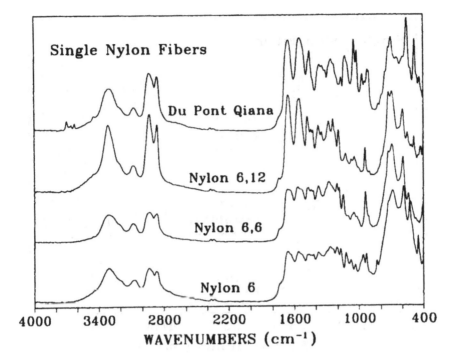

FIGURE 17.   FT-IR/PAS spectra of four common types of nylon single fibers.

collection but were identified as nylon 6,6 when computer-searched against the library. The lower spectrum of Figure 18 is the nylon 6,6 spectrum from the library. The peak positions of the two upper spectra are well matched to those of the lower library spectrum. The two upper spectra also show that variations are observed in relative peak intensities and shapes due to differences in fiber diameter and cross-sectional geometry. These variations provide additional information beyond just polymer type and could be useful in establishing that two fibers in a forensic investigation are from a common source.

The FBI collection consists of approximately 50 fibers and includes all of the common types of polymers. Over 20 FT-IR/PAS spectra have been searched successfully against an FT-IR/PAS spectral library of the FBI fibers without any false identifications.

2. Single Coal Particles

Microparticles are supported on very fine tungsten needles for FT-IR/PAS analysis. Manipulation of microparticles by tungsten needles is a common practice in optical microscopy[17] and is often done by hand under a microscope by experienced microscopists. Particles usually attach to the needle tip by electrostatic attraction. In some instances when humidity is high, particles are best attached using a speck of electrically conducting colloidal graphite glue that can be obtained from distributors of microscopy supplies. The glue is made to flow using isopropanol, then the particle is rapidly touched before the glue dries. This latter method of attachment allows for both FT-IR and SEM analysis of microsamples after mounting.

Microparticles are manipulated and mounted for FT-IR/PAS analysis using items shown in the right foreground and background of Figure 14. The sample holder in the photograph contains a desiccant holder, tungsten needle socket, and polished conical insert that are used

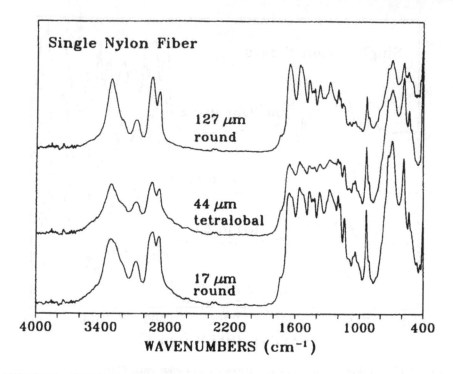

FIGURE 18. FT-IR/PAS spectra of single nylon 6,6 fibers with different cross-sectional geometries and fiber diameters.

in FT-IR/PAS particle analysis. In the right foreground and background are a tungsten needle storage holder and a micromanipulator for the needles, respectively. The micromanipulator attaches to a laboratory microscope as shown in Figure 19 and allows precise control of the needle point.

The best FT-IR/PAS spectra of microparticles are obtained with low mirror velocities, a helium gas atmosphere, and adequate time for the desiccant to work. It is possible to measure spectra of single particles as small as 25 µm in diameter with a 2X reduction mirror to concentrate the infrared beam intensity on the sample and an OPD FT-IR mirror velocity of 0.05 cm/s. FT-IR/PAS analysis of smaller particles should be possible with higher concentration of the beam intensity on the sample and lower mirror velocity.

Spectra of three single coal particles are plotted in Figure 20 with a macrosample spectrum of the same coal at the top of the figure. The coal particle size is between 125 and 150 µm. The particle spectra show that there are a number of compositional differences in coal particles in this size range related to mineral, hydroxyl, and aromatic/aliphatic concentrations.

C. Quantitative Analyses

The application of factor analysis for processing FT-IR/PAS spectra enables quantitative analyses to be readily performed with standard error of prediction (SEP)[18, 19] values below 1%, in spite of the truncation of strong absorbance bands that occurs due to photoacoustic signal saturation. Both principal components regression (PCR)[20] and partial least squares (PLS)[20] factor analysis routines tolerate nonlinearities in spectra well and allow concentrations of

FIGURE 19.   Micromanipulator attached to a laboratory microscope. The manipulator allows precise control of the tungsten needle which is used to pick up particles for analysis.

multicomponent systems to be determined as well as other physical and chemical properties of materials. The quality of the analysis of the FT-IR/PAS technique is sensitive to the same considerations as other FT-IR sampling methods. (For additional information, see Chapter 8.) These items should be considered when a new quantitative method is being developed for a specific application:

1. The number of learning set samples used and the differences between their physical or chemical property values and those of the unknowns
2. The provision for several learning set samples with physical or chemical property values above and below those of the unknowns
3. The accuracy of the learning set
4. The spectral range or ranges used in the factor analysis
5. The spectral resolution of spectra
6. The SNR of spectra
7. The system response stability over time and the reproducibility from sample to sample

The first six considerations are very important to the accuracy of quantitation but are not specific to the sampling method used and will not be discussed in detail here. The last consideration depends the FT-IR, the sampling accessory, the samples, and the background spectrum. The following conditions must remain constant or be accounted for in quantitative FT-IR/PAS analyses:

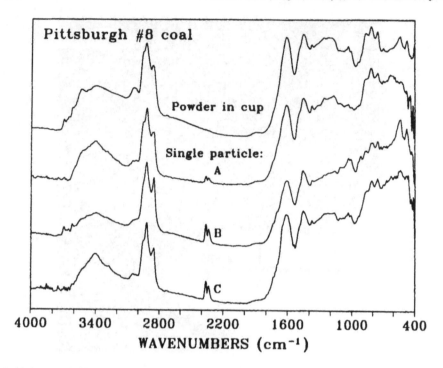

FIGURE 20. FT-IR/PAS spectra of coal in powder (upper) and single particle form (A, B, and C). Variations in the mineral content of the particles are indicated by bands in the 400 to 1000 cm⁻¹ and 3600 to 3700⁻¹ cm regions.

1. FT-IR mirror velocity
2. FT-IR and PAS detector amplifier gain settings
3. PAS detector optical alignment
4. FT-IR source intensity
5. FT-IR interferometer alignment
6. Helium gas concentration in the PAS detector
7. Sample volume
8. Sample morphology
9. Sample matrix
10. Microphone sensitivity
11. Carbon black standard response

The FT-IR mirror velocity must be set at the same value when spectra of the calibration standards (learning set) and of unknowns are acquired. If the mirror velocity setting is not maintained, changes in the PAS sampling depth will occur which are analogous to changing sample thickness or concentration in transmission spectroscopy.

All amplifier gain settings should be at the same values when sample spectra are acquired in order to have consistent measurement conditions.

Good quantitative analyses also require that the detector positioning in the FT-IR is reproducible when the accessory is put in and taken out of the FT-IR and that its position remains fixed when the accessory is installed. Properly designed kinematic mount and baseplate registration accomplish these requirements.

TABLE 4. Analysis Conditions for Determinations of Vinyl Acetate in
Polyethylene Copolymers and Ash in Coal

Condition	Vinyl acetate	Ash
Number of scans	128	64
OPD mirror velocity (cm/s)	0.10	0.05
Resolution (cm−1)	8	8
Number of standards	13	11
Concentration range	8.81–51%	6.0–31.6%
Type of standards	Pellets of different sizes	-200 mesh powder
Method of analyzing standards	Titration	ASTM ash analysis
Type of factor analysis	PLS[a]	PCR[b]
Spectral ranges of factor analysis	578–1596 cm^{-1} 1630–1900cm^{-1} 2479–3150 cm^{-1}	720–1818 cm^{-1}
Spectrum standardization	Area of CH band 2750–3120	none
Number of data sets averaged	3	1
SEP	0.55% (mass)	0.44% (mass)

Note: Standard errors of prediction (SEP) are given for cross-correlations of the standard set.

[a] Partial least squares.
[b] Principal components regression.

   Scale variations are caused in FT-IR/PAS spectra if changes occur in source intensity, interferometer alignment, helium concentration, sample volume, morphology, matrix, microphone sensitivity, and carbon black response. In some instances, if proper care is taken, these changes can be held small enough to yield satisfactory quantitative analyses. But in most instances, adding a standardization procedure that addresses all of these potential changes is the best way to insure that quantitative results will have the degree of reproducibility necessary for a particular application.

   FT-IR/PAS spectra are usually best standardized by one of two methods. The first is applicable to situations where the spectral changes over the full analyte range are small relative to the whole spectrum. In this case, the spectra can be scaled so that the area under the whole spectrum from 400 to 4000 cm^{-1} is held constant.

   The second method exploits the fact that highly saturated bands in FT-IR/PAS spectra vary much slower with changes in analyte concentration than do unsaturated bands. Consequently, FT-IR/PAS spectra can be standardized by scaling spectra so that a particular intense band's height or area remains constant.

   Both of the standardization methods appear to work well with factor analysis routines. Standardization methods should be tested and adjusted, if necessary, after a cross-correlation test of the calibration spectra. This test involves using each calibration standard, in turn, as an unknown and the others as the learning set, and then calculating the standard error of prediction.[18, 19]

   In the following two examples, quantitative analyses are performed in one case with and in the other without standardization of spectra.

1. Vinyl Acetate Concentration in Polyethylene Copolymer Pellets
   Vinyl acetate concentrations in pellets of varying sizes are determined by FT-IR/PAS and a PLS factor analysis routine (Spectra-Calc, Galactic Industries Corp., Salem, NH) without

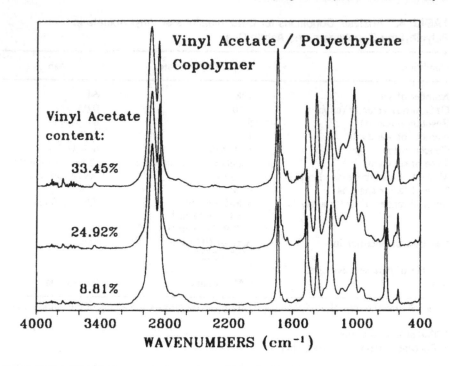

FIGURE 21.   FT-IR/PAS spectra of three compositions of vinyl acetate/polyethylene copolymers.

sample preparation. The direct analysis of pellets is complicated by variations in sample volume and morphology. This variability and the other potential changes listed above are accounted for by using the area under the C–H band (2750 to 3120 cm⁻¹) for standardization of spectra. The analysis conditions are given in Table 4. A cross-correlation analysis of the standards set resulted in a SEP value of 0.55% (mass). The reproducibility of the PLS analysis was tested by twice running spectra on pellets of three concentrations. The second run involved reloading the same pellets in the sample cup followed by the purge and seal operation, and the spectrum acquisition. The average values of the vinyl acetate concentrations were 5.89, 9.49, and 16.27%, and the differences between the two analysis values for each concentration were 0.12, 0.21, and 0.00%, respectively. The manufacturer's stated concentrations for the samples were 5.84, 9.36, and 14.90%. All of the data are consistent with the cross-correlation analysis SEP except for the highest concentration value reported by the manufacturer.

Figure 21 shows typical FT-IR/PAS spectra of two vinyl acetate-polyethylene copolymer samples that were in the learning set. The cross-correlation analysis results are plotted in Figure 22.

2. Ash Concentration in Coal

In industry, coal is cleaned by various processes in order to reduce the concentration of ash which causes boiler fouling and environmental problems. The froth flotation cleaning process produces float material with reduced ash concentration and a sink material with high ash content. FT-IR/PAS can be used to rapidly check the ash concentration in a coal cleaning operation using standards that are synthesized from float and sink material. The analysis conditions are given in Table 4. The coal samples were in fine powder form and were

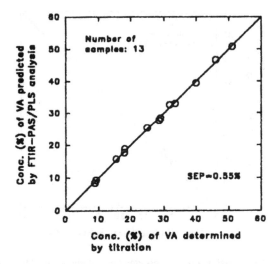

FIGURE 22. Cross-correlation plot of vinyl acetate concentration determined by titration against predicted concentration by FT-IR/PAS/PLS analysis for pellet vinyl acetate/polyethylene copolymer samples.

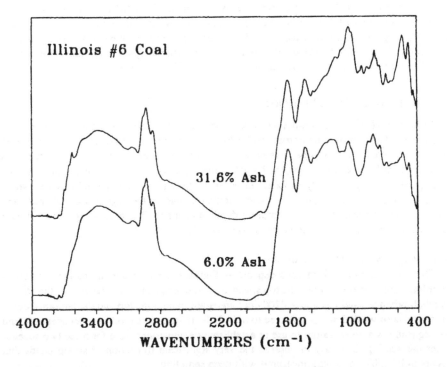

FIGURE 23. FT-IR/PAS spectra of two typical samples of float (6.0% ash) and sink (31.6% ash) material from cleaning Illinois #6 coal.

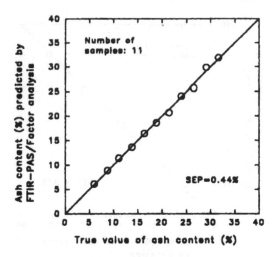

FIGURE 24. Cross-correlational plot of ash concentration determined by ASTM ash analysis against predicted concentration determined by FT-IR/PAS/factor analysis of cleaned Illinois #6 coal.

formulated to have different ash concentrations by mixing calculated weights of float and sink material. Portions of the mixtures were measured out for FT-IR/PAS analysis by volume simply by filling small stainless steel sample cups as shown in Figure 5 and sliding a spatula across the cup rim to remove excess powder. Figure 23 shows two typical spectra of the float (6% ash) and sink (31.6% ash) material that were used in the analyses. The results of the cross-correlation analysis are plotted in Figure 24. A SEP of 0.44% (mass) was obtained.

D. Variation of Sampling Depth

In Section II, signal generation considerations related to sampling depth variation were discussed and an approximate expression for the sampling depth of FT-IR/PAS measurements was introduced. In this section, examples are presented where spectra are acquired with different sampling depths by varying FT-IR mirror velocity. The first example deals with a homogeneous sample of polycarbonate in order to show that some spectral changes which occur with varying sampling depth are due solely to changes in the degree of absorbance band truncation that appears when the sampling depth is changed. The other examples deal with samples which do have varying compositions with depth.

1. Homogeneous Polycarbonate

The top and bottom polycarbonate spectra of Figure 25 are measured at shallow- and deep-sampling depths, respectively. The top spectrum is reduced so that the heights of the starred small absorbance band just above 2200 cm^{-1} are equal in both the top and the bottom spectra. This band is so weak that it is expected to have little, if any, truncation in either spectrum and consequently is a good feature to use in establishing a consistent scale for these two spectra. After this scaling, it is easy to observe the very significant truncation of strong bands that occurs in FT-IR/PAS spectra measured with deep sampling.

The middle spectrum of Figure 25 is expanded so that the strongest band denoted by the star just above 1200 cm^{-1} has the same height as this band in the bottom spectrum. Many of the absorbance band heights in the middle spectrum are now observed to exceed those in the

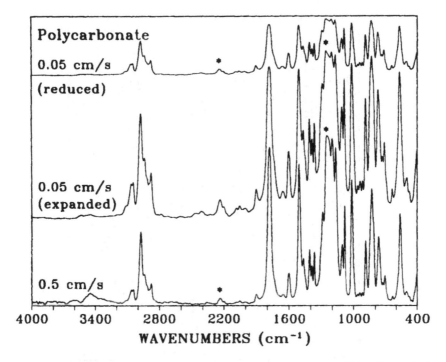

FIGURE 25. FT-IR/PAS spectra of polycarbonate measured at OPD mirror velocities of 0.05 cm/s (deep-sampling depth) and 0.5 cm/s (shallow sampling depth). The upper spectrum has been reduced so that the starred weak band height just above 2200 cm¹ is equal to that of the 0.5 cm/s spectrum in order to show the truncation of strong bands that occurs with deep sampling. The middle spectrum is scaled so that the height of the starred strong band height just above 1200 cm⁻¹ is equal to that of the 0.5 cm/s spectrum.

bottom spectrum as, for example, the band at 1600 cm⁻¹. This observation should not, however, be interpreted as an increase in species concentrations with depth. It is instead due to inappropriate scaling, for comparative purposes, of one spectrum relative to the other.

2. Mylar-Coated Polycarbonate

The next example is a sample with a depth-varying composition consisting of a 2.5 μm thick coating of mylar on a polycarbonate substrate. The three top spectra of Figure 26 show that as sampling depth is increased, more and more of the substrate spectra appear. As the FT-IR mirror velocity is decreased from 1.5 to 0.5 cm/s and finally to 0.05 cm/s, the polycarbonate band at 1775 cm⁻¹ is seen to grow relative to the mylar band at 1725 cm⁻¹. Wherever there are bands of little or no spectral overlap, similar behavior is observed allowing differentiation between a number of substrate and coating features.

3. Cure of an Acrylic Coating

The sampling depth variability of FT-IR/PAS measurements is especially useful in analysis of coatings where it is desirable to enhance the coating spectrum relative to the substrate spectrum. A typical example is monitoring of cure due to ultraviolet-induced polymerization of an acrylic coating on polycarbonate. The spectra of Figure 27 show the cure monitoring as indicated by reduction of the cure band height with increasing polymerization. All of the

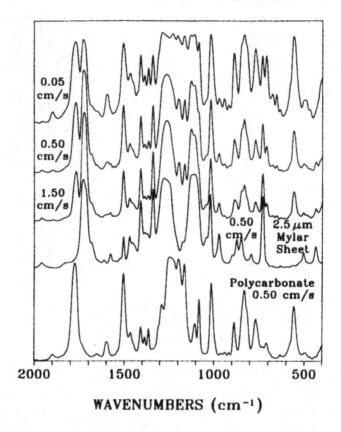

WAVENUMBERS (cm⁻¹)

FIGURE 26.   FT-IR/PAS spectra of a 2.5 μm mylar coating on a polycarbonate substrate measured with increasingly deep-sampling depths going from mirror velocities of 1.50, 0.50, and 0.05 cm/s. Spectra of the mylar sheet and polycarbonate alone are shown for comparison with the upper spectra. Note the increasing prominence of the polycarbonate substrate band at 1775 cm⁻¹ as the sampling depth increases.

spectra in Figure 27 are scaled so that the polycarbonate peak heights at 1600 cm⁻¹ are equal. Considerably more contrast is observed in the cure bands measured by shallow-sampling than by deep-sampling. Consequently, the more coating-specific, shallow-sampling FT-IR/PAS analysis provides the most sensitive monitor of the degree of coating cure.

4. Plastic-Coated Paper

Plastic-coated paper is another example of a layered material. The material used in this example has a 35 μm thick coating and was analyzed by both FT-IR/PAS and DRIFTS in order to probe deeper into this rather thickly coated material than can be achieved by FT-IR/PAS alone. A MTEC multisampling option (pictured in Figure 28) was used to measure the DRIFTS spectrum. This option allows DRIFTS, PAS, and transmittance spectra to be obtained rapidly by interchanging sampling heads shown in the foreground of Figure 28 without changing FT-IR accessories.

Spectra of the paper obtained with shallow- and deep-sampling PAS and DRIFTS are shown in sequence going from bottom to top in Figure 29. The shallow-sampling photoacoustic

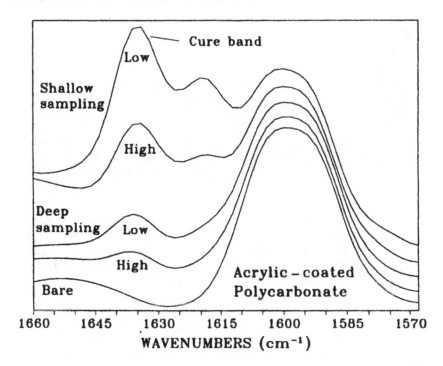

FIGURE 27. FT-IR/PAS spectra of an acrylic-coated polycarbonate with high and low degrees of cure that are measured with shallow and deep sampling. The intensity of the cure band decreases going from a low to a high degree of cure. Note that shallow sampling is most sensitive to the cure process because it enhances the coating vs. the substrate spectrum. All spectra are scaled so that the substrate band at 1600 cm⁻¹ are of equal height.

spectrum measured at an OPD velocity of 0.5 cm/s is dominated by the absorbance bands of the 35 μm thick coating. It is possible in this case to essentially isolate the plastic spectrum by shallow-sampling alone, without spectral subtraction due to the coating's thickness and thermal properties.

The deep-sampling spectrum, measured at an OPD velocity of 0.05 cm/s, reveals additional absorbance bands of the paper such as the feature at 900 cm⁻¹. The DRIFTS spectrum, plotted in Kubelka-Munk units, probes still deeper into the paper and shows absorbance bands, including the one at 3700 cm⁻¹ that are not observed in the photoacoustic spectra. The DRIFTS spectrum also contains artifacts such as the dip at approximately 1260 cm⁻¹. This feature is due to absorbance by the plastic coating and should be a peak rather than a dip. In spite of some artifacts, the information in the DRIFTS spectrum gives spectral information from deeper within the sample than FT-IR/PAS probes at a mirror velocity of 0.05 cm/s. New step-scan FT-IR interferometers provide considerably lower mirror velocities which will extend FT-IR/PAS sampling depth by over a factor of three. This advance will yield sampling depths comparable to or greater than DRIFTS on many classes of samples.

5. Chemically Treated Surfaces of Polystyrene

A final example in this topic deals with infrared analysis of chemically altered surfaces of polystyrene spheres which are approximately 0.5 mm in diameter. Both FT-IR/PAS and

FIGURE 28. MTEC multisampling options with interchangeable sampling heads shown in the foreground (left to right) for diffuse reflectance (DRIFTS), photoacoustic (PAS), and transmission sampling. The sampling heads for DRIFTS and transmission operation have carbon black absorber elements incorporated within to sense the fraction of the infrared beam that is diffusely reflected or transmitted, respectively.

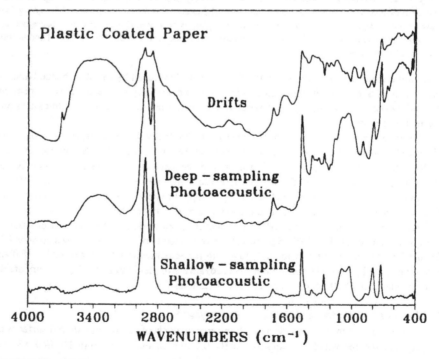

FIGURE 29. FT-IR spectra of a plastic-coated spectra measured by PAS and DRIFTS sampling. The sampling depth of the spectra increases from the bottom to the top of the figure.

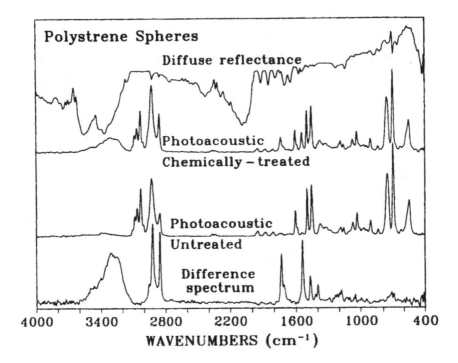

FIGURE 30. FT-IR spectra of polystyrene spheres with and without chemically treated surfaces. In this case, the very deep-sampling depth of DRIFTS results in grossly distorted spectra. FT-IR/PAS allows the spectrum due to the chemical surface treatment to be observed after spectral subtraction. The samples were provided by Prof. R. A. Kellner of the Institute of Analytical Chemistry, Technical University of Vienna.

DRIFTS spectra are plotted in Figure 30. The DRIFTS spectrum shows very expanded weak absorbance bands, inverted or derivatized strong bands, and some absent bands. The FT-IR/PAS spectra measured at an OPD mirror velocity of 0.5 cm/s show both polystyrene bands and bands due to the chemical treatment. In this instance, it was not possible to isolate the chemically treated layer from the underlying polystyrene. Consequently, spectral subtraction using a spectrum of untreated polystyrene was necessary in order to isolate the absorbance bands due to the chemical treatment.

E. Carbon-Filled Materials

Materials with significant concentrations of carbon in fiber or powder form are highly opaque and consequently are difficult to analyze by infrared spectroscopy. These materials, however, are of considerable industrial importance, and their infrared spectra provide very useful information for research and development and for production operations.

Baseline slope is often present in spectra of very strongly absorbing samples. It is magnified by the need to expand the ordinate scale of spectra to a high degree in order to see absorbance bands above a strong background absorption due, for instance, to carbon black.

There are two basic mechanisms that lead to baseline slope with strongly absorbing samples. The first is due to light scattering within the sample which causes the baseline to slope

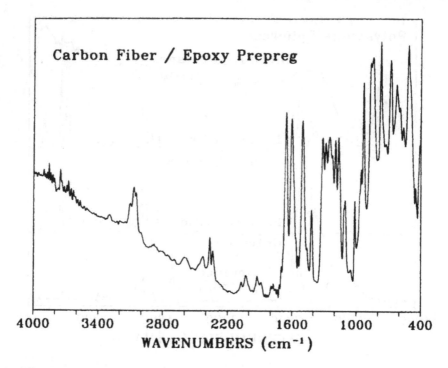

FIGURE 31. FT-IR/PAS spectrum of a carbon fiber/epoxy prepreg used in composite materials. The spectrum has a sloping baseline due to light scattering within the sample. Absorbance bands appear in the spectrum between 1400 to 1900 cm^{-1} and 3500 to 3900 cm^{-1} due to moisture in the sample chamber. If more time is allowed for purging and for a desiccant to work, these bands can be eliminated.

upwards with increasing wavenumber. The second is due to differences in thermal response as a function of infrared beam modulation frequency between the sample and the reference standard used to ratio spectra. For instance, there is usually less baseline slope with a solid slab sample geometry if the reference standard has the same geometry. Consequently, a slab of glassy carbon or graphite is a preferable reference standard to carbon black in either a thin coating or powder form if a slab of material is being analyzed.

1. Composite Material Prepreg

Carbon fiber/epoxy prepregs are too opaque for transmission spectroscopy but yield a good FT-IR/PAS spectrum of the epoxy component as is shown in Figure 31. This spectrum was obtained on a 7 x 7 mm square of material cut with scissors. The plot shows a baseline that slopes upward with increasing wavenumber which is characteristic of light scattering within the sample.

2. Automobile Tire

Figure 32 shows a FT-IR/PAS spectrum of a typical automobile tire with a high carbon-black concentration. Tire samples are among the most difficult materials from which to obtain infrared spectra. The spectrum of Figure 32 was obtained on a slab of tire cut out with a razor blade. For sample and reference spectra 20,000 FT-IR scans were co-added at an OPD mirror velocity of 0.5 cm/s and 8 cm^{-1} resolution.

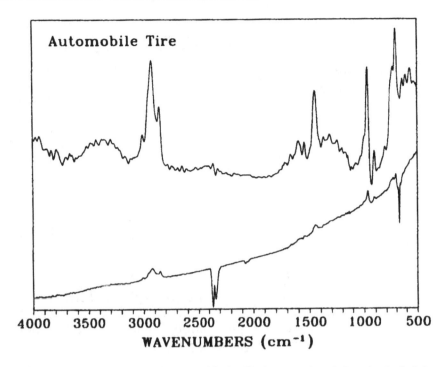

FIGURE 32. FT-IR/PAS spectra of an automobile tire. The lower spectrum is the ratio of a single beam tire spectrum to a single beam glassy carbon spectrum. The upper spectrum is the result of baseline flattening, subtraction of CO_2 bands and blanking of another gas feature between 2000 and 2100 cm^{-1}, and a 19 point smoothing.

F. Polymer Films

1. Elimination of Interference Fringes by FT-IR/PAS

Interference fringe bands are often observed in infrared transmission spectra of polymer films. These bands interfere with and obscure important small features in spectra associated with various types of additives. FT-IR/PAS spectra are free of interference artifacts due to the differences in signal generation between an absorption vs. transmission-based measurement.

Figure 33 shows spectra measured by the conventional transmission method and by FT-IR/PAS of the same polyethylene film. The spectra were measured directly on the film material after punching out a disk of the film with a cork borer. The FT-IR/PAS spectrum has been plotted as a transmission spectrum using the procedure discussed in Section V.A.1. Many of the small- and medium-sized features of the conventional transmission spectrum are so seriously distorted that they are not discernable for a qualitative analysis and that observable features are not useful for quantitative analyses. For instance, it is not possible to observe the small band at 2330 cm^{-1} in the transmission spectrum, and the larger band at 1100 cm^{-1} is not suitable for quantitative analysis due to the fringe bands.

2. Polarized Beam Measurements on Oriented Films

In production, polymer fibers are oriented by drawing in order to improve tensile strength. Polarized FT-IR/PAS measurements are useful in monitoring the orientation process. A

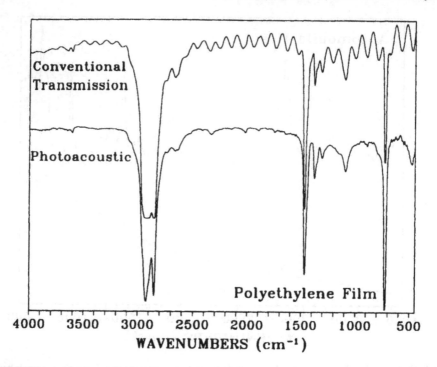

FIGURE 33. Conventional transmission and FT-IR/PAS spectra of a 32 μm thick polyethylene film. The PAS spectrum has been converted to transmittance for the purpose of comparison. Note the absence of interference fringes in the PAS spectrum.

polarizer is placed in front of the photoacoustic detector so that the infrared beam incident on the sample in the detector sample holder is polarized. It may be useful to initially rotate the polarizer until the signal is maximized with a carbon-black reference in the detector. This procedure will ensure that the polarizations imposed by the FT-IR optics and the polarizer are in coincidence resulting in maximal intensity in the polarized beam.

After this adjustment, the polarizer should be left fixed and the sample should be rotated in the sample cup. Figure 34 shows spectra of a PET film with the infrared beam polarized parallel and perpendicular to the direction of draw. A number of spectral changes are observed as a function of orientation particularly in the 800 to 900 cm^{-1} and 1300 to 1400 cm^{-1} ranges. The absence of interference fringes in FT-IR/PAS spectra is obviously also an advantage in this application.

3. Polymer Coatings on Metal Containers

Polymer films are often used as a barrier coating on beverage and food containers. Figure 35 shows a FT-IR/PAS spectrum of a polymer coating on the interior of a beverage can. The spectrum was measured directly on a small specimen cut with scissors from the can. The surface morphology of such a specimen is of no consequence in a FT-IR/PAS measurement, whereas surface flatness is critical in a specular reflectance measurement in that the reflected fraction must be focused on the detector.

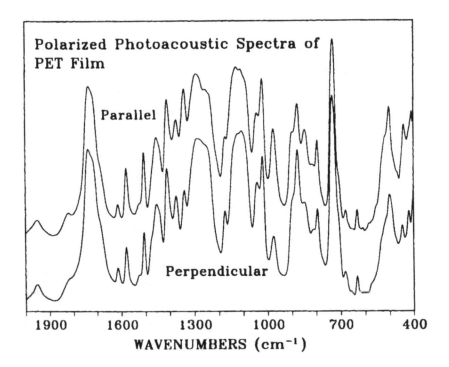

FIGURE 34. FT-IR/PAS polarized light spectra of a 31-μm thick PET film which has been drawn in production to induce molecular orientation. Spectra were measured with the direction of draw-oriented parallel and perpendicular to the polarization direction of the infrared beam.

G. Semisolids, Gels, and Liquids

FT-IR/PAS can be applied to semisolid, gel, and liquid samples when considerations of light scattering or contact with infrared window materials makes transmission or ATR methods impractical. In these instances, only enough sample should be used to cover roughly a 5-mm diameter area centered in a stainless steel cup or other vessel. If the sample is aqueous, a desiccant should be used under the sample cup as discussed in Section III.A. Lower mirror velocities will help in reducing the intensity of vapor spectra which may be superimposed on the sample spectrum. If necessary, spectral subtraction can be used to remove the vapor spectrum. A spectrum of the vapor alone can be obtained by covering the sample with a small disk of aluminum foil and thereby masking the sample from the infrared beam. The vapor will still be present above the foil and will produce only a vapor spectrum.

Figure 36 shows a spectrum of an aqueous collagen gel, which is primarily water obtained with a desiccant but without spectral subtraction to remove the small amount of moisture and CO_2 spectra that are present. The spectrum clearly shows the absorbance bands of the non-water species.

Care should be taken when vapor species are associated with samples to avoid vapors that might damage or contaminate the photoacoustic detector.

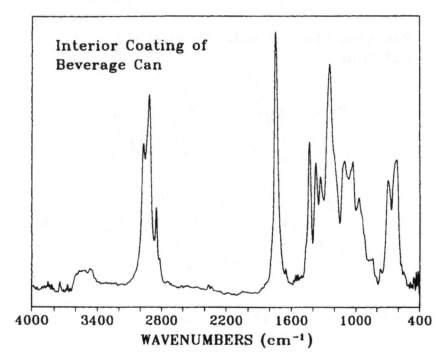

FIGURE 35. Infrared spectrum of the internal coating of a beverage can measured by FT-IR/PAS.

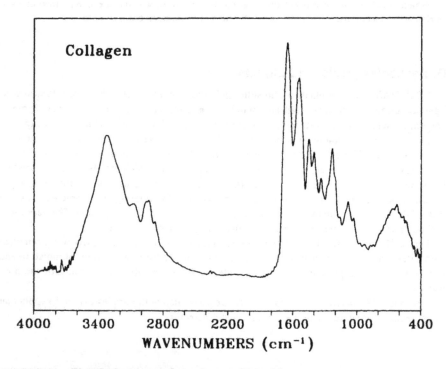

FIGURE 36. FT-IR/PAS spectrum of an aqueous collagen gel.

VI. CONCLUSION

PAS provides the FT-IR spectroscopist with a rapid nondestructive means of directly obtaining spectra of materials without traditional sample preparation to reduce opacity. The PAS method is applicable to all types of samples in macro and micro forms. FT-IR/PAS spectra have the same absorbance peak wavenumber locations as classic transmission spectra but usually have truncation of strong absorbance bands due to photoacoustic signal saturation. The presence of band truncation, however, has not been found in applications explored to date to limit the capability of FT-IR/PAS in either qualitative or quantitative determinations based on commercial FT-IR search and factor analysis software, respectively.

FT-IR/PAS has the unique capability of being able to vary sampling depth by changing the modulation frequency of the FT-IR beam. This capability allows, for instance, measurement of spectra with either high surface specificity to analyze a coating, or with bulk specificity to observe the absorbance bands of a substrate.

Other important aspects of FT-IR/PAS measurements include operation over all spectral regions, absence of interference fringes, elimination of microsample pressing and aperturing, and capability to switch between sampling modes by interchanging photoacoustic detector sampling heads for DRIFTS, PAS, and transmission analyses.

ACKNOWLEDGMENTS

This work was funded by MTEC Photoacoustics, Inc.; by the Center for Advanced Technology Development, which is operated by Iowa State University for the U.S. Department of Commerce under Grant No. ITA 87-02; and by Ames Laboratory, which is operated by Iowa State University for the U.S. Department of Energy under Contract No. W-7405-ENG-82, and supported by the Assistant Secretary for Fossil Energy.

We wish to thank the following people for supplying samples used in this work: J.B. Callis (University of Washington), L. Bright (DuPont), H.L.C. Meuzelaar (University of Utah), M.W. Tungol and E.G. Bartick (FBI), P. Milne (University of Miami), P.J. Codella (General Electric), B.J. Slomka and W.H. Buttermore (Ames Laboratory), and R. Kellner (Technical University of Vienna).

REFERENCES

1. Colthup, N.B., Daly, L.H., and Wiberley, S.E., *Introcution to Infrared and Raman Spectroscopy*, 3rd ed., Academic Press, New York, 1990.
2. *Laboratory Methods in Infrared Spectroscopy*, 2nd ed., Miller, R.G.J. and Stace, B.C., Eds., Heyden and Sons, London, 1979.
3. *Practical FT-IR Spectroscopy: Industrial and Laboratory Chemical Analysis*, Ferraro, J.R. and Krishnan, K. Eds., Academic Press, New York, 1990.
4. Osborne, B.G. and Fearn, T., *Near Infrared Spectroscopy in Food Analysis*, Longman Scientific and Technical, Essex, U.K., 1986.
5. Graham, J.A., Grim, W.M., III, and Fateley, W.G., Fourier Transform Infrared Photoacoustic Spectroscopy of Condensed-Phase Samples, in *Fourier Transform Infrared Spectroscopy*, Vol. 4, Ferraro, J.R. and Basile, L.J. Eds., Academic Press, Orlando, 1985.
6. Griffiths, P.R. and de Haseth, J.A., *Fourier Transform Infrared Spectrometry*, John Wiley & Sons, New York, 1986, 45.

7. Crocombe, R.A. and Compton, S.V., *The Design, Performance and Applications of a Dynamically-Aligned Step-Scan Interferometer, FTS/IR Notes No. 82*, Bio-Rad, Digilab Division, Cambridge, MA, 1991.
8. Michaelian, K.H., *Appl. Spectrosc.*, 43, 185, 1989.
9. McDonald, F.A., *Am. J. Phys.*, 48, 41, 1980.
10. Carslaw, H.S. and Jaeger, J.C. *Conduction of Heat in Solids*, Clarendon, Oxford, 1959.
11. Dittmar, R.M., Chao, J.L., and Palmer, R.A., *Appl. Spectrosc.*, 45, 1104, 1991.
12. Oelichmann, J. and Freseuius, Z., *Anal. Chem.*, 333, 353, 1989.
13. Kinsler, L.E. and Frey, A.R., *Fundamentals of Acoustics*, 2nd ed., John Wiley & Sons, New York, 1950, 186.
14. Griffiths, P.R. and de Haseth, J.A., *Fourier Transform Infrared Spectrometry*, John Wiley & Sons, New York, 1986.
15. *Perkin-Elmer Polymer Library*, D.O. Hummel, Perkin-Elmer, Norwalk.
16. Tungol, M.W., Bartick, E.G., and Montaser, A., *Appl. Spectrosc.*, 44, 543, 1990.
17. McCrone, W.C. and Delly, J.G., *The Particle Atlas: An Encyclopedia of Techniques for Small Particle Identification*, 2nd ed., Ann Arbor Science, Ann Arbor, 1973–1980.
18. Haaland, D.M. and Thomas, E.V., *Anal. Chem.*, 60, 1202, 1988.
19. Haaland, D.M. and Thomas, E.V., *Anal. Chem.*, 60, 1193, 1988.
20. Malinowski, E.R. and Hovery, D.G., *Factor Analysis in Chemistry*, John Wiley & Sons, New York, 1980.

Chapter 6

THE VERSATILE SAMPLING METHODS OF INFRARED MICROSPECTROSCOPY

Patricia L. Lang and Lisa J. Richwine

CONTENTS

I. Introduction 146
II. Types of Experiments Done with a Typical IR Microscope 147
   A. Experimental 147
   B. Transmission Collection Mode 147
   C. Reflection Collection Mode 148
III. Experimental Factors and Sampling Considerations 153
   A. Stereoscopic Examination and Manipulation 153
   B. Sampling Away Unwanted Optical Effects 153
      1. Total Absorption 153
      2. Diffraction 155
      3. Scattering 158
      4. Reflection 158
      5. Refraction 159
   C. Sample Purity and Homogeneity 159
   D. Accessory Alignment 160
IV. Parameters to Report when Publishing 161
V. Other Special Techniques 161
   A. Emission Microspectroscopy 161
   B. Grazing Angle and Attenuated Total Reflection 161
   C. Chromatographic Coupling 161
Acknowledgments 162
References 162

0-8493-4203-1/93/$0.00+$.50
© 1993 by CRC Press Inc.

I. INTRODUCTION

Infrared (IR) microspectroscopy could be defined as the coupling of a microscope to an infrared spectrometer. This coupling allows one to focus IR radiation onto the sample, to collect and image transmitted or reflected IR radiation from the sample onto a detector, and to view and position the sample.[1] The essential features required to perform these aforementioned functions include an optical microscopic system for viewing the sample and reflective optics for imaging the IR radiation. In addition, the system contains one or more remote apertures used for defining the sample area of interest. Specific details concerning the design of infrared microscope accessories have been adequately reviewed elsewhere.[1]

Perhaps another definition of IR microspectroscopy is the study of how infrared radiation interacts with microscopic particulates. Indeed, diffraction, refraction, reflection, and absorption effects (which result from the interaction of IR radiation with microscopic samples) routinely play a much more important role in microspectroscopy than in its macroscopic counterpart. Consequently, by virtue of size, the microscopic sample is an integral part of the microspectrometer.[2] One must therefore consider how to minimize these effects by altering the sample or by changing instrumental parameters in order to obtain the most representative, meaningful spectra possible.

Regardless of the definition, IR microspectroscopy provides the capability to obtain spectra on extremely small samples, as small as about 10 μm in diameter. Another significant advantage, however, is that it can allow one to acquire spectra on microscopic and macroscopic samples "as is". For instance, substances which are expensive and/or which react with IR transparent salts, such as certain solid proteins of biological interest, can be easily sampled without preparation of a salt pellet using the microscope.[3] Polarization studies can be performed on single fibers or small single crystals, rather than large, artificially prepared films.[4-6] The infrared spectrum of an intact, red blood cell in saline solution can be obtained with the infrared microscope, facilitating the study of carbon monoxide binding under conditions which are not unlike those found in the body.[7]

Another advantage to IR microspectroscopy is that the heterogeneity of microscopic and macroscopic samples can be examined, or even profiled, using the microscope accessory. For example, hydrocarbon fluid inclusions inside quartz and halite crystals can be analyzed using infrared microspectroscopy, thus aiding in the study of the geological history of these minerals.[8] The identification of the various layers of fiber optics or polymer laminates can be obtained by cross-sectioning the materials and focusing on individual layers with the IR microscope.[9,10] The profiling can be automated by the use of a computer-controlled x,y mapping stage which can be stepped in small increments. A functional group image in one or two dimensions can be obtained, and the two-dimensional images can be represented as axonometric or contour plots.[11]

Infrared microspectroscopy is a versatile technique and the microscope a versatile accessory; both permit a wide range of samples to be studied.[12] In this chapter, the basic experiments which can be done with an IR microscope will be examined, along with the unwanted optical effects from microscopic samples, how they affect each experiment, and how to correct them by proper sampling. Then, sample purity and homogeneity will be addressed. Finally, additional instrumental parameters affecting performance and other special sampling techniques will be mentioned.

II. TYPES OF EXPERIMENTS DONE WITH A TYPICAL IR MICROSCOPE

There are essentially four types of experiments which can be performed using the basic IR microscope accessory: transmission, specular and diffuse reflection, and reflection-absorption. For illustrative purposes, a *macroscopic* sample of a nylon tie (used in fastening tubing to laboratory glassware) is prepared for microspectroscopy, and spectra obtained by transmission, specular and diffuse reflection, and reflection-absorption are compared.

A. Experimental

Infrared spectra of the nylon tie were obtained at 4 cm^{-1} using a Perkin-Elmer 1760-X FT-IR spectrometer coupled with a Spectra Tech-IR Plan I. A narrow band mercury-cadmium-telluride (MCT) detector with a detector area of 0.25 mm X 0.25 mm and an air-cooled ceramic source were used. A Norton-Beer apodization function was used in the Fourier transform. For each spectrum, 100 signal-averaged scans were acquired, unless otherwise noted.

The type of aperture used, round or rectangular, was chosen to match the sample geometry as closely as possible in order to maximize energy throughput. In addition, the apertures were always closed inside the edges of each sample's image in order to minimize stray light. The specific details are described in the sections which follow.

B. Transmission Collection Mode

The most common applications of IR spectroscopy and microspectroscopy involve measuring IR radiation which has been transmitted through a sample. To illustrate this type of experiment, a nylon tie was shaved over a microscope slide using a razor blade. The shaving was then flattened by rolling the tip of a metal probe over it, applying enough pressure to flatten the sample but not to break it apart. The shavings were then placed on top of a potassium bromide (KBr) salt plate which was mounted over a hole in an aluminum plate cut approximately to the size of a microscope slide.

Next, the sample was placed on the microscope stage, centered, and brought into focus using the stage adjustment. The adjustable, rectangular aperture located between the sample and detector was brought into focus using the substage condenser. Then that aperture and the one located between the source and the sample were both adjusted around the sample's image, which was about 100 μm by 100 μm. After the single-beam spectrum of this sample was obtained, the sample was moved out of view and a single-beam spectrum was obtained at the same aperture settings on the bare salt window. The latter spectrum served as the background spectrum. The percent transmittance (T) spectrum of the nylon tie obtained in this manner is shown in Figure 1.

The ideal sample for transmission microspectroscopy is one which does little to alter the optical path of incoming and outgoing IR radiation. This requires that the diameter of the sample be as large as possible so as to minimize the diffraction of light. Optimistically, one desires a sample with a diameter larger than the longest wavelength of light being used, and this corresponds to about 20 μm for mid-IR measurements. Additionally, the sample's shape should be as smooth and flat as possible, so that it does not scatter or refocus the light. Finally,

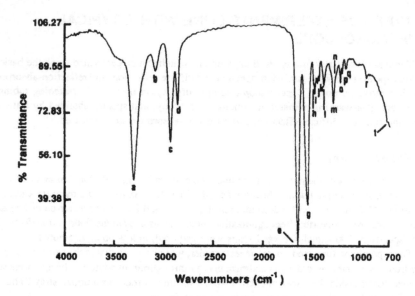

FIGURE 1. Spectrum of a nylon tie obtained by transmission. Letters refer to the peak position comparison made in Table 1.

the ideal transmission sample is thin enough to yield a spectrum in which the most intense bands would lie above 10% T for a redundantly apertured or physically masked sample. For many organic samples, this corresponds to a thickness of approximately 10 μm.

C. Reflection Collection Mode

When the microscope is placed in the reflection collection mode, three experiments can be performed, depending on the nature of the sample.

The first experiment that can be performed is the reflection-absorption (RA) experiment. In this experiment, the sample is in close contact with a reflective surface. The incident IR radiation passes through the sample, reflects from the substrate, and passes back through the sample again. The incident radiation also reflects at the surface of the sample, but if the sample is of the proper thickness, the contribution from this reflection to the overall reflection from the sample will be small.

To illustrate this experiment, the nylon tie was dissolved in phenol, and the solution was slowly deposited on top of an aluminum foil surface using a microliter pipet. The foil, with its shiny side up and its grain running parallel to the long side of the slide, had been rubber-cemented to a glass microscope slide. The area of the applied solution was about 1 mm in diameter. The solution was then allowed to thoroughly dry under a heat lamp to remove the phenol. A single beam spectrum of a 100 μm diameter spot of the nylon film using a fixed, circular aperture was obtained. Then, a single beam spectrum of an area of the foil with no sample observed was obtained using the same aperture in place. The sample spectrum was divided by the background spectrum, and the ratioed spectrum is shown in Figure 2. The weak bands at 756 and 1593 cm⁻¹ indicate that the phenol was not totally removed.

One can observe a spectrum with similar spectral features to that in Figure 1. The similarities are expected since the absorption-reflection experiment is fundamentally a

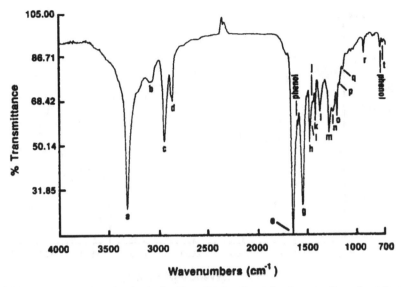

FIGURE 2. Spectrum of a nylon tie obtained by reflection-absorption. Letters refer to the peak position comparison made in Table 1.

transmission experiment. The frequencies, relative intensities, and band shapes in Figures 1 and 2 are very much alike, although the absorption bands are, in general, sharper in Figure 2 than those observed in Figure 1. Considering that fact, along with the observation that some of the bands are split, the nylon film cast from phenol is probably a more crystalline form of nylon than that prepared for the transmission experiment.

Since RA is basically the same phenomenon as transmission, the criteria for an ideal sample is the same. However, since RA is approximately a double-pass experiment, the sample pathlength would need to be about half that for the same sample measured in transmission for good spectral quality, provided that the Cassegrainian objective delivers light at an average incident angle of about 25° from normal. This pathlength corresponds to about 5 μm for organic materials. The thickness of a film, d, on a metal substrate required to yield a transmittance spectrum obtained by a RA experiment of the same intensity as the transmittance spectrum obtained by a transmission experiment done on a 10 μm film is given by the following equation, where a is the angle of incidence.[13]

$$2d = 10 \cos a \tag{1}$$

The second experiment that can be performed is the diffuse reflection experiment. In this experiment, the direction of reflected light is random with respect to the incoming beam. To illustrate this experiment, a piece of fine grit emery paper (the particle diameters ranging from 40 to 80 μm) was brushed across the surface of the nylon tie. The emery paper, along with the fine bits of nylon which adhered to it, served as the sample. With no apertures in place, the spectrum of the sample was obtained and ratioed against a spectrum of unused emery paper. The spectra were obtained on an area which was about 250 μm X 250 μm, approximately the effective size of the IR beam at the focal point of the microscope.

Diffuse reflection is a different physical phenomenon than absorption, and consequently, the diffuse reflectance spectrum, shown in Figure 3, is not equivalent to the transmittance

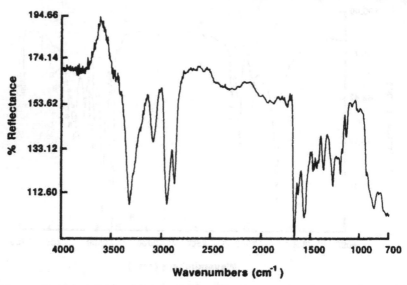

FIGURE 3. Spectrum of a nylon tie obtained by diffuse reflection (520 scans). Letters refer to the peak position comparison made in Table 1.

spectra shown in Figures 1 and 2. The diffuse reflectance spectrum, R'_∞ is defined as the ratio of the single-beam diffuse reflectance spectrum of the sample to the single-beam diffuse reflectance spectrum of a nonabsorbing standard.[14]

$$R'_\infty = \frac{R'_\infty(\text{sample})}{R'_\infty(\text{std})} \qquad (2)$$

The Kubelka-Munk theory shows that there is a linear relationship between the diffuse reflectance function, $f(R'_\infty)$, known as the Kubelka-Munk function, and c, the concentration.[15,16]

$$f(R'_\infty) = \frac{(1 - R'_\infty)^2}{2R'_\infty} = \frac{c}{k} \qquad (3)$$

where $k = s/2.303\varepsilon$, s = scattering coefficient, and ε = molar absorptivity.

Consequently, the Kubelka-Munk spectrum, $f(R'_\infty)$ would be expected to be similar to an absorbance spectrum, and the antilog of the negative Kubelka-Munk spectrum, shown in Figure 4, would be expected to be similar to a transmittance spectrum as, indeed, it is. The band frequencies match extremely well to those of Figures 1 and 2 as do the relative intensities of the bands.

The ideal sample for a diffuse reflection experiment is one in which the particle size is relatively uniform. The particles need to be larger than the wavelength used, otherwise diffraction, rather than diffuse reflection, will occur on a large scale. Fuller and Griffiths[14] have shown that the IR diffuse reflectance spectrum varies with particle size and that when potassium chloride (KCl) is used as the nonabsorbing substrate, the KCl particles (which range between 75 to 90 μm) are the most ideal size because they show little variation in reflectance over the mid-IR region. Particles larger than 90 μm show a large change in reflectance over

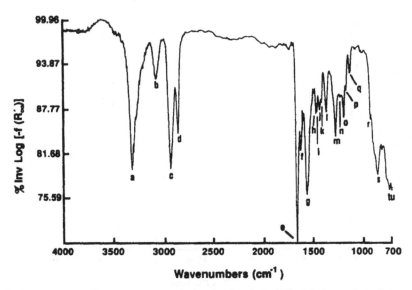

FIGURE 4. Diffuse reflectance spectrum of nylon tie after the Kubelka-Munk transform. Letters refer to the peak position comparison made in Table 1.

the same region. Particulate samples which are mixed with a nonabsorbing substrate give rise to the best diffuse reflectance spectra. However, since the microscope optics are not maximized for the diffuse reflection experiment, one usually needs to obtain the spectrum using the widest apertures possible, so that the most energy possible is used for the measurement. (See Chapter 4 for more details on diffuse reflection.)

The third experiment that can be performed in the reflection collection mode is the specular reflection experiment. This experiment requires very little sample preparation. The tip of the nylon tie was cut off and simply placed on a surface to support it. A rectangular aperture was masked down around the image of the tie and the spectrum recorded on a sample area of about 200 μm X 200 μm. This spectrum was ratioed against a front-coated gold mirror background using the same aperture setting.

Specular reflection is also a different physical phenomenon than absorption. The radiation is reflected from the surface of the sample at an angle which is the equivalent to the angle of incidence. Radiation can, of course, also be refracted and transmitted, but in the reflection mode, this component is not collected. One could have reflection at the bottom surface as well, but if the sample is of proper thickness, this component will not contribute much.

In the specular reflectance spectrum of the nylon tie shown in Figure 5, one can observe several first derivative-like spectral features. These bands, known as restrahlen bands, have arisen because of the refractive index dispersion which occurs in the region of an absorption band. In other words, a plot of the complex refractive index vs. wavenumber shows discontinuity in the region where the medium absorbs radiation. Since reflection is strongly dependent upon refractive index behavior, the reflectance spectrum will display this large change in refractive index, as well.

A mathematical process, known as the Kramers-Kronig (KK) transform, can convert these derivative features to a spectrum that describes extinction coefficient as a function of wavenumber. This transformation is useful because absorbance as a function of wavenumber can be easily calculated from the extinction coefficient spectrum.[17]

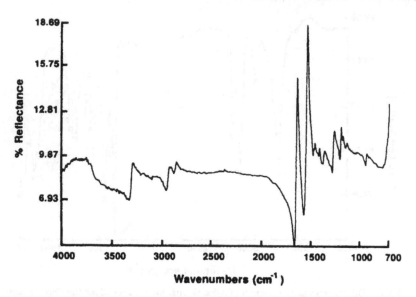

FIGURE 5. Spectrum of a nylon tie obtained by specular reflection. Letters refer to the peak position comparison made in Table 1.

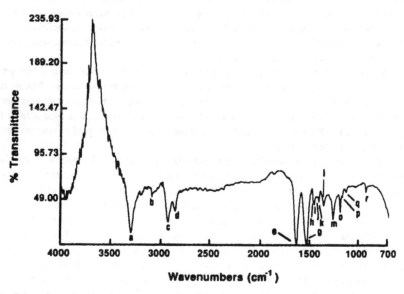

FIGURE 6. Kramers-Kronig calculated percent transmittance spectrum of nylon tie.

The KK transform of the specular reflectance spectrum is shown in Figure 6. The calculated band frequencies and relative intensities match extremely well with those obtained by the transmission experiment. One can observe two spectral anomalies: the first is that the shape of the overtone at 3091 cm^{-1} is distorted. The second, and most obvious, is the large negative feature observed at about 3600 cm^{-1}. The distortions indicate that the sample was not an ideal specular reflector, which is smooth and optically thick. In fact, the tie had some grain associated with it and, therefore, probably gave rise to a significant amount of diffuse reflection as well.

Transmission, reflection-absorption, diffuse reflection, and specular reflection are all types of experiments which can be performed easily using the modern microscope accessory. The type of experiment performed will depend largely on the sample under study. Ordinarily, IR spectra using the microscope accessory will be obtained on samples which are smaller and less well-defined geometrically than the nylon tie samples used. Consequently, the transmission experiment will be, practically speaking, the most widely used and most successful experiment. Routinely, many microscopic samples do not happen to be on a reflective substrate, and many do not lend themselves to being dissolved and applied to one. In addition, many microscopic particulates are not smooth and highly reflective, or lend themselves to being dispersed evenly in a nonabsorbing substance. Many, however, *can* be made thin enough so that a transmission experiment can be performed. Nonetheless, the nontransmission experiments offer a variety of alternative ways to obtain spectra when appropriate. One should be aware that there can be differences in the spectral information obtained by each experiment, as the detailed listing of peak positions in Table 1 illustrates.

The type of experiment performed using the microscope accessory will also depend on the analytical questions which need to be answered.

III. EXPERIMENTAL FACTORS AND SAMPLING CONSIDERATIONS

A. Stereoscopic Examination and Manipulation

Examination of the sample under the stereobinocular microscope is typically the first step of sample preparation. The visible analysis of the sample allows one to decide whether spectra should be obtained using the infrared microscope accessory in the reflection or transmission collection mode. The next step may be the manipulation of the sample in some fashion, removing a microscopic particle from a matrix and mounting it on a salt window, for example. Such manipulations can easily be performed using the stereoscope because it provides an erect image and a long working-distance. Tools needed for transport manipulations include fine-pointed probes and microforceps, the latter being very useful for fiber manipulation. Fine-pointed probes may be made from 24 to 26 gauge tungsten wire by chemical or electrolytic etching and then held in a pin vise.[18,19] Microliter pipets are useful for applying small amounts of liquids or dilute solutions to a salt window. The liquid or solution is applied in a small area of the window, and the application is repeated over the same area if a higher concentration is needed.

B. Sampling Away Unwanted Optical Effects

1. Total Absorption

Most microscopic sampling is performed on solids, particles, or fibers which, although they are small in diameter, are too thick for one to obtain representative spectra without reducing their pathlength. In macroscopic infrared spectroscopy, this consideration often may not pose a problem, since a solid sample is usually prepared for infrared spectroscopy by dilution with solvent, salt, or mineral oil.

TABLE 1. Peak Locations in Wavenumbers (cm⁻¹)

Peak label	Reflection-absorption	Transmission	Diffuse reflection	Specular
a	3303	3302	3316	3304
b	3080	3082	3083	3091
c	2938	2933	2938	2936
d	2861	2860	2862	2861
e	1636	1641	1660	1644
f			1626	
g	1541	1543	1563, 1553	1544
h	1475	1475	1478	1475
i	1466	1465	1467	1465
j	1440	1439	1441	1437
k	1417	1420	1422	1419
l	1372	1372	1373	1372
m	1277	1275	1278	1275
n	1240	1243 sh	1237 w	
o	1200	1201	1202	1200
p	1182	1181	1183	1181
q	1151	1144	1146	1142
r	935	935	935	935
s			854	
t	731	726	731	
u			708	

Total absorption is readily apparent with traditional spectra, however, by the observation of absorptions which hover close to the 0% T line. With spectra obtained using an infrared microscope, these features may not be present, especially if one has used a single remote aperture to mask the sample image.

Spectrum A in Figure 7 is that of a 30-μm diameter Dacron fiber, and Spectrum B is the same fiber flattened to about twice that of its original diameter.[1] The spectra were obtained using a single aperture placed in a focal plane between the source and the sample. The same aperture settings and background were used in obtaining both spectra. In the spectrum of the flattened fiber, the most intense bands are at about 18% T. In the spectrum of the unflattened fiber, the most intense bands are at an apparent 50% T, even though the sample is much thicker. One can observe that the bands are much broader as well in the unflattened fiber spectrum. A large amount of stray radiation which has reached the detector (due to diffraction which will be discussed in the following section) is the cause of the false transmittance reading. The unflattened sample is actually too thick to provide the details needed in a representative spectrum of the sample.

Reducing the pathlength of a microscopic solid sample can be done by flattening with a probe tip, probe heel, or any tool which can withstand the pressures applied. One common method for reducing the pathlength is rolling a stainless steel bearing mounted on an axle handle (commonly known as a roller) across the sample.[1] If the sample is fairly pliable, rolling may be done with the sample directly placed on a sturdy salt window, such as BaF_2 or ZnSe. More resilient samples can be rolled on a harder surface, then peeled or scraped off and transferred to a salt window. Squeezing the sample between two salt windows or in specially designed cells, such as a diamond anvil cell, is useful for flattening more resilient samples.[1]

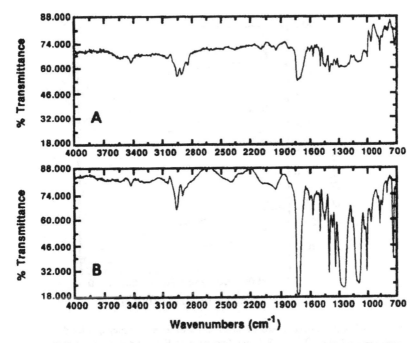

FIGURE 7. Spectra of (A) a 30-mm diameter Dacron fiber as received and (B) the fiber flattened to about twice its original diameter. (From Katon, J.E., Sommer, A.J., and Lang, P.L., *Appl. Spectros. Rev.*, 25(3,4), 173, 1989, 1990. With permission.)

Slicing the sample is an alternative method of reducing the pathlength which may keep the crystalline structure more intact. A slice of the material may be obtained by holding it firmly between two clean glass slides, so that a small bit of the sample protrudes from the edge of the slide. Then a razor blade or scalpel is run along the edge of the slide until a section of an appropriate thickness is cut. Samples can be more precisely cross-sectioned using a microtome. Some samples are firm enough to be microtomed directly while some must be supported in an embedding material such as paraffin wax, β-pinene wax, acrylic resin, or epoxy resin. The resins are more difficult than the waxes to remove after sectioning and can cause spectral interferences.[19]

2. Diffraction

When the sample size is on the order of the wavelength of the light being utilized for a measurement, diffraction starts to play a major role. Diffraction can occur at the adjustable aperture(s) used to mask the sample image and at the sample's edges.

Turner[20] and Chase[4] have each pointed out that when radiation is bent around an aperture, a baseline distortion with a loss of transmission at the longer wavelengths is observed. Spectra obtained by Turner illustrating this are shown in Figure 8. The spectra shown are those obtained through circular apertures with an effective diameter of 29, 25, 21, 17, 12.5, and 8.3 μm which have each been ratioed against the spectrum obtained for a 33 μm (effective) diameter aperture. Both the loss of energy as the aperture becomes smaller, and the sloping baseline at longer wavelengths are apparent. However, this distortion can be ratioed out of the spectrum by using the identical aperture settings in the acquisition of the background as used in the acquisition of the sample.

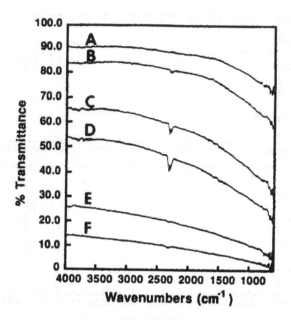

FIGURE 8. Diffraction effect caused by small apertures. Energy throughput for decreasing aperture diameters (29, 25, 21, 17, 12.5 and 8.3 mm, Spectra A to F, respectively) vs. throughput for 33 mm aperture. (From Turner, P.H., *Anal. Proc.*, 23, 268, 1986. With permission.)

Diffraction at the slit edges can cause a redistribution of the infrared radiation to areas located outside that defined by the apertures. Spectrum A in Figure 9 shows a reference infrared spectrum of a cellulose acetate film recorded with a single 8 μm X 240 μm aperture located before the sample.[21] The film aperture width was such that the entire area was filled. Spectrum B was obtained when the edge of the film was lined up with the edge of the image of the aperture. Spectrum C was obtained after the edge of the film had been stepped 40 μm away from the edge of the image of the aperture. The latter two spectra indicate that radiation is present outside areas defined by the aperture. This has some important practical ramifications. First, the single-beam reference spectrum needs to be obtained on a location on the salt window which is far enough away from the sample so it does not contribute to the reference. In this example, 40 μm was not far enough for the 8 x 240 μm sample. Multiple samples should be spaced adequate distances, perhaps 50 μm, apart on the salt window as well. Second, when sampling multilayered polymers, layers outside that defined by the aperture will contribute to the spectrum of the defined layer.

Upon placing a sample in the system, light is bent around the sample's edges to some extent. Spectrum B in Figure 10 compares the spectrum of a cellulose acetate film with the spectrum (Spectrum A) of a 15 μm x 15 μm slice of cellulose acetate. Both samples are the same thickness with the single remote aperture defining a 15 μm square area. The film is much larger in area than the area defined by the aperture, so any spectral effects from diffraction around the edges of the film should not be observed.[22]

They are apparent in the 15 μm square slice of cellulose acetate, however. The peak intensity of any absorption band differs significantly between the spectra. The carbonyl stretch at about 1750 cm[-1], for example, has a peak intensity at 28% T vs. 36% T in the 15 μm square acetate slice. In addition, the C–O stretch at about 1225 cm[-1] has an 18% T peak intensity in the film spectrum vs. 38% T in the spectrum of the 15 μm square slice. Light which has been

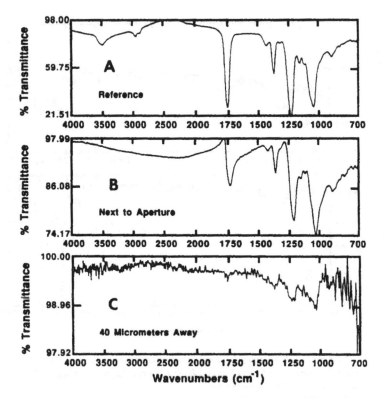

FIGURE 9. Spectra of a cellulose acetate film recorded within a 8 mm x 240 mm diameter aperture, of the same film but with its edge located next to and 40 mm away from the edge of the aperture image (Spectra A, B, and C, respectively.) (From Sommer, A.J. and Katon, J.E., *Appl. Spectros.*, 45(10), 1633, 1991. With permission.)

bent around but not passed through the sample has reached the detector, and this stray light increases with increasing wavelength in agreement with diffraction theory.

Varying changes in the relative intensities of absorption bands may also be a result of diffraction. As sample size approaches the diffraction limit and as aperture sizes get smaller, the loss of energy may be so severe at the longer wavelengths that absorption bands cannot be distinguished from the baseline. Indeed, the lower frequency absorptions in the spectrum of the acetate slice appear fairly "washed out". This affects the relative intensity of particularly the methyl bend at 1375 cm^{-1} and the C–O stretch at 1225 cm^{-1}, which are drastically different in the two spectra.

Diffraction effects can be minimized in several ways. First, the widest possible apertures should be used. The ideal aperture setting is one in which the apertures are closed within the sample's image but not closed enough to reduce energy throughput drastically.[21] Consequently, a sample with the largest possible area should be used. The sampling technique of flattening to reduce pathlength will also help to reduce diffraction effects by enlarging the sample area. A physical mask in the sample plane will eliminate the stray light from diffraction which results from the use of remote apertures. Finally, using the same aperture setting in the background scan as in the sample scan will serve to ratio out the baseline slope resulting from diffraction effects from the apertures.

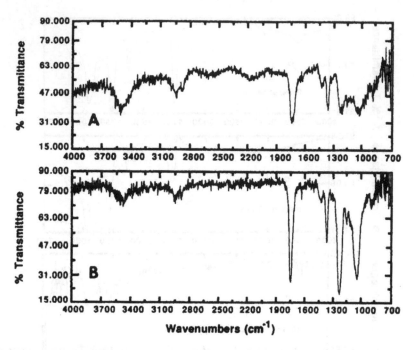

FIGURE 10. Spectra of (A) a 15 mm x 15 mm slice of cellulose acetate and of (B) a cellulose acetate film. (From Lang, P.L., Sommer, A.J., and Katon, J.E., *Particles on Surfaces 2*, Mittal, K.L., Ed., Plenum Press, New York, 1989, 143. With permission.)

3. Scattering

Microscopic samples which are opaque or finely powdered scatter the infrared radiation in the same way as macroscopic samples. Consequently, the remedies are much the same as well, except on a much smaller scale. For example, a microscopic particle which is too small to be handled by a "conventional" micropellet accessory can be placed next to a small crystal of KBr or a similar salt.[23] Then pressure is applied by using the end of a sturdy probe or roller until the salt and sample form a somewhat homogeneous, clear plate. Potassium iodide (KI) is preferred over KBr by some analysts, since it is a softer material and presses easily by hand. The process also serves to dilute the solid sample as well. Wopenka and Swan[24] have found that scattering by opaque particles can be reduced simply by pressing them into freshly cleaved KBr windows.

4. Reflection

Cells used to contain or flatten microscopic samples can give rise to interference fringes which obscure more interesting spectral information. This problem arises in the use of pressure cells, such as the diamond anvil cell. To avoid the fringes, the following technique can be used if the sample is not too elastic. After the sample is flattened in the cell, one can usually observe that it sticks to one half of the cell but not both. The portion of the cell which is free of the sample is then removed. The sample scan is obtained through the sample and one half of the diamond cell and is ratioed to a background obtained on a bare part of the same cell half. In this way, one of the two faces which produce the interference fringes is removed.[25] The technique also allows for a more effective ratioing out of the diamond absorptions, since the diamond pathlength is essentially cut in half.

5. Refraction

Samples or substrates which have a high refractive index or which are very thick can give rise to focal shifts. The shift is equal to $(n - 1)t/n$, where n is the refractive index and t is the thickness.[26] This shift can be corrected by adjusting the focal system of the microscope, either by visually observing the sharpness of the aperture edge or by observing the energy throughput in the energy mode of the spectrometer. The consequences of not making this adjustment or using a microscope which is not capable of this adjustment (one with a fixed focal system) may be a drastic reduction in SNR when using thick or high refractive index substrates.

C. Sample Purity and Homogeneity

Sample purity is a major concern in microspectroscopy since contaminants are not adequately diluted by the sample on a microscopic scale.[27] If fact, because of the small sizes involved it is not uncommon for an analyst to obtain spectra on particles unrelated to the problem under investigation. It is therefore helpful to know the type, nature, source, and spectral features of the contaminants which one might encounter.

One of the major components of office/lab dust is cellulose. Originating primarily from paper products, cellulose can be in either particle or fiber form. The IR spectra of cellulose dust may exhibit kaolin absorptions in addition to cellulose absorptions, since kaolin is a widely used paper additive. The kaolin bands usually observed are those at 3697 and 3620 cm^{-1}; the lower frequency kaolin absorptions are typically masked by cellulose absorptions. One may also observe lignin absorptions in the IR spectra of cellulose from unbleached paper products.

Polyamides are another ubiquitous component of office/lab dust. Present as flat, flaky particulates, most of the polyamide dusts originate from human skin. The epithelial cellular features can be confirmed by microscopic examination under 400X magnification.

Both natural and synthetic polyamide fibers are occasionally found in office/lab dust, too. Obviously originating from human or animal hair and synthetic textiles, respectively, natural and synthetic polyamides can usually be easily differentiated from their infrared spectral features.[28]

Other contaminants can originate from the surface where samples are transported, stored, or prepared. For example, formic acid salts have been found on "pre-cleaned" microscope slides.[29] The IR spectra of formic acid salts have C–H stretching absorptions at relatively low frequencies, such as 2730, 2830, and 2930 cm^{-1}, as well as intense absorptions due to carbonyl stretching at about 1600 and 1360 cm^{-1}. Therefore, it is wise to physically scrape clean the working surface of any glass microscope slide with a razor blade prior to its use.

Minor impurities in solvents can be another source of contamination.[19] After the solvent evaporates, the impurity may significantly contribute to spectrum. The use of pure reagents and the frequent use of blanks help one to minimize and recognize the presence of these types of contaminants.

Next, because samples may not be homogeneous on a small scale, it is easy to obtain spectra on particulates which are not representative of the sample under study. Consider the three spectra in the region 1600 to 700 cm^{-1} which were obtained about 1 mm apart on a single bleached cotton fiber shown in Figure 11.[30] The spectra are obviously different in this region, and any one of these spectra would not give an accurate representation of the fiber. In addition, the possibility that the fiber was part of a larger, inhomogeneous sample must also be considered. For instance, obtaining a spectrum of a single fiber from an "unknown" fabric to be identified may lead to inaccurate results if the fabric was a blend.

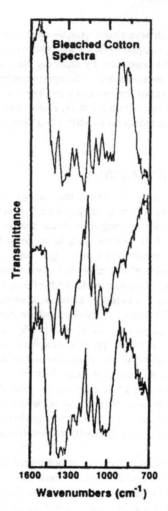

FIGURE 11. Spectra obtained approximately 1 mm apart on a bleached cotton fiber. (From Lang, P.L., Katon, J.E., O'Keefe, J.F., and Schiering, D.W., *Microchem. J.*, 34, 319, 1986. With permission.)

D. Accessory Alignment

It is very important that the infrared microscope be properly aligned. Levy[31] has published a simple experiment to find where the center of infrared energy is relative to the visible field. If the IR center is not coincident with the visible beam-path, one can realign the microscope by focusing a small pinhole on the microscope stage and by installing one or both remote apertures — of equivalent size *when* installed in the focal plane(s). When these apertures are visually aligned, then the infrared energy is maximized through these apertures according to the microscope manufacturer's instructions.[19]

IV. PARAMETERS TO REPORT WHEN PUBLISHING

The American Society of Testing and Materials has suggested that when publishing the results of an infrared microspectroscopic study, the following parameters should be noted: the area of the specimen being analyzed; the size and type of the detector element; the mode (reflection or transmission) in which the spectra were obtained; the specimen geometry and method of preparation; the shape, location, and type of masking apertures used.[19] In addition, the spectral resolution, the data collection time, the nature of the background spectrum, and any computer manipulation of the spectrum should be noted.

V. OTHER SPECIAL TECHNIQUES

A. Emission Microspectroscopy

Emission done through a microscope accessory has been reported.[32-34] According to Handke and Harrick,[32] the collection of a sample's emitted IR radiation over very large solid angles improves the SNR significantly and allows the emission spectra from nonflat and rough sample surfaces to be obtained. Lauer[33] has used emission microspectroscopy to study the surface temperatures at the interface between a moving disk and silicon head.

The most obvious advantage to emission microspectroscopy is that high-temperature studies can be performed on small quantities of materials. Indeed, the emission spectrum of a single 12 μm diameter Kevlar filament has been obtained using a microscope attachment.[34]

B. Grazing Angle and Attenuated Total Reflection

Some specialized infrared microscope objectives offer the analyst additional methods for obtaining spectra on otherwise difficult samples to handle by infrared microspectroscopy. These objectives include the grazing angle objective and the attenuated total reflection (ATR) objective which have recently been introduced by Spectra-Tech, Inc.

The grazing angle objective offers the capability to examine and obtain spectra on extremely thin coatings on reflective surfaces in which the area may be as small as 25 μm x 25 μm. Although detailed specifications have been described elsewhere, the objective allows for surveying the area of interest in near normal light before switching to grazing angles (37 to 85°) for additional viewing and infrared analysis.[35] One important consideration is that purge conditions must be rigorously controlled if IR spectra of extremely thin films are to be obtained.

The ATR objective contains an ATR crystal designed for a single reflection at a nominal 45° angle of incidence. The objective, which provides an optic for locating the sample, allows one to obtain spectra readily on highly absorbing materials of small size with little or no sample preparation.[36]

C. Chromatographic Coupling

Infrared microspectrometers have been interfaced to various types of chromatography.[37-41] In the trapping mode, chromatographic coupling with the IR microspectrometer has the advantage that the effective infrared beam size can be approximately matched to the sample area. In Pentoney, Shafer, and Griffiths's[37] supercritical fluid study, where the sample area was typically less than 200 μm in diameter, the end of the gas restrictor was positioned 40 μm above the surface of a zinc selenide (ZnSe) window which was cooled by Peltier coolers. The window was held stationary during all depositions and moved to a new position after the elution of each peak. It was determined that window temperature was an important parameter, one which defined the sensitivity of the method.

The microscope has been used to assist in on-line chromatographic detection using an ultramicro cylindrical internal reflectance (CIRCLE) cell.[40,41] This cell, made by Spectra-Tech, Inc. and specially designed for an IR Plan microscope, utilizes the Cassegrainian condenser and objective of the microscope to focus light into and collect light exiting the crystal. The results of these studies indicate, however, that the detection limit for the ultramicro cell is generally poorer that of the micro CIRCLE cell, both in the flow injection analysis and reverse phase high-pressure ligquid chromatography systems studied.

ACKNOWLEDGMENTS

Spectra-Tech, Inc. is gratefully acknowledged for funding, in part, the IR Plan Infrared Microscope. Other funding was provided by the Ball State University College of Sciences and Humanities, the Office of Research, and the Department of Chemistry.

REFERENCES

1. Katon, J.E.; Sommer, A.J., and Lang, P.L., *Appl. Spectros. Rev.*, 25(3,4) 173, 1989, 1990.
2. Messerschmidt, R.G., *Spectroscopy*, 2(2), 16, 1986.
3. Okada, K., Ozaki, Y., Kawauchi, K., and Muraishi, S., *Appl. Spectros.*, 44(8), 1412, 1990.
4. Chase, D.B., Infrared microscopy: a single fiber technique, in *The Design, Sample Handling and Applications of Infrared Microscopes*, (ASTM, STP 949), Roush, P.B., Ed., American Society for Testing and Materials, Philadelphia, 1987, 4.
5. Barer, R., Cole, A.R.H., and Thompson, H.W., *Nature*, 6, 198, 1949.
6. Johnston, C.T., Agnew, S.F, and Bish, D.L., *Clays and Clay Minerals*, 38(6), 573, 1990.
7. Messerschmidt, R.G. and Reffner, J.A., *Microbeam Analysis*, 1988, 215.
8. Barres, O., Burneau, A., Dubessy, J., and Pagel, M., *Appl. Spectros.*, 41(6), 1000, 1987.
9. Harthcock, M.A., Lentz, L.A., Davis, B.L., and Krishnan, K., *Appl. Spectros.*, 40(2), 210, 1986.
10. Carl, R.T., *Microbeam Analysis*, 1988, 223.
11. Harthcock, M.A. and Atkin, S.C., Infrared microspectroscopy: development and applications of imaging capabilities, in *Infrared Microspectroscopy: Theory and Applications*, Messerschmidt, R. and Harthcock, M., Eds., Marcel Dekker, New York, 1988, 21.
12. Bergin, F.J., *Appl. Spectros.*, 43(3), 511, 1989.
13. Griffiths, P.R. and de Haseth, J.A., *Fourier Transform Infrared Spectrometry*, John Wiley & Sons, New York, 1986, 189.
14. Fuller, M.P. and Griffiths, P. R., *Anal. Chem.*, 50(13), 1906, 1978.
15. Kubelka, P. and Munk, F., *Z. Tech. Phys.*, 12, 593, 1931.
16. Kubelka, P. and Munk, F., *J. Opt. Soc. Am.*, 38, 448, 1948.
17. Krishnan, K., Applications of the Kramers-Kronig Dispersion Relations to the Analysis of FT-IR Specular Reflectance Spectra, FTIR/IR Notes 51, Biorad Digilab Division, Cambridge, MA, August, 1987.

18. Humecki, H.J., Specimen preparation for microinfrared analysis, in *The Design, Sample Handling and Applications of Infrared Microscopes* (ASTM, STP 949), Roush, P. B., Ed., American Society for Testing and Materials, Philadelphia, 1987, 39.

19. Standard practice for general techniques of infrared microanalysis, American Society for Testing and Materials E334, Philadelphia, 1990.

20. Turner, P.H., *Anal. Proc.*, 23, 268, 1986.

21. Sommer, A.J. and Katon, J.E., *Appl. Spectros.*, 45(10), 1633, 1991.

22. Lang, P.L., Sommer, A.J., and Katon, J.E., Infrared and Raman microspectroscopy: an overview of their use in the identification of microscopic particulates, in *Particles on Surfaces 2*, Mittal, K.L., Ed., Plenum Press, New York, 1989, 143.

23. Oma, M.V., Lang, P.L., Katon, J.E., Mathews, T.F., and Nelson, R.S., Applications of infrared microspectroscopy to art historical questions about medieval manuscripts, in *Archaeological Chemistry IV*, Allen, R.O., Ed., American Chemical Society, Washington, 1989, 265.

24. Wopenka, B. and Swan, P., *Mikrochim. Acta*, I, 183, 1988.

25. Lin-Vien, D., Bland, B.J., and Spence, V.J., *Appl. Spectros.*, 44(7), 1227, 1990.

26. Messerschmidt, R. G., Photometric considerations in the design and use of infrared microscope accessories, in *The Design, Sample Handling and Applications of Infrared Microscopes* (ASTM, STP 949), Roush, P.B., Ed., American Society for Testing and Materials, Philadelphia, 1987, 12.

27. Lang, P.L., Katon, J.E., and Bonanno, A.S., *Appl. Spectros.*, 42(2), 313, 1988.

28. Scott, R.M. and Ramsey, J.N., *Microbeam Analysis*, 239, 1982.

29. Sommer, A.J. and Katon, J.E., *Microbeam Analysis*, 207, 1988.

30. Lang, P.L., Katon, J.E., O'Keefe, J.F., and Schiering, D.W., *Microchem. J.*, 34, 319, 1986.

31. Levy, F., *Anal. Chem.*, 60, 1623, 1988.

32. Handke, M. and Harrick, N.J., *Appl. Spectros.*, 40(3), 401, 1986.

33. Lauer, J.L., Vollenweider, J., and Vogel, P., Infrared Emission Spectroscopy of Operating Magnetic Recording Devices, Paper No. 334, Pittsburgh Conference and Exposition, Atlantic City, NJ, 1986.

34. DeBlase, F.J. and Compton, S., *Appl. Spectros.*, 45(4), 611, 1991.

35. Reffner, J.A., *Appl. Spectros. Mater. Sci.*, 1437, 89, 1991.

36. ATR Microscopy, Spectra-Tech Product Data Sheet, PD-8, Spectra-Tech, Inc., Stamford, CT.

37. Pentoney, S.L., Shafer, K.H., and Griffiths, P.R., *J. Chromatogr. Sci.*, 24, 230, 1986.

38. Haefner, A.M., Norton, K.L., Griffiths, P.R. Bourne, S., and Curbelo, R., *Anal. Chem.*, 60, 2441, 1988.

39. Fraser, D.J.J., Norton, K.L., and Griffiths, P.R., HPLC/FTIR measurements by transmission, reflection-absorption and diffuse reflection microscopy, in *Infrared Microspectroscopy: Theory and Applications*, Messerschmidt, R. and Harthcock, M., Eds., Marcel Dekker, New York, 1988, 197.

40. McKittrick, P.T., Danielson, N.D., and Katon, J.E., *J. Liq. Chromatogr.*, 14(2), 377, 1991.

41. McKittrick, P.T., Danielson, N.D., and Katon, J.E., *Microchem. J.*, 44, 1, 1991.

Chapter 7

GENERAL METHODS OF SAMPLE PREPARATION FOR INFRARED HYPHENATED TECHNIQUES

Gregory L. McClure

CONTENTS

I. Introduction 166
II. Overview of FT-IR Combined Techniques 166
   A. FT-IR Microscopy and Visible Light Microscopy 166
   B. TG/FT-IR and Its Variations 166
   C. Dense Fluid Phase Chromatographies 167
   D. Gas Chromatography 168
III. Specific Sample Preparation Techniques for GC/FT-IR 169
   A. General Strategy of GC Sample Introduction 169
   B. Solid Samples 170
      1. Supercritical Fluid Extraction (SFE) 170
      2. Pyrolysis and Thermolysis by Thermogravimetry (TGA) 173
      3. Thermal Desorption 174
   C. Liquid Samples 174
      1. Preparation Strategies 174
      2. Liquid Sample Introduction Devices 176
         a. Split Injection 176
         b. Splitless Injection 177
         c. Cold On-Column Injection 178
   D. Gaseous Samples 179
      1. Gas Streams 179
      2. Direct (Static, Equilibrium) Headspace 179
      3. Indirect (Dynamic) Headspace 181
IV. Applications Illustrating Diverse Methods of Sample Preparation for GC/FT-IR 181
   A. Distillation 181
   B. Solvent Extraction 186
   C. Direct (Static, Equilibrium) Headspace Applications 189
   D. Indirect (Dynamic) Headspace Applications 194
   E. Supercritical Fluid Extraction (SFE) 198
   F. TG/FT-IR and TG/GC/FT-IR Applications 202
   G. Artifacts Commonly Encountered in GC/FT-IR 207
V. Summary and Conclusions 211
Acknowledgments 212
References 212

0-8493-4203-1/93/$0.00+$.50
© 1993 by CRC Press Inc.

I. INTRODUCTION

Determination of molecular identities and quantities in complex mixtures may be the most common problem in chemistry. One may reasonably argue that, throughout the history of chemistry, the total time and effort spent on analyzing results of synthetic experiments has generally exceeded the time spent on the actual synthesis itself. In most instances the result of synthetic chemical operations is not a single product, but a distribution of multiple products. In other applications, such as perfumery, the construction of complicated molecular mixtures is the purposeful objective. Because of the difficulty and importance of the task, multiple analytical techniques are combined to form what are often called hyphenated techniques. Chromatography provides a means of separating the mixture into individual compounds, while spectrometry and other physical techniques provide a means of identifying (and optionally quantitating) the specific compounds. This work will include an overview of techniques of molecular mixture analysis which involve FT-IR as the spectrometric detector. Sample preparation for gas chromatography/Fourier-Transform-Infrared (GC/FT-IR) will be discussed in detail, since the latter is the most developed and practiced of the techniques which combine chromatography and FT-IR.

II. OVERVIEW OF FT-IR COMBINED TECHNIQUES

A. FT-IR/Microscopy and Visible Light Microscopy

Combined analytical techniques utilizing FT-IR have been employed for solid, liquid, and gas phase samples. For solids, variations in composition over the spatial dimensions of the sample can be examined by FT-IR/microscopy. By contrast, liquids and gases need a device such as a chromatographic column to develop a dimension of compositional variation in the sample over both space and time, i.e., the chromatogram. FT-IR/microscopy has been developed so extensively that it merits separate coverage (see Chapter 6).

B. TG/FT-IR and Its Variations

Examination of spectra of solid samples by microscopy is not always sufficient to provide enough information to characterize the sample fully. Further information can often be gained by studying the thermal properties and decomposition products of the sample with thermogravimetric FT-IR (TG/FT-IR). This technique is becoming increasingly important for the analysis of solids and occasionally liquids, particularly in the analysis of polymers, fossil fuels, and inorganics.[1-7] Thermogravimetric analysis (TGA) is used to measure the fraction of sample weight lost as a function of temperature and/or time. The loss of sample weight implies that some portion of the sample is evolved into the surrounding purge gas stream. Conducting the purge gas into an infrared sampling cell often enables the identification of the evolved gases and provides insight into the sample thermochemistry.

The major factor in satisfactory sample preparation for TG/FT-IR is control of particle size. For example, massive polymers and geologic materials must be ground, homogenized, and subsampled. It is desirable to reduce the particle diameter below 1 mm or less, if possible. Particle size reduction below 1 μm is probably not productive, and very small particles often present problems with static charge accumulation. However, with large particles it is difficult to achieve sufficient temperature homogeneity across the particle radius without very slow heating-rates. Temperature inhomogeneity of the sample results in broadening of the sample weight-loss event. Such broadening may lead to loss of resolution between multiple separate weight-loss events. Infrared is a concentration-sensitive detector, rather than a mass-sensitive detector. Consequently, event broadening leads to more gradual gas evolution rates in grams per second. Because the systems are operated at fixed purge rates (ml/min) to promote TGA balance stability, the result of event broadening is an undesirable reduction in absorbance signal levels from the evolved gases because of lower dynamic concentrations (mg/ml) in the infrared cell.

For some studies, TGA has been performed initially as a guide for temperature programming of the controlled sample pyrolysis. Subsequent spectroscopic investigation may be carried out separately with sample heating by a tube furnace. This experimental variation is referred to as evolved gas analysis (EGA).[8] Other experimental variations include techniques such as SMATCH/FT-IR, which is useful in studying rapid gas evolutions of reactive compounds, such as explosives.[9] SMATCH is an acronym for Simultaneous Mass Temperature CHange.

C. Dense Fluid Phase Chromatographies

High pressure liquid chromatography (HPLC), supercritical fluid chromatography (SFC), and thin layer chromatography (TLC) use mobile phases which are generally more dense and viscous than those of gas chromatography. HPLC, SFC, and TLC can be grouped together as dense fluid-phase chromatographies. White[10] has described the more widely investigated combinations of dense fluid chromatography with FT-IR, including high pressure liquid chromatography (HPLC/FT-IR), supercritical fluid chromatography (SFC/FT-IR), and thin layer chromatography (TLC/FT-IR). The general FT-IR text by Griffiths and deHaseth[11] includes a discussion of possibilities for HPLC/FT-IR interfaces. The combination of FT-IR with gel permeation chromatography (GPC/FT-IR) has recently been used to advantage in polymer analysis by Nishikida.[12] A review by Fujimoto and Jinno[13] summarizes many of the latest developments in chromatographic detection with FT-IR.

The combination of HPLC, SFC, and TLC with infrared spectrometry faces a major obstacle to facile molecular analysis. The mobile phases of HPLC and SFC and stationary phases of TLC typically absorb strongly over a large part of the mid-infrared region (4000 to 400 cm^{-1}). Measurements of detected infrared energy will often be more strongly affected by artifact of the chromatographic technique than by the signal of the eluted component of interest. Examples are very strong or opaque solvent bands and fringes produced by the high relative refractive index of ZnS or ZnSe cell windows. (Note: Infrared windows made of ZnS or ZnSe are preferred in some applications because of their chemical inertness, particularly when measuring spectra of aqueous samples.)

The significant probability of interfering absorbances limits the flexibility and range of applications for dense-fluid phase chromatographies directly coupled to FT-IR detectors. As a consequence of such inconveniences, development of dense-fluid phase chromatographies coupled to FT-IR has not developed as extensively as the more facile GC/FT-IR. In a recent text monograph on SFC edited by Jinno, the various aspects of SFC/FT-IR are reviewed for both the deposition scheme with mobile phase removal and the "on-the-flow" scheme which incorporates a high-pressure cell.[14, 15]

The one notable exception to the comments made above involves the use of supercritical xenon as a mobile phase for SFC/FT-IR. as has been decribed by Healy, Jenkins, and co-workers.[16-18] The rare gas xenon has, strictly speaking, no infrared (vibrational) spectrum, because it is monatomic. However, its large electron cloud appears to be quite polarizable, and xenon can be very useful as a supercritical mobile phase. The supercritical temperature and pressure of xenon are 16.6°C and 58 atm, respectively.

In any case, SFC and HPLC with solvent and modifier removal remain areas of high potential. In terms of solvent removal with analyte deposition, the adaptation by deHaseth[19] of the "MAGIC" system described for HPLC/MS appears to be a promising approach. MAGIC is an acronym for Monodisperse Aerosol Generator Interface Combining.

Traditionally, sample preparation for HPLC and SFC have generally amounted to "dissolve and filter" operations, the main objective being removal of particulates to avoid problems with the inlet sampling valve. More recently, supercritical fluid extraction (SFE) has been shown to provide an excellent sample preparation tool for HPLC and SFE, especially for the SFE/SFC combination.[20]

TLC is less susceptible to the particulate problem, because insolubles will remain at the origin on the plate. Another advantage of TLC is that the resultant plate can be scanned for extended periods to improve the spectral signal-to-noise ratio. TLC/FT-IR using the convenience of X-Y mapping stages has been described.[21]

D. Gas Chromatography

By far the greatest wealth of analytical experience and successful application of combining a form of chromatography with infrared spectrometry (to date) has occurred with GC/FT-IR. Heres[22] has devoted an entire volume to theory and applications of GC/FT-IR. Detailed discussions of the topic are offered by White[10] and also Griffiths and deHaseth.[11]

The extent to which GC/FT-IR has been developed and successfully applied has resulted from the combination of a number of factors. First, carrier gases used as the mobile phase in gas chromatography (typically helium, hydrogen, and nitrogen) do not absorb in the infrared region. The infrared spectra of the eluted components can be recorded directly in the gas phase without major interference from the operational elements of the chromatographic technique. In addition, the availability and convergence of three major technical developments has been a key factor. These were the rapid scan interferometer, the mercury cadmium telluride (MCT) detector, and the gold-coated glass "lightpipe".[23,24]

Early interferometers designed for astronomy were not rapid scanning instruments. During the decade of the 1970s a combination of electronic and mechanical design improvements contributed to the development of an interferometer which could retain reasonable optical stability while scanning rapidly enough to provide data rates compatible with the speed of gas chromatography. This achievement included the ability also to record and store the potentially large data files which could be generated with such instruments.

The gold-coated lightpipe amounts to a small-volume gas cell which could be constructed to match expected elution volumes of gas chromatographic peaks. While the geometric optical constraints of lightpipes result in strong attenuation of the light throughput of FT-IR instruments, this loss in energy ultimately provides an indirect benefit. The optical system can be designed such that the residual energy transmitted though the lightpipe roughly matches the useful operating range of an MCT detector. Ironically, the lightpipe serves to protect the MCT from saturation by excess light and also trades the light-energy loss for a potential absorbance signal gain of transiting eluted components. The MCT detector is more sensitive and has a faster response time than most other mid-infrared (4000 to 600 cm^{-1}) detectors.

Early attempts to combine GC and IR were less direct and more cumbersome to execute properly. Fractions eluted from packed GC columns were condensed in glass melting-point tubes held at the exit of a column or thermal conductivity detector.[25,26] The condensed material was then transferred to a sodium chloride window as a film and inserted in the beam of a dispersive infrared spectrometer. After several minutes scan time, a spectrum of the fraction may have been obtained if the sample had not been lost by evaporation. This was GC-IR at the milligram level. While it was useful for some applications, it did not provide sufficiently low detection limits or high separation efficiency for many applications such as fragrances and petrochemical samples.

The major growth in interest in combining gas chromatography and infrared spectrometry occurred when the high resolution GC (HRGC) associated with capillary column technology was coupled to FT-IR in the mid to late 1970s and early 1980s.[27, 28] Analytical performance improved and detection limits decreased below the microgram level and into the nanogram range. In addition to the much improved hardware performance, the software also developed to provide "on-the-fly" HRGC/FT-IR. This development enabled the analyst to see an infrared spectrum of the material almost immediately as each component exited the column. The experienced operator could then identify many materials by visual inspection of the component spectra as they appeared on the computer screen.

Detection limits below the nanogram level are now claimed for deposition methods of HRGC/FT-IR.[29] These systems are designed to condense eluted components into small areas on refrigerated surfaces for investigation by transmittance or reflectance infrared techniques. These systems are more expensive and complicated than those of the lightpipe variety. In addition, molecular interaction plays a role in making solid phase spectra different from vapor phase spectra. This phenomenon tends to complicate the operation of library searching to identify the components of a chromatographed mixture. Nevertheless, among commercially available GC/FT-IR system designs to date, the decomposition systems generally offer lower detection limits than the lightpipe systems.

A third factor must be recognized which has contributed to promote the success of GC/FT-IR. Because it is an analytical technique combining two technologies, GC/FT-IR has benefited from technical advances in each of the separate parent analytical fields. Sample introduction in GC has been the subject of extensive study and creative invention.[30, 31] One result has been the development of numerous and diverse sample introduction devices for GC. The next section will discuss general aspects of the major types of GC sample introduction, and the section following will illustrate these by application to a variety of samples.

III. SPECIFIC SAMPLE PREPARATION TECHNIQUES FOR GC/FT-IR

A. General Strategy of GC Sample Introduction

Numerous techniques have been developed to introduce a sample into a gas chromatograph. The sample may be solid, liquid, or gas at ambient conditions. Usually multiple options exist for handling samples of any phase. The unifying objective of all sample introduction methods is control of the transfer of volatile materials from the sample container onto the stationary phase coated on the inside walls of a column. Minimizing extra peak width contributed by the sample inlet process is an important overall goal. This problem is known as "injector broadening". This often means attempting to localize the sample (on the stationary phase) at the front or head of the column in as short a length as is practical.

The majority of samples analyzed by GC are liquids introduced by syringe into an "injector" designed to volatilize the sample and convey the vapor cloud (or some fraction of it, in the case of split injection) to the GC column for separation of the components. Traditionally, the standard method of isolation of volatiles from nonvolatiles in organic chemical preparations of liquid phase samples has been distillation, often at pressures below 1 atm. In addition, preconcentration of volatiles and separation from complex sample matrix materials has been typically accomplished by extraction with organic solvents.[32-34] The mainstay of apparatus for extraction of solids and immobilized materials has been the device described by

FIGURE 1.   Diagram of Soxhlet extraction apparatus. (Used with permission of Kontes Glass, Inc.)

Soxhlet[35] over a century ago, and it has changed little since the original publication. A modern version of the Soxhlet extractor is shown in Figure 1.

The fundamental objective of both extraction and distillation is control, both inclusive and exclusive in nature. The inclusive control is centered around providing a proper amount of sample for the chromatographic separation without requiring direct introduction of the sample volume in which the volatiles were originally dispersed. The exclusive control involves isolation of the desired set of compounds in the sample from those which are not of interest and would needlessly complicate the separation effort. Practically, it is best to minimize introduction of nonvolatile and low-volatility organics into the injector, because they often slowly degrade. The result is inlet system contamination and possibly interference with subsequent chromatography. Because this problem is not totally avoidable, it is customary to use glass liners in injectors. Contaminated glass liners can be easily cleaned or replaced.

B. Solid Samples

Samples that are solids under ambient conditions may be analyzed by GC/FT-IR if they are volatile at GC operating temperatures, usually 250 to 300°C, but possibly as high as 400 to 450°C with special equipment. Otherwise, solid samples are not directly suitable for GC unless some form of fractionation or pretreatment is used. Headspace sampling is a form of fractionation and will be discussed as part of the section on gaseous samples. Other important methods of handling solid samples are SFE, TGA, and pyrolysis.

1. Supercritical Fluid Extraction (SFE)

In the past few years, there has been exceptionally high interest in the development of SFE as an inlet method for GC. SFE is a technique with high potential for growth and is currently the subject of active research and development for analytical applications. A supercritical fluid

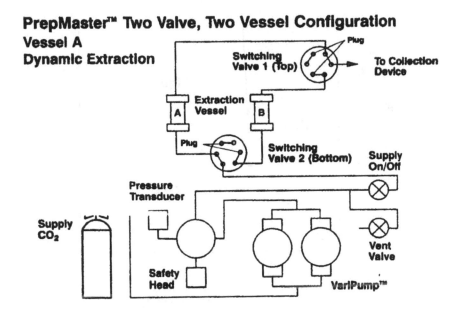

FIGURE 2. Block diagram of a supercritical fluid extractor (SFE) as in Perkin-Elmer PrepMaster.

extractor is basically a device to transport fluid in a supercritical state through a vessel containing a solid material. Some components in the sample may dissolve in the supercritical fluid and be transferred out of the extraction vessel through porous metal fritted disks into some form of collection device. The block diagram of one design of a supercritical fluid extractor is shown in Figure 2.

SFE bears some resemblance to solvent extraction except that a supercritical fluid is used rather than a liquid. A supercritical fluid is a material that is held above its critical temperature and at a pressure above its critical pressure. In simple terms, a material in such a state is too diffusive to fit the description of a liquid and too dense and too good a solvent to conform to what is usually considered a gas. Rather, supercritical fluids display an unusual combination of properties somewhere between that of a liquid and that of a gas. In terms of their densities and solvating powers, they resemble liquids. In terms of their diffusivities and ability to penetrate sample matrices, they resemble gases.[36]

The unusual behavior of supercritical fluids was first reported about 170 years ago.[37] Somewhat later, evidence was presented to support the idea of a specific critical point in a phase diagram.[38] Figure 3 shows a phase diagram of carbon dioxide, which is similar to that of many materials. The supercritical state has been used by chemical engineers for applications such as the decaffeination of coffee.[39] Despite its long industrial history, supercritical fluid extraction has only recently begun to be exploited for its analytical utility.

One major factor for the driving interest in SFE is the impending requirement to replace chlorocarbon and chlorofluorocarbon extraction solvents for health and environmental reasons.[40] Use of organic solvents for extractions requires eventual handling and disposal of used solvents. The rising costs of purchase and disposal of reagent chemical solvents loom as ominous and forceful economic incentives for development of replacement methods which do not impose such difficulties. SFE methods will typically consume carbon dioxide (CO_2) and small amounts of organic solvents (e.g., methanol) as reagents. Consequently, SFE methods pose very little problem in terms of chemical waste disposal.

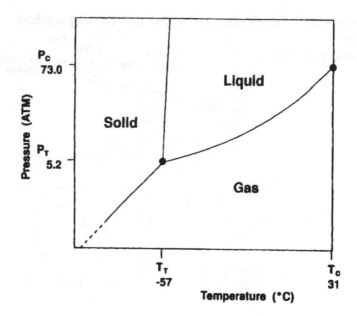

FIGURE 3.   Phase diagram of carbon dioxide showing supercritical region.

SFE methods are accompanied by several technical, as well as environmental, advantages. Among other factors, supercritical extractions are often carried out in shorter periods and with higher efficiency than with Soxhlet extractions. Moreover, in many cases, SFE can impose less thermal stress on the native analytical sample and final extracts. Therefore, SFE is less likely to produce artifacts of sample preparation than the Soxhlet technique.

An important phenomenon in supercritical fluids is the strongly increasing solubility of most materials with increasing pressure and fluid density above the critical point. Hannay and Hogarth[41] reported this phenomenon in 1882. They observed that ionic salts such as cobalt chloride could be reversibly dissolved and reprecipitated in ethanol by raising and lowering the pressure, respectively, of the supercritical ethanol at constant temperature.

At relatively low pressures (e.g., 100 to 200 atm, 1500 to 3000 psig), CO_2 is a good solvent for aliphatic hydrocarbons, with solvent properties similar to those of hexane. At higher pressures (e.g., 400 to 500atm, 6000 to 7500 psig), CO_2 displays much greater solvent power and has been compared to dichloromethane.[42]

Although a variety of materials have been investigated as supercritical solvents, the material most favored for analytical applications is CO_2. The critical temperature is 31.3°C, and the supercritical pressure is 72.9 atm. Both of these conditions may be easily and safely achieved with laboratory scale equipment. In contrast, supercritical water, an exceptionally powerful solvent, will probably not become prominent for analytical laboratory applications in the near future, since the supercritical temperature is over 370°C.[43] Even small leaks in apparatus containing supercritical water could pose a safety risk.

SFE with CO_2 has the limitation that highly polar materials are not extracted effectively. This limitation mainly refers to those materials which are readily soluble in aqueous media. However, the solvent properties of CO_2 for more polar materials can be significantly enhanced by addition of moderate amounts of polar organic solvents, such as methanol. Other modifier agents include acetonitrile, formic acid, 1,4-dioxane, methylene chloride, and numerous others.[44] Nitrous oxide (N_2O) is the second most used supercritical extraction for analytical applications. Pure N_2O behaves as a more polar solvent than pure CO_2, and N_2O can be used in place of CO_2 modifier combinations in some applications.[45] The reader should be cautioned that there have been reports of explosions using supercritical N_2O. The dangers of SFE with N_2O are not yet well-understood, and this issue remains the subject of debate and further investigation.

SFE effluents can be sampled "on-line" by direct connection to a GC injector, or "off-line" with intermediate isolation of the extract before analysis. The on-line approach has the advantages that it is fast and sensitive. It provides a means of reducing the risk of loss of the more volatile components of a sample between the extraction vessel and the GC column. The on-line experiment is generally performed by conducting the SFE effluent in a steel capillary tube, through the septum of a split/splitless GC injector operated in split mode.[46] The steel tube is crimped at the exit end to limit the flow rate. The heat from the injector barrel compensates for the Joule-Thompson cooling which accompanies decompression of the CO_2. For best results with volatile compounds, it is useful to reduce the GC oven temperature below ambient during the dosing step.

The simplest method of off-line collection in SFE involves decompression of the supercritical phase through a length of fused silica capillary into a collection vial partially filled with a solvent. Typical solvents are methanol and dichloromethane. This method of collection, although simple, quick, and easy, suffers from two disadvantages. First, as the SFE effluent bubbles into the solvent, the more volatile components are likely to be lost by the sparging effect of the escaping CO_2 gas. Second, the SFE effluent cools as it decompresses. Both the lower pressures and temperatures tend to make the extracted materials less soluble in the extracting fluid near the decompression region. The result is that extracted materials may separate out as solid particulates and plug the restrictor. The plugging may be reversed in many cases by warming the restrictor with a hot-air blower or passing the effluent through a heated block as it enters the restrictor.

For off-line collection, a better strategy seems to be decompression through a heated restrictor into an adsorption trap cooled to subambient conditions. Extracted materials can be transferred quantitatively to a collection vial by warming the adsorption trap slightly above ambient and rinsing it with a small volume of organic solvent. With this arrangement of collection components, losses of volatiles are reduced and restrictor plugging is minimized.[47]

SFE is likely to become the sample preparation method of choice for important environmental analyses, such as total petroleum hydrocarbons (TPHs) in soil and water, polycyclic aromatic hydrocarbons (PAHs), polychlorinated biphenyls (PCBs), and chlorinated pesticides. The exploration of the range of useful applications for SFE continues to be the subject of active developmental efforts. Generally, use of this technique in analytical work is not yet at the routine level. Other areas which are likely to benefit from SFE are analyses of food materials, pharmaceuticals, and petroleum and petrochemical products.

2. Pyrolysis and Thermolysis by Thermogravimetry (TGA)

Both pyrolysis probes and TGAs serve to force the liberation of volatile fractions of the sample into a vapor phase which can be transported by the carrier gas stream. Pyrolysis probes are designed to add a large amount of heat to a sample quickly, such that the majority of the material can be efficiently vaporized into a small volume and moved onto the head of the column. More recently, programmable pyrolytic probes have become commercially available.[48] TGAs and programmable pyrolitic probes enable precise control of sample temperature. The objective is to enable differentiation and identification of materials evolved in separate weight-loss events during the temperature programmed run.

The fact that infrared is a nondestructive technique confers an advantage to TG/FT-IR analysis, in that it is possible and also convenient to collect fractions from the exit of the infrared cell on adsorbent traps.[49] The adsorbent traps can be sequentially analyzed by thermal desorption into a GC/FT-IR system.[50] This arrangement amounts to TG/GC/FT-IR, and it contains an extra dimension of information. This dimension is observed as knowledge of the distribution of products evolved in each thermally induced weight-loss event. Further extension of this approach has also been reported with the inclusion of mass spectrometry to provide TG/GC/FT-IR/MS.[51]

3. Thermal Desorption

Materials such as Tenax-TA (Polyphenylene oxide), XAD-2, and other thermostable polymeric materials can be used to scrub dilute gas streams of volatile organic molecules at ambient temperatures. Such traps reversibly release many adsorbed materials when heated to temperatures in the 100 to 400°C range. While these are nominally solid samples, their analytical utility arises from their convenience as templated gas and vapor samples. This area will be reserved for discussion under dynamic headspace techniques.

C. Liquid Samples

1. Preparation Strategies

The use of distillation and extraction ∾ ιιn tools for preparing liquid samples for GC was mentioned at the beginning of this secti . A number of interesting devices have evolved to serve preparative needs. Traditionally, volatiles have been extracted with light hydrocarbons, such as pentane. To concentrate the extract, a Kuderna-Danish device, shown in Figure 4, provides for relatively efficient solvent evaporation. A sample concentrate is obtained with minimal loss of volatile components.[32] This device has been used extensively in the flavor and fragrance industry.

The most rudimentary extraction device is perhaps the so-called "separatory funnel", with which stepwise extraction can be carried out with multiple aliquots of solvents. Some extension of the concepts of the separatory funnel have led to a device known as the continuous liquid-liquid extractor. Recently, an apparatus was described which combined the continuous liquid-liquid extractor with a Kuderna-Danish concentrator in one unit.[52]

Likens and Nickerson[53] described an apparatus designed to concentrate volatile organic compounds by an extraction which occurred during condensation of the distillate from aqueous and organic solvents on the same condensation surface. The Likens-Nickerson apparatus for simultaneous distillation extraction (SDE) has been adapted for micro-scale preparations by Sandra and coworkers.[54, 55] The micro-SDE apparatus requires minor variations, depending on the density of the organic solvent relative to that of the aqueous phase being extracted, as shown in Figure 5. Apparatus A in Figure 5 is intended for use with extraction solvents more dense than water while apparatus B is for solvents less dense than water. Volatiles from both the aqueous and organic distillation flasks are distilled onto the cold finger, where intimate mixing of the phases leads to redistribution of the volatiles, preferentially into the organic phase. The condensate drops into a separation zone, where the lighter and heavier phases return to their respective distillation flasks for recycle. The overall process affords a concentrated, small volume organic phase extract from a dilute, larger volume aqueous phase, with a minimum loss of volatiles during the operation.

In view of the development of convenient, commercially available equipment for SFE (discussed earlier) and the multiple advantages of this technique, SFE can be expected to

FIGURE 4.   Diagram of Kuderna-Danish concentrator. (Used with permission of Kontes Glass, Inc.)

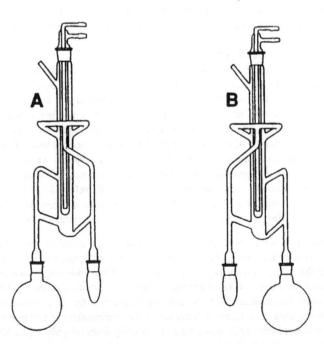

FIGURE 5.   Diagram of microscale Likens-Nickerson apparatus for simultaneous distillation extraction (SDE). (A) Apparatus for extraction of aqueous phases with organic solvents more dense than water. (B) Apparatus for extraction of aqueous phases with organic solvents less dense than water. (Figure adapted from References 43 and 44, used with permission.)

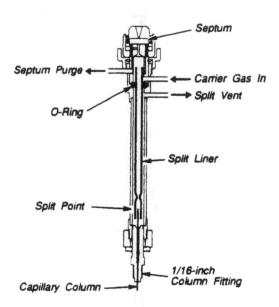

FIGURE 6.   Diagram of split/splitless injector for gas chromatography.

become a standard method in many applications, replacing older operations such as distillation, vacuum distillation, extraction, continuous liquid-liquid extraction, and SDE.

2. Liquid Sample Introduction Devices

By far the simplest and most common form of sample introduction in GC involves the
injection by syringe into a heated chamber fitted with a carrier gas inlet and at least one outlet
connected to a column. This roughly defines what is now called an "injector". Such a device
can be operated in a number of ways, depending on the nature and properties of the sample
being analyzed. Packed columns have high sample capacities and relatively low separation
efficiencies. Packed column GC injectors are relatively simple devices which fit the minimalist
definition given above. Capillary columns have comparatively low sample capacities, but
exceptionally better separation efficiencies than packed columns. For capillary column GC, a
more sophisticated injector is appropriate, and this is usually referred to as a "split-splitless"
injector, which is shown in Figure 6. This device may be used alternatively in either split or
splitless injection modes, depending on the nature and requirements of the sample.

a. Split Injection   With split injection, the sample vapor cloud in the injector barrel is
divided into two streams. One goes into the column for analysis, and the other fraction is
directed out of the injector into a trap or purge vent. Split injection has the advantage that the
total flow through the injector is rapid enough to move the sample onto the head of the column
as a narrow band of material, and the injector clears rapidly of residual sample vapor.
Introduction of the sample as a narrow band favors the achievement of better resolution of the
components on the column. Split ratios may vary widely from roughly 1:1 to 500:1, but they
generally range from 10:1 to 100:1 in most applications.

A secondary advantage of split injection is that it reduces the amount of material loaded
onto the column. This plays a role when a sample is concentrated (e.g., an essential oil,
petroleum distillate, etc.), and it is considered undesirable to dilute the sample. Splitting the

sample in the injector facilitates a compromise between introducing the minimum amount of sample that can be easily manipulated and the maximum amount the column can handle without significant overloading.

The complaint often lodged against split injectors involves what is called "mass discrimination". This problem arises from the fact that the split ratio for the carrier gas is not always the same as the split ratio for all components in the sample. The problem often shows up in the heavier components which are not volatilized at the same rate in a heated injector as the lighter compounds. While mass discrimination poses a problem for accurate quantitation by GC, it is not usually as large a concern for GC/FT-IR, which is most often used as a qualitative tool. However, this problem should be mentioned, since the future course of GC/FT-IR development may include a greater emphasis on quantitation.

b. Splitless Injection   When the concentration of the components of interest in the sample is relatively low, introduction of more sample onto the column can be accomplished by injecting more material or closing the split vent for a fixed period at the start of the run. Since there is a limit to the amount of sample an injector can efficiently vaporize, injecting more is not always a useful approach. However, if the split vent is closed for a period (on the order of 0.1 to 1.0 min), substantially more of the material injected will be forced by the carrier gas pressure onto the column rather than out the split vent. In such cases, it is useful to hold the GC oven at a relatively low initial temperature in order that the injected materials become trapped in the stationary phase on the head of the column as they leave the heated injector zone. For some applications, the oven is purposely cooled below ambient with liquid nitrogen or CO_2 to ensure that injected volatile matter is condensed in a very narrow band. The use of such coolants to concentrate components on the head of the column is called cryofocusing. For many components with boiling points above 100°C and molecular weights above 100 Da, special GC oven coolants may not be necessary. The trapping effect of the stationary phase itself may be sufficient to provide adequate performance, assuming the GC oven is cooled to close to ambient (35 to 40°C) during sample introduction. In addition, the use of bonded phase, thick films (1 to 5 μm) also facilitates this effect. The thick-film columns have a greater sample capacity than thin film (0.1 to 0.5 μm) columns of the same internal diameter, and it takes more sample mass to overload thick-film columns at the same column temperature. A column is overloaded when more material is forced on to the column than can be absorbed by the film of the stationary phase. The resulting peaks have very poor shape and tend to broaden substantially in base width, particularly on the leading edge. This effect is known as "pretailing" and is direct evidence that a greater amount of material is present in the peak than can be readily accommodated by the film of stationary phase immobilized on the column wall. This is illustrated by the Gram-Schmidt chromatogram obtained from GC/FT-IR analysis of a sample of dill seed oil, shown in Figure 7. Table 1 lists the major components identified by GC/FT-IR.

The Gram-Schmidt procedure allows the rapid calculation of changes in portions of the interferograms obtained during the GC/FT-IR on-the-fly data collection. The result of the calculations plotted vs. time of acquisition is generally equated with absorbance changes due to peaks passing through the lightpipe. This amounts to a general infrared chromatogram, similar to the total ion chromatogram (TIC) used in MS. The Gram-Schmidt procedure has the advantage that calculation can be made on the interferograms as they are acquired, without the delay which results from Fourier transformation to obtain spectral domain data.[56]

Note that overload of the column does not necessarily imply anything negative about the quality of the infrared spectra obtained from the ill-shaped peaks. At the same time, it is generally preferable to avoid attempting spectral comparisons on the basis of strongly absorb-

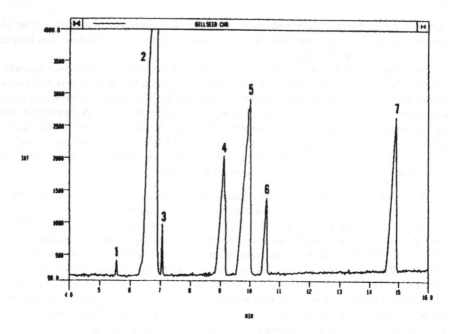

FIGURE 7. Gram-Schmidt chromatogram of dill seed oil, illustrating pretailing of overloaded peaks.

TABLE 1. GC/FT-IR Analysis of Commercial Dill Seed Oil (Major Components)

Peak #	Identification (GC/FT-IR)
1	β-Pinene (Nopinene)
2	Limonene (overlap)
3	Cymene (overlap)
4	γ-Terpinene
5	Dihydrocarvone
6	Carvone
7	Thymol
8	Apiol (dill type)

Note: Column: 25 m, 0.32 mm i.d., 1 μm film, (5% phenyl-methylsilicone)

ing spectra. The optical configurations of lightpipe GC/FT-IR systems are typically highly convergent, and absorbance linearity is usually not reliable over a very large range. Absorbance linearity with concentration (Beer's law) fundamentally assumes that light rays entering the sample are parallel (collinated beam) not convergent (focused beam). Consequently, the relative absorbances of bands in strong and weak spectra will typically not be constant in GC/FT-IR data derived from light pipe systems.

c. Cold On-Column Injection   There are GC samples which are sensitive to the sudden shock of flash vaporization in a split/splitless injector. To avoid degradation of sensitive materials upon injection, a different approach to the use of an injector is preferred. With cold on-column injection, the sample is injected directly into a piece of deactivated, fused silica or capillary column tubing with a special syringe. The sample is allowed to diffuse into the stationary phase along the length of the head of the column before temperature programming both the injector and column oven. The low chemical activity of the stationary phase and the reduction in heat shock applied to the sample components caused by sudden (flash) vaporization can be quite useful in avoiding degradation in the injector for certain sensitive materials.

D. Gaseous Samples

One might expect that gas phase samples would be most easily handled in a technique named "gas" chromatography. However, gases readily diffuse or expand to fill their containers, and gaseous samples are always subject to loss by degradation or incomplete containment after sampling and also to loss on transfer to an instrument. Consequently, the major strategy of gaseous sample introduction in GC centers around control and concentration of the components onto an intermediate trap or the head of the column. For gas samples, this requires techniques that are somewhat different from those used with condensed phase samples.

1. Gas Streams

Gas sampling from flowing streams usually requires special sampling valves and sampling loops. Industrially, much of this type of work is still done with packed columns and single channel detectors, such as flame ionization (FID) and thermal conductivity detectors (TCD), instead of spectrometer detectors. Typically, the composition of such streams is known fairly well, and the analytical interest is centered around qualitative or quantitative variations.

FT-IR analysis with long path gas cells has been used for combustion gas and process waste-gas analysis. However, such streams often contain reactive and unstable species such as nitrogen oxides, sulfur oxides, formaldehyde, etc. While such reactive species may be analyzed directly as mixtures, they would not necessarily provide equivalent analytical data if passed through a GC column.

Some process streams contain low levels of catalyst activity modifiers in addition to the major products and reactants. For example, 1,2-dichloroethane has been used to control catalyst activity of some silver catalysts used for epoxidation of ethylene. As in this example, catalyst modifiers tend to be used in very dilute concentrations (e.g., ppm to ppb range), while reactants and products are present at higher levels (e.g., 5 to 50% range). In the example sited, a problem arises because the large ethylene peak tails and obscures the dichloroethane peak. FT-IR is not particularly selective for chlorocarbons relative to olefins. Consequently, GC/FT-IR would not be a satisfactory technique for applications such as this.

2. Direct (Static, Equilibrium) Headspace

Direct (static) headspace analysis has been commercially available as a GC accessory for over a decade. Among its earliest uses was the determination of blood alcohol levels. Subsequently, direct headspace was used routinely to determine residual monomer in polymer, such as vinyl chloride in polyvinyl chloride (PVC).

The success of headspace sampling for GC depends on partitioning of volatile components between an immobilized solid- or liquid-phase sample and the confined gas volume above it. The term "immobilized phase" means that some fraction of the analytical sample is restricted from access to the inlet of the GC. In the case of direct headspace, the immobilization is provided simply by gravity, in that no part of the sample introduction system provides any connection to the condensed phase component of the analytical sample. The steps in headspace

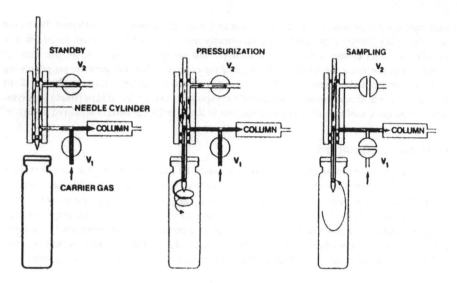

FIGURE 8.   Diagram of balanced pressure headspace sampling, as in Perkin-Elmer HS-40 Automated Headspace Accessory: STANDBY — incubation and equilibration of the sample at a set experimental temperature; PRESSURIZATION — pressurization of the sample vial to a constant, reproducible pressure; SAMPLING — dosing of the sample headspace from the vial on to the column.

sampling diagrammed in Figure 8 generally include the following: (A) STANDBY — incubation and equilibration of the sample at a set experimental temperature; (B) PRESSURIZATION — pressurization of the sample vial with carrier gas to a constant, reproducible pressure; and (C) SAMPLING — dosing of the headspace from the vial onto the column. Several augmentations of the procedure include steps for backflush, multiple headspace extraction (MHE), and others.

For quantitative applications, equilibration of concentrations between the gas and condensed portions of the sample is important for achieving analytical reproducibility. Great quantitative accuracy and precision can be obtained with direct headspace sampling, if proper care is given to reproducing equilibration and sampling times, temperatures, pressures, and flows involved in the analysis.[57, 58]

When the objective is qualitative analysis, using direct headspace sampling with GC/FT-IR, it is appropriate to alter sampling parameters to suit the needs of the experiment. Operationally, this means using longer dosing times (30 to 60 s) to transfer volatiles from the pressurized sample vial to the column. By comparison, dosing times for quantitative headspace GC may fall in the 5 to 15 s range. GC/FT-IR is typically carried out with 0.32 mm i.d. rather than 0.25 mm i.d. capillary columns. Sample vial decompression and mass transfer is more rapid with the wider bore column for the same dosing period and column linear-gas velocity. Larger sample loadings will probably be desired for GC/FT-IR than GC/FID or GC/MS, and the use of 0.32 mm i.d. columns, rather than 0.25 mm i.d., will facilitate this objective.

The advantage of direct headspace is that it provides a rapid, facile method of separating components of analytical interest for GC analysis from other materials in the sample that are unsuitable for introduction into a GC column. The difficulty with direct headspace is that more volatile materials tend to dominate the headspace volume. For example, two components of equal concentration in the condensed-phase sample may show very disparate representation in the headspace chromatogram if one has a significantly higher boiling point, or if one is significantly more polar and associates strongly with components in the condensed phase of the sample. In some cases, increasing the incubation temperature will drive more of the less volatile component into the vapor phase. Higher temperatures will also normally raise the total headspace pressure in the sample vials, which are designed with a safety limit on the contained pressure. Increasing the temperature will also increase the vapor pressure of the more volatile components, sometimes to the point that the safety vent releases and the headspace sample is released and lost. If a suitable compromise incubation temperature cannot be found, the analyst may seek to actively remove the more volatile material by multiple headspace injections of the same sample. As the more volatile components are depleted from the sample, the incubation temperature can be raised to increase the concentration of the less volatile materials in the condensed phase of the sample.

3. Indirect (Dynamic) Headspace

The critical advantage of dynamic (indirect) headspace rests in the fact that the volume of the sample introduced into the GC is decoupled from (i.e., essentially independent of) the original volume of the sample. In practice, dilute analytes in a gas sample are transported dynamically (i.e., with positive net flow) through a chamber containing a solid sorbent which tends to strongly retain them at the trapping temperature. The analytes are subsequently transferred from the trap as a concentrated evolved vapor by elevating the trap temperature (thermal desorption), or by extraction with a solvent (solvent desorption). Because the desorption volume can generally be so much smaller than the volume of the native sample, this technique offers relatively low detection limits and is excellent for trace analysis. A block diagram of an automated thermal desorption system for dynamic headspace analysis is shown in Figure 9. The valve position is set for desorption of the experimental sample tube into the internal trap in Figure 9A. The valve position is switched in Figure 9B to provide for desorption of the internal trap onto the GC column.

The method of water analysis for volatile pollutants known as "purge and trap" is fundamentally a dynamic headspace method in which the initial sample is a liquid rather than a gas or vapor.[59,60] The first operation of purge and trap analysis is sparging the volatiles from the aqueous phase with an inert gas onto a polymer trap. The success of this technique relies heavily on the favorable qualities of the polymer trap, which is usually filled with a heat-stable material such as Tenax-TA or XAD-2 resin. These polymers are highly aromatic structures which may be heated between ambient and 300°C for many cycles without major degradation (in the absence of oxygen). In addition, these materials have a relatively low affinity for water, and organic materials tend to be selectively trapped.

Thermal desorption can be operated on-line, directly to the GC column inlet, or off-line to an intermediate collection vessel.[61, 62] Off-line collection is more likely to expose the sample to greater risk of loss of volatile components. However, it has the advantage that a surplus of sample may be prepared and reserved for subsequent analysis by additional techniques such as GC/MS or nuclear magnetic resonance (NMR).

IV. APPLICATIONS ILLUSTRATING DIVERSE METHODS OF SAMPLE PREPARATION FOR GC/FT-IR

A. Distillation

Liquid samples produced by distillation are usually easy to introduce into the GC, since the work of isolating the volatiles from the nonvolatiles has been inherently handled by the

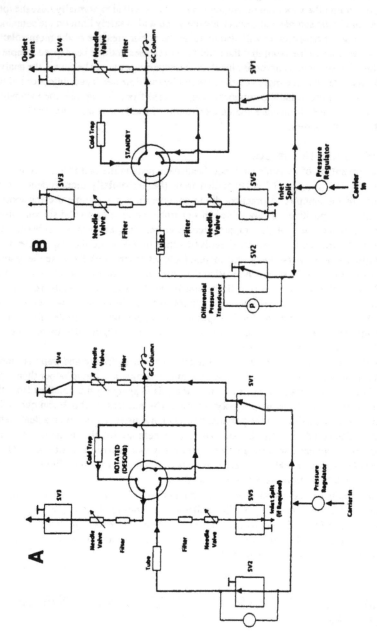

FIGURE 9.   Diagram of automated thermal desorption system for dynamic headspace analysis as in Perkin-Elmer ATD-400: (A) operating mode for desorption of collection trap into the cryotrap and (B) operating mode for transfer of sample from cryotrap to the GC column.

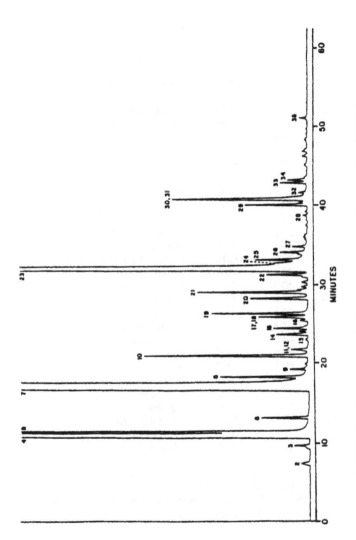

FIGURE 10.   Chromatogram of solvent mixture recovered by conventional distillation.

TABLE 2.   GC/FT-IR Analysis of Paint Thinner

Peak #	Time (min.)	Area % (FID)	Identification (GC/FT-IR)
1	8.8	——	Water
2	9.8	0.18	Methanol
3	12.7	0.18	Ethanol
4	14.3	32.83	Acetone
5	14.9	2.77	Isopropanol
6	17.3	0.39	*DICHLOROMETHANE
7	19.9	39.58	2-Butanone (MEK)
8	23.9	0.58	Ethyl acetate
9	25.2	0.14	1-Propanol
10	27.4	1.20	*1,2-DICHLOROETHANE
11	28.5	0.14	Tetrahydrofuran
12	28.5	——	1-Butanol
13	29.7	0.03	Alkane
14	31.4	0.24	Alkane
15	31.9	0.24	Alkane
16	33.2	0.06	Alkane
17	33.8	0.41	*TRICHLOROETHYLENE
18	33.8	——	Aliphatic alcohol
19	34.4	0.55	Alkane
20	36.9	0.40	4-Methyl-2-pentanone (MIBK)
21	37.9	0.66	Alkane
22	40.9	0.29	Isobutyl acetate
23	41.6	15.78	Toluene
24	42.6	0.10	Alkane
25	43.3	0.32	Mesityl oxide
26	44.6	0.15	n-Butyl acetate
27	45.5	0.07	Alkane
28	50.3	0.03	Alkane
29	52.3	0.37	Ethylbenzene
30	53.2	1.15	m-Xylene
31	53.2	——	p-Xylene
32	54.3	0.13	Aliphatic ketone
33	55.9	0.17	Tetralin
34	56.3	0.10	Alkane
35	66.7	0.04	Alkane

Note: Column: 100 m, 0.32 mm i.d., 5 mm film (methylsilicone)
       Program: 4 min hold at 60°C, ramp 4°C/min to 260°C, 30
       min hold at 260°C.

distillation process. In contrast, samples isolated by extraction may well contain some non-volatile components.

Figure 10 shows the chromatogram of a distilled solvent mixture. This sample was obtained by distillation of a collection of used solvents. The objective of this investigation was to evaluate the potential of the distilled product for possible reuse by a major American railroad as a recovered paint solvent. However, for several reasons, the requirement was imposed by the user that the recovered solvent contain no chlorinated solvents. It was the objective of the

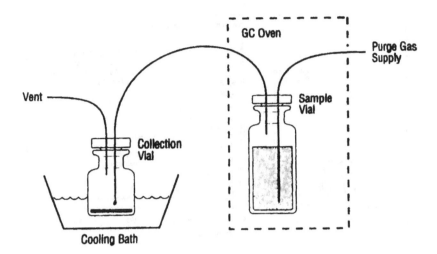

FIGURE 11. Diagram of microdistillation apparatus constructed from headspace vials and fused silica capillary tubing.

GC/FT-IR analysis to detect and identify any chlorinated solvents present in the clear, colorless distillate. Table 2 shows that GC/FT-IR was able to determine that multiple chlorinated compounds had found their way into the recovered solvent mixture.[63] Other components in the chromatogram were either specifically identified or at least classified as to compound type, e.g., alkane, etc. Based on these results, this batch of recovered solvent was not approved for further use as a paint thinner.

One efficient, inexpensive, and convenient way for isolating volatile materials from solid matrices involves the use of a combination of two septum-capped vials and fused silica capillary tubing, as is diagrammed in Figure 11. It can be easily constructed from parts in common use for GC. This preparation process has many aspects in common with direct and indirect headspace sampling, the purge and trap technique, as well as conventional distillation operations, particularly so-called "dry distillation". The technique is simple to implement by an operator of a GC/FT-IR system, because the oven and carrier gas of the GC play the role of the thermal and pneumatic drivers, respectively, for transfer of materials.[64]

A solid sample is filled and sealed into a septum-capped vial. Particularly useful are the vials usually used in automated headspace analysis. Lengths of fused silica capillary tubing (0.32 mm i.d. tubing is typically used) are the conduits to introduce carrier gas into the vial containing the solid sample and to conduct volatile materials from the solid sample vial to the cooled collection vial. The success of this system is based on the fact that fused silica tubing can be inserted through a headspace vial septum and that the septum forms a reasonably tight seal around the fused silica tubing. Losses due to leakage are negligible, and the system runs at low operating pressures, as long as the system is monitored to ensure the transfer tubing is not plugged by solid sublimate from the sample. With a hand-held magnifying glass, one can see through the polyimide coating on the fused silica tubing and observe the droplets of distillate as they are carried over to the collection vial. Insertion of the silica tubing through the septa is easier if a pilot hole is established by piercing the septum with a standard GC syringe needle before the fused silica tubing is forced through. Establishing a pilot hole reduces the force necessary to penetrate the septum and minimizes the chance of plugging the fused silica with septum material. Pilot holes should also be established in the collection vial:

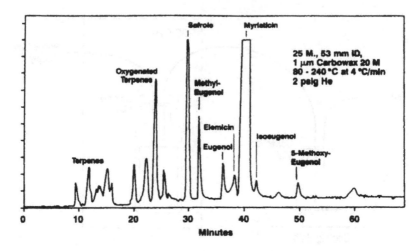

FIGURE 12. Gram-Schmidt chromatogram of nutmeg distillate, prepared using the microdistillation apparatus of Figure 11.

one for receiving the capillary carrying the distillate from the solid sample vial and one to serve as a vent port for carrier gas.

The solid sample vial is placed in the GC oven; the carrier gas-source capillary is introduced through one guide hole, and the transfer capillary is introduced through the other. The exit end of the transfer capillary is fed through a port in the GC oven wall to the cooling bath outside of the GC. Most chromatographs have potential exit ports designed in the oven wall for sampling valves, subambient operating accessories, and transfer lines to external detectors such as MS, FT-IR, etc.

This distillation system enjoys the inherent advantage of excellent control of both the distillation temperature and the purge-gas flow rate, which are afforded as standard facilities of a gas chromatograph.

Figure 12 shows the results of chromatography with a polar Carbowax capillary column of the distillate obtained from ground nutmeg, based on preparation with the device described above. GC/FT-IR analysis of this material shows that the distillate components are terpenes, oxygenated terpenes, aromatics, sesquiterpenes, and some fatty acids. The major aromatic component is myristicin (mol wt 192), which is obtained along with other aromatics, such as elemicin, safrole, eugenol, methyleugenol, and 5-methoxyeugenol (4-desmethylelemicin).[64] The structures of these di- and trisubstituted propenylbenzenes are shown in Figures 13 and 14, respectively.

In a similar experiment, the composition of volatiles in parsley seeds was investigated by GC/FT-IR, based on a sample prepared with the distillation device described above. The Gram-Schmidt chromatogram of the parsley seed distillate, along with those of similarly prepared distillates of sassafras root bark and nutmeg volatiles, are shown together in Figure 15. Table 3 lists the various peaks identified, including the propenylbenzene derivatives encountered.[65] In both the nutmeg and parsley seed volatiles, myristicin (peak 5) is one of the major aromatic components. The final large peak, marked 11, is due to parsley apiol (mol wt 222). The experimental spectra of myristicin and parsley apiol are shown in Figures 16A and 17A, respectively.

Between myristicin and apiol in the parsley-seed distillate chromatogram there are several sesquiterpene peaks and one due to another aromatic compound, peak 9. The infrared spectrum

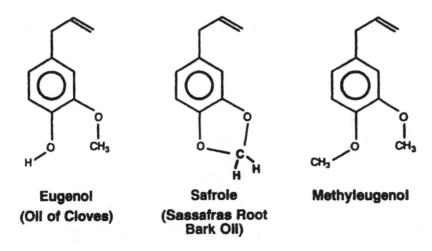

FIGURE 13. Structures of disubstituted propenylbenzene compounds found in nutmeg distillate.

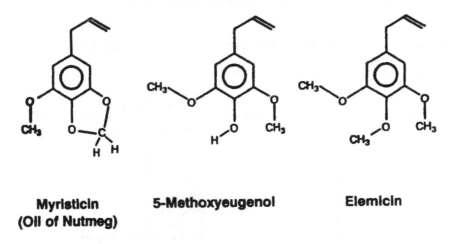

FIGURE 14. Structures of trisubstituted propenylbenzene compounds found in nutmeg distillate.

of this sample component of peak 9 is shown in Figure 16B. The comparison of this spectrum with those of myristicin and apiol indicate that, in this material, there is no evidence of an aliphatic methylene C–H stretch of the methylenedioxy group around 2775 cm⁻¹. However, there are peaks at 910, 990, and 3080 cm⁻¹, which are characteristic of a terminal vinyl (–CH=CH₂) of the allyl substituent found in myristicin, apiol, eugenol, safrole, and other allylbenzene homologues. In addition, the strongest bands are observed in the region of aromatic C–O stretch. Based on these results, it was hypothesized that the unknown aromatic was most probably the 2,3,4,5-tetramethoxyallylbenzene. This hypothesis was subsequently confirmed by supporting GC/MS data.[66]

Figure 17B contains the spectrum of dill apiol, which corresponds to Peak 8 in Figure 7, discussed earlier. The GC/FT-IR spectra of the two apiol isomers are easily differentiable on the basis of these reference spectra. The structures of the three tetrasubstituted propenylbenzenes, dill apiol, parsley apiol, and the tetramethoxy compound are shown in Figure 18.

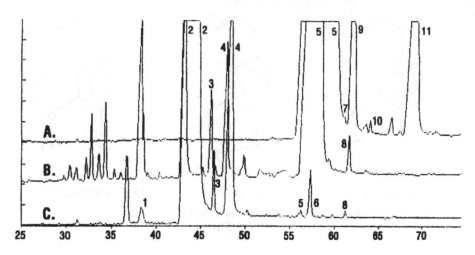

FIGURE 15. Comparison of Gram-Schmidt chromatograms of volatiles isolated by microdistillation of natural materials: (A) parsley seeds, (B) nutmeg, and (C) sassafras root bark.

TABLE 3. GC/FT-IR Analysis of Propenylphenols in Plant Materials by Microdistillation

Peak #	Identification (GC/FT-IR)
1	Estragole
2	Safrole
3	Eugenol
4	Methyleugenol
5	Myristicin
6	Elemicin
7	Sesquiterpene
8	5-Methoxyeugenol
9	Tetramethoxyallylbenzene
10	Sesquiterpene
11	Apiol (parsley type)

Note: Seeds of parsley (*Petrosalinum sativum*), nutmeg (*Myristica fragrans*), root bark of sassafras (*Sassafras albidum*)

Column : 50 Meter, 0.32 mm i.d., 5 micron film (methylsilicone)

B. Solvent Extraction

Solvent extraction, in addition to and in combination with distillation, has probably been the most used technique of the classical laboratory synthetic procedures adapted to the service of GC sample preparation. Liquid solvent extraction may become much less used in the future because it may be replaced by supercritical fluid extraction. Nevertheless, for rapid, uncomplicated, small scale isolation of organic materials from aqueous solutions or adsorbed surface deposits, it may continue to be used, particularly for preliminary investigations.

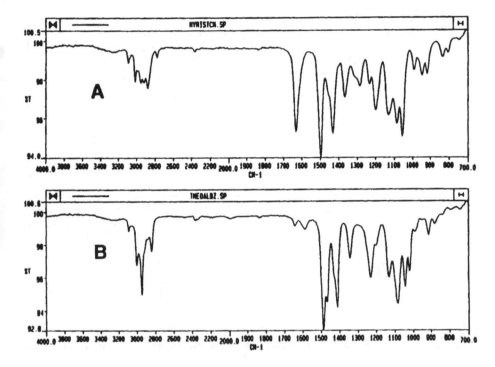

FIGURE 16. GC/FT-IR spectra of propenylbenzene compounds found in parsley seed oil: (A) myristicin, peak 5 and (B) 2,3,4,5-tetramethoxyallylbenzene, peak 9.

In practice, solvent extraction involves the production of a solution phase sample suitable for introduction, usually by syringe, into a GC. In practice, this often amounts to extraction of aqueous phases by organic solvents, although many variations exist, such as in the following example.

The Gram-Schmidt chromatogram in Figure 19 shows the result of hexane extraction of volatiles directly from blooms of domestic sage or herb sage (*Salvia officinalis*). To obtain the data shown, the fragrant light blue flowers of the herb sage were simply surface-rinsed with the hydrocarbon solvent. The extract was analyzed by GC/FT-IR; the identified components are listed in Table 4. The results obtained are not inconsistent with published studies on sage oil.[67] However, sage oil is usually obtained from leaves rather than from blooms, and some variation from the leaf is not unexpected.

C. Direct (Static, Equilibrium) Headspace Applications

Direct headspace sampling is also called static or equilibrium headspace sampling. It is likely to be the method of choice for GC/FT-IR when two conditions are met. First, the possible presence of nonvolatiles in the sample may rule out direct injection. Second, at incubation temperatures that the sample can withstand without decomposition, the components of analytical interest have appreciable vapor pressure in the vapor phase volume, i.e., the headspace, above a confined sample.

Figure 20 shows the headspace GC/FT-IR chromatogram of a analgesic ointment. Some ointments are formulated with an oil base or grease-like material as a carrier for the active

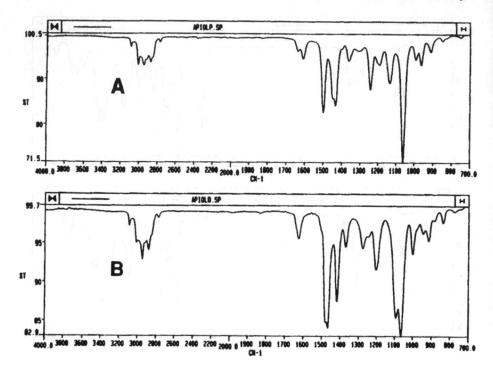

FIGURE 17. Comparison of GC/FT-IR spectra of apiol isomers: (A) parsley apiol spectrum from peak 11 in Figure 6, and (B) dill apiol spectrum from peak 7 in Figure 7.

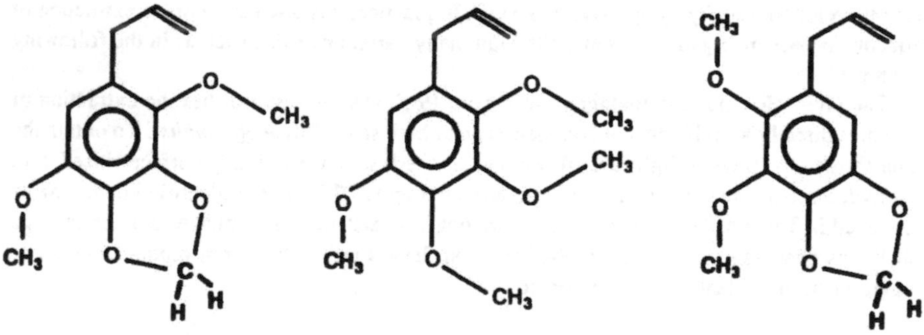

FIGURE 18. Structures of tetrasubstituted propenylbenzenes found collectively in parsley and dill seed oils.

ingredients. Advertising claims purported that this ointment was "greaseless". However, as a precaution against injection of any possible nonvolatiles, headspace sampling was performed to investigate what volatile components might be present in the ointment besides those declared on the label. In fact, the two major peaks in the chromatogram were confirmed by the

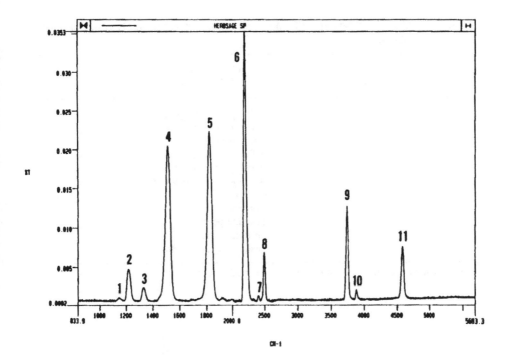

FIGURE 19. Gram-Schmidt chromatogram of volatiles isolated from domestic sage *(Salvia officinalis)* by solvent extraction.

GC/FT-IR data to be menthol and methyl salicylate as claimed on the packaging. The smaller peaks early in the chromatogram are due to water and methanol.[68]

Headspace sampling is useful for materials with sufficient vapor pressure to transfer nanogram to microgram amounts of compounds to the GC column. However, large molecules and polar materials may not have sufficient vapor pressure for headspace sampling. One solution is derivatization of the materials in the native sample to produce relatively more volatile compounds.[69] This conversion to the derivatized homologue can often be carried out with convenience directly in the headspace vial. Figure 21 shows the chromatogram of the volatile reaction products of sodium methoxide and a synthetic lubricant intended for use in high temperature applications. Such lubricants are often composed of fatty acid esters of poly-ol materials such as pentaerythritol. From the infrared spectrum of the bulk material, it might be possible to determine what poly-ol formed the basis of this material. On the other hand, the spectral differences among medium and longer chain fatty acids are relatively slight, which makes it difficult to distinguish what fatty acids are incorporated in the ester. On the other hand, vapor phase infrared spectra of fatty acid methyl esters are easier to differentiate. The addition of sodium methoxide in methanol to a headspace vial containing the ester lubricant leads to an *in situ* transesterification to methyl esters of the respective acids.[70] However, a better procedure was devised using neat sodium methoxide mixed with the ester and equili-

TABLE 4.   GC/FT-IR Analysis of Sage (*Salvia officinalis*) from Solvent Extraction
of Floral Volatiles.

Peak #	Identification (GC/FT-IR)
1	α-Thujene (3-thujene)
2	α-Pinene (2-pinene)
3	Camphene
4	β-Pinene (nopinene)
5	Eucalyptol (1,8-cineole)
6	α-Thujone (3-thujanone)
7	Camphor
8	Borneol
9	Sesquiterpene
10	Sesquiterpene
11	Sesquiterpene alcohol

Note: Column: 50 m, 0.32 mm i.d., 5 μm film (methylsilicone)

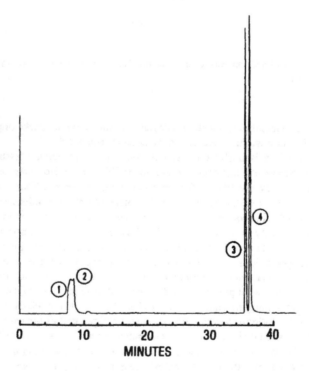

FIGURE 20.   Gram-Schmidt chromatogram of commercial analgesic ointment analyzed by direct
headspace GC/FT-IR. (1) water, (2) methanol, (3) menthol, (4) methyl salicylate.

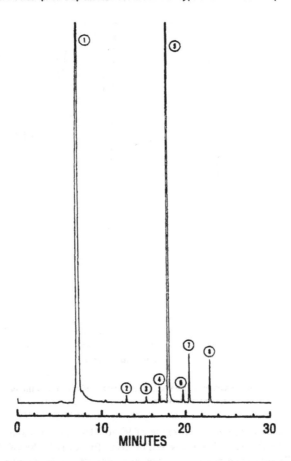

FIGURE 21. Gram-Schmidt chromatogram of transesterification products from high temperature lubricant analyzed by derivitization headspace GC/FT-IR.

brated in the headspace vial. This improvement avoided the problem with excess vial pressure due to methanol vapor and the swamping of the chromatogram by the oversized solvent peak. The transesterification rate is likely to be higher in a homogeneous phase with a large excess of methanol solvent. However, more than sufficient conversion is achieved by the simpler, solvent-free headspace direct derivatization. Additionally, methyl esters are relatively volatile and are easily separated by GC. Figure 22 shows the spectrum of the major methyl ester peak from Figure 21. This material was easily identified as methyl heptanoate.[68]

Reaction chemistry may occur in the headspace vial without the addition of derivatizing reagents. Some samples generate headspace components by thermolysis of materials at experimental incubation temperatures, even though these components are not free under ambient condition. Figure 23 shows the chromatogram obtained by headspace GC/FT-IR of Gumbo Filé (the ground leaves of the sassafras plant). This material is commonly called for in soup and gumbo recipes in the popular Louisiana Cajun cuisine.[71] As the incubation temperature of

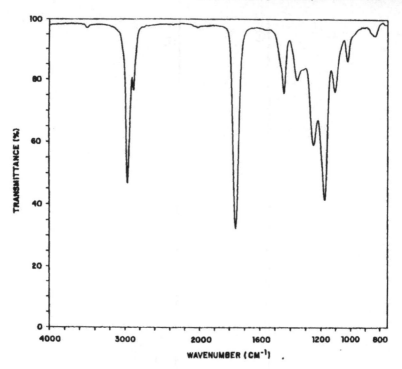

FIGURE 22. GC/FT-IR spectrum of major methyl heptanoate, identified as the major ester obtained by *in situ* transesterifcation of high temperature lubricant.

these sassafras leaves is raised to the range of 90 to 110°C, a substantial increase in vial pressure is observed. Along with some water, the large first peak in the GC/FT-IR chromatogram can be easily identified as CO_2, evolved in quantity from the sample.[72]

Along with CO_2 evolution, a group of other small molecules are evolved along with multiple members of the terpene family. The heated sample produces a markedly characteristic flavorful aroma, if allowed to depressurize through a spare syringe needle. Among the group of small molecules evolved is a set of several aliphatic aldehydes. The evolution of these aldehydes was not anticipated. However, these aldehydes can be related to corresponding amino acids, possibly arising from oxidative decarboxylation of amino acids or their bioprecursors, as shown in Figure 24. Figure 25 shows the correspondence between the aldehydes observed and their respective associated amino acids.

D. Indirect (Dynamic) Headspace Applications

Indirect or dynamic headspace sampling involves passing components in a vapor phase sample into an intermediate adsorption trap prior to their subsequent desorption. The use of an adsorption trap has the advantage of increasing the effective amount of sample available for chromatographic analysis, because large volumes of a dilute vapor may be passed through the trap. The adsorbed compounds are usually removed from the trap by either solvent or thermal desorption. In either case, dynamic (indirect) headspace sampling method provides a way for analysis of volatiles free of nonvolatiles in a native condensed phase sample. For gas phase samples, it provides a way to make the volume of sample introduced to the GC a parameter

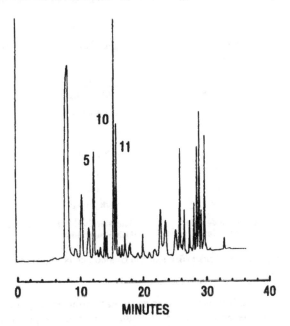

FIGURE 23. Gram-Schmidt chromatogram of gumbo filé (ground leaves of sassafras, *Sassafras albidum*) analyzed by thermolysis headspace GC/FT-IR.

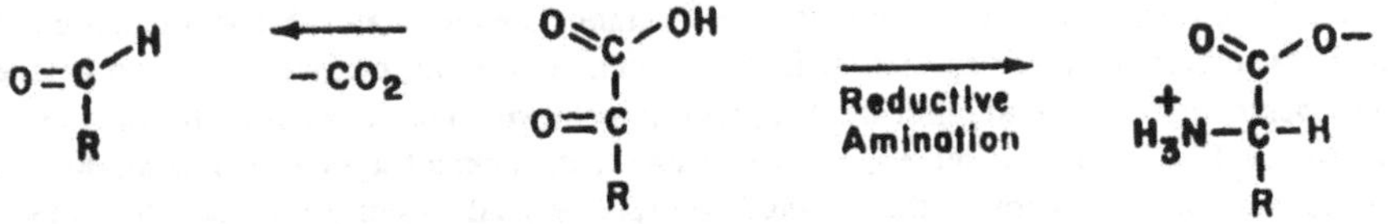

$$O=\overset{H}{\underset{R}{\overset{|}{C}}} \quad \xleftarrow{\quad -CO_2\quad} \quad \overset{O=\overset{OH}{\overset{\diagup}{C}}}{\underset{R}{\underset{|}{C}}}=O \quad \xrightarrow[\text{Amination}]{\text{Reductive}} \quad \overset{O=\overset{O-}{\overset{\diagup}{C}}}{H_3\overset{+}{N}-\underset{R}{\overset{|}{C}}-H}$$

FIGURE 24. Reaction pathways of potential amino acid precursors that could lead to aldehyde formation.

independent of the original volume in which the sample components were dispersed. This last point accents the major differentiating factor between direct and indirect headspace. When equilibrium concentrations of analytes are too low for practical operation of direct headspace sampling, indirect sampling with adsorption traps may be appropriate, in order to achieve the required detection limits.

The analysis of floral fragrances illustrates the utility of indirect headspace sampling for analysis of dilute gaseous samples. Classical methods for preparing floral fragrances include extraction with fat, steam distillation, and vacuum distillation, among others. However, these methods risk loss of more volatile components, degradation of reactive materials, and artifact component production. For preparative scale operations, such as isolation of rose, jasmine, and other natural fragrances for perfumery, the tolerance for compositional variation due to preparative technique may be greater. However, for critical analytical purposes, these older methods are often unsuitable.

The accurate determination of tropical orchid fragrance composition became an important step in the study of the cooperative behavior between orchids and several species of Central

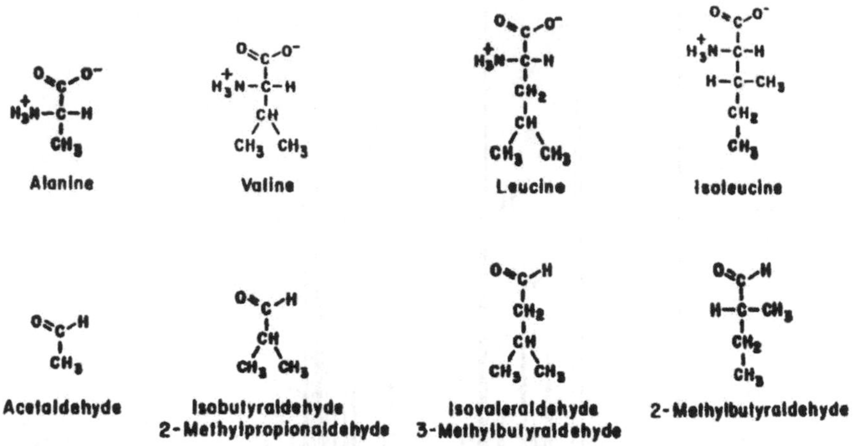

Alanine   Valine   Leucine   Isoleucine

Acetaldehyde   Isobutyraldehyde   Isovaleraldehyde   2-Methylbutyraldehyde
                  2-Methylpropionaldehyde 3-Methylbutyraldehyde

FIGURE 25. Correspondence of aldehydes found in Gumbo Filé headspace and their possible amino acid precursors.

and South American bees. Various bee species are each preferentially attracted to particular corresponding orchid species on the basis of the fragrance composition of the orchid's inflorescence (flower or flower cluster). These are male bees, not female or worker bees, and the fragrance is not ingested but collected in hind tibial organs on the rear legs, roughly the bee's "knee". This unusual phenomenon involving orchids and bee's knees is termed "the Euglossine Syndrome".[73]

Not all components of the evolved orchid fragrance are attractants. Some are pronounced repellants to particular bee species. It is the combination of various attractant and repellent components that appears to tailor the fragrance to achieve a high selectivity for only certain bee species. Because the orchid appears to be able to develop a fragrance composition which is effectively species specific, the targeted bee species inadvertently becomes the agent for transfer of pollen granules only from that respective orchid species. By this scheme, the orchid maintains its genetic isolation in a biological environment rich in opportunities for loss of species identity loss by hybridization. It is thought that genetic isolation of the particular bee species is served by its interaction with a specific orchid species. Genetic isolation of the bee species is accomplished by pheromones, specific sex attractants.

The major analytical question attached to investigation of the Euglossine syndrome centers around determining the composition of fragrances of a number of tropical orchids and attempting to relate that information to the composition of materials collected from the hind tibial organs and other target zones of the respective bee species that pollinate that orchid.

Certain tropical orchids produce fragrance materials only during specific hours of the day. To obtain enough of the fragrance components to provide material for multiple analytical techniques and to allow for archival storage of part of the sample, the dilute materials were sampled by the off-line indirect headspace method. At the appropriate point in its development, the inflorescence (bloom) of the orchid was surrounded by a clear plastic housing. Air was drawn by a pump, through the housing, and into a glass tube containing Tenax-TA to trap the fragrance components. The air inlet to the housing was "scrubbed" by a large charcoal trap. Visible light exposure was timed to match that which the orchid would experience under natural growing conditions. The trapped fragrance components were desorbed with gentle heating under high vacuum into a trap cooled with liquid nitrogen. After the desorption step, the recovered fragrance materials were rinsed into a collection vial with a small volume of pentane and stored in a freezer until needed for analysis.

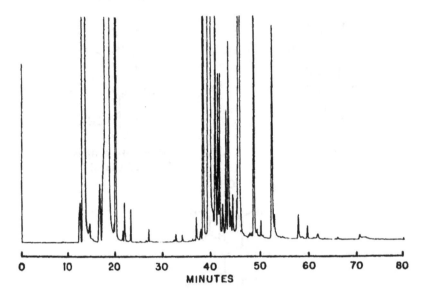

FIGURE 26. Chromatogram of *Stanhopea anfracta* orchid fragrance, prepared by dynamic headspace sampling.

Figure 26 shows the chromatogram of the fragrance from the *Stanhopea anfracta* orchid. This orchid generates a complicated fragrance with a large variety of components. The GC/MS analysis of this fragrance suggested that the compound represented by the major peak around 40 min was an unsaturated terpene alcohol with an empirical formula of $C_{10}H_{16}O$. However, there are numerous natural products with this empirical formula. Based on other additional data from other sources, it was suspected that the material in question was a substance with the common name of ipsdienol. The common name is related to the fact that ipsdienol is a sex attractant of the ips bark beetle, a scourge of the ponderosa pine in the western U.S.[74] Figure 27 shows the structures of ipsdienol and two related structures which were initially considered as possible candidates for the identity of the substance responsible for the peak at 40 min. The comparison of the GC/FT-IR spectrum of the peak and that of a synthetic reference standard of ipsdienol, as shown in Figure 28, provides a band for band match in the two materials. Additionally, the retention time of the peak in the orchid fragrance exactly matches that of the ipsdienol standard. These results represented the first occasion of the identification of ipsdienol as a component in the fragrance of a tropical orchid.[75]

Figure 29 shows the chromatogram of a fragrance sample isolated from the orchid *Catesetum longifolium*. Peak 9 was easily identified by both GC/FT-IR and GC/MS as carvone ($C_{10}H_{14}O$), a common terpene ketone. The GC/MS analysis suggested that peak 10 was carvone with an addtional oxygen ($C_{10}H_{14}O_2$) and that it was probably bonded on the ring as epoxide. Figure 30 shows the structure of carvone and that of two stereochemical isomers of epoxycarvone. It was impossible to determine which isomer was the orchid epoxycarvone strictly on the basis of an initial GC/MS work.

Epoxycarvone was prepared by laboratory synthesis and produced unequal amounts of two ring epoxide products. On the basis of detailed analysis of the products by NMR, it was concluded that the major epoxycarvone isomer formed in the laboratory synthesis was the *cis* isomer.[76] In this case, the terms *cis* and *trans* refer to the orientation of the methyl and isopropenyl groups, on the same or opposite sides of the plane of the ring respectively. In the GC/FT-IR analysis of the synthetic epoxycarvone reaction mixture, the major product, the cis

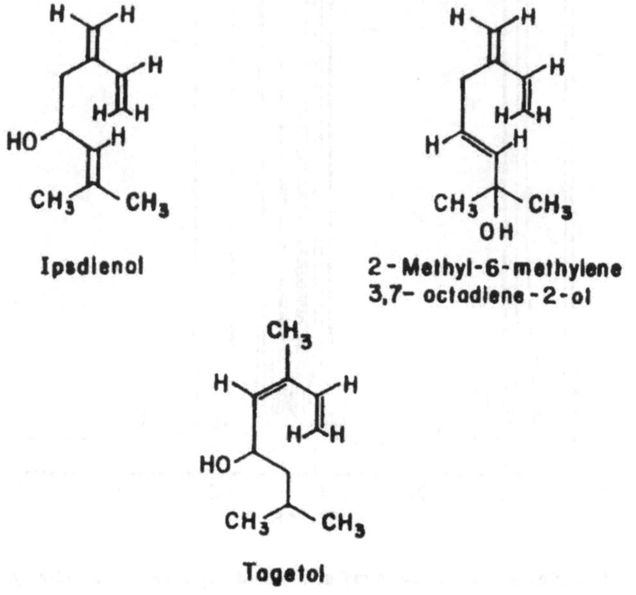

Ipsdienol

2 - Methyl-6-methylene
3,7- octadiene -2-ol

Tagetol

FIGURE 27.   Structures of ipsdienol and related terpene alcohols.

isomer eluted in 26.88 min and the carbonyl band maximum absorbance was found at 1728 cm[-1]. The minor product, the trans isomer, eluted in 26.27 min, and the carbonyl band maximum absorbance was found at 1732 cm[-1]. Other differences were noted in the IR spectra of the two isomers in the C–O stretching region around 1200 to 1000 cm[-1]. The GC/FT-IR spectra of the *cis* and *trans* isomers of epoxycarvone are shown in Figure 31.[75]

Figure 32 shows the GC/FT-IR spectra of carvone and epoxycarvone found in the orchid fragrance. The epoxycarvone isomer in the orchid fragrance in Figure 32B has a GC/FT-IR spectrum identical with that of the *trans* isomer of Figure 31A, with carbonyl-band maximum absorbance at 1732 cm[-1] and the same band pattern in the C–O stretching region. The GC/FT-IR data conclusively confirm what had been suspected on the basis of GC retention time data and the NMR analysis of related compounds of known absolute configuration.[76]

E. Supercritical Fluid Extraction (SFE)

One of the key aspects of SFE is pressure dependence of solubility. Not only are most compounds more soluble in supercritical fluids at higher pressures, but there appears to be a pressure threshold for solubility of certain compounds. This behavior is easy to demonstrate with commonly available materials, such as a commercially available product known as Constant Comment Tea™. To carry out SFE on this sample, the tea mixture was removed from the paper bag enclosure. The yellow pieces of lemon peel were physically removed, and the residual dark fragments were sealed into a steel extraction vessel. The sample was put through an initial extraction under mild conditions (120 atm, 50°C, 20 min). The extract was decompressed through a steel flow-restrictor into a trap filled with glass beads cooled to -30°C. The trap was warmed to 40°C and eluted with 1.5 ml of methanol. Results of GC/FT-IR analysis on mild SFE condition sample are shown by the Gram-Schmidt chromatogram in Figure 33B. The two major peaks observed at 8.6 and 13.9 min arise from limonene and eugenol, respectively. The spectrum of the eugenol peak is shown in Figure 33A.

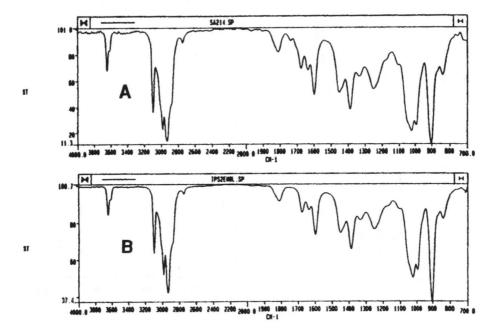

FIGURE 28. Investigation of composition of *Stanhopea anfracta* orchid fragrance: (A) spectrum of major peak at 40.2 min in chromatogram in Figure 25 and (B) spectrum of ipsdienol synthetic reference standard.

The identical sample was subjected to another extraction under more strenuous conditions (450 atm, 70°C, 20 min). The Gram-Schmidt chromatogram in Figure 34B shows the results of extraction under more strenuous SFE conditions and includes a third peak which was not in evidence in the extract obtained with mild conditions. The GC/FT-IR spectrum of the third peak, shown in Figure 34A, is that of caffeine, a relatively polar molecule which was not extracted under the initial mild conditions.

As mentioned in earlier sections, SFE may replace older methods of sample preparation, both at the analytical and synthetic scale of operation. The use of SFE for extraction of volatile flavor components can be demonstrated by the extraction of the herb known as sweet basil (*Ocymum basilicum*) under moderate conditions (250 atm, 60°C, 20 min). The Gram-Schmidt chromatogram of basil obtained by on-line SFE/GC/FT-IR is shown in Figure 35. The heated GC injector helps overcome the cooling caused by CO_2 decompression and promotes volatilization of the extracted materials onto the initially unheated column. After the CO_2 has purged from the system, temperature programming of the GC oven is initiated. The zoom box in the lower window of Figure 35 shows the expanded chromatogram in the region which contains the volatile basil fragrance compounds.

Figure 36 shows the spectra of peaks 12 and 15 identified as *cis*- and *trans*-methyl cinnamate. Figure 37 shows the spectra of peaks 6 and 14 identified as linalool and methyleugenol. Peaks 6, 12, and 14 were observed but not specifically identified in a previous study of basil volatiles prepared by SFE.[46] Figure 38 shows the spectrum of estragole and chavicol from peaks 8 and 9, respectively. The structures of estragole, chavicol, and chavicol acetate are shown in Figure 39. The presence of chavicol acetate in basil is suspected, but has not yet been established by GC/FT-IR. Table 5 lists the compounds (excluding the sesquiterpenes) identified in this study.

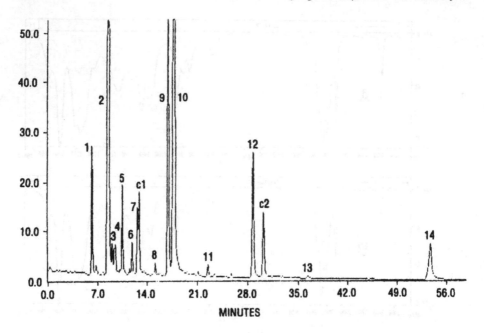

FIGURE 29. Chromatogram of the *Catesetum longifolium* orchid fragrance, prepared by dynamic headspace sampling.

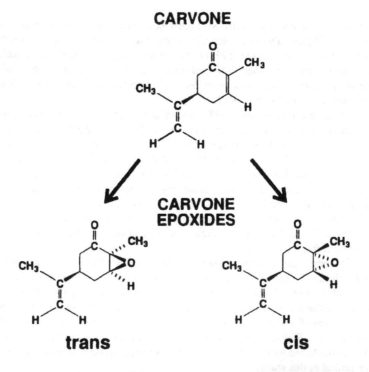

FIGURE 30. Structures of carvone and epoxycarvone isomers, potentially in the fragrance of *Catesetum longifolium*.

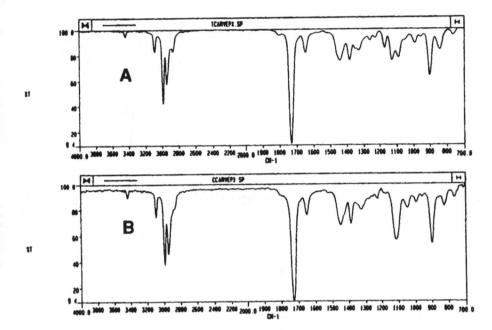

FIGURE 31. GC/FT-IR spectra of synthetic epoxycarvones: (A) *trans*-epoxycarvone, 26.27 min, $v_{(C=O)}$ = 1732 cm^{-1} and (B) *cis*-epoxycarvone, 26.88 min, $v_{(C=O)}$ = 1728 cm^{-1}.

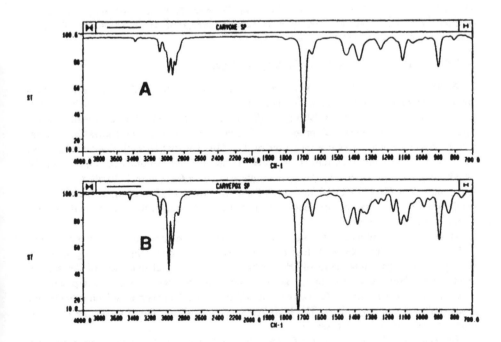

FIGURE 32. GC/FT-IR spectra of carvone and epoxycarvone from *Catesetum longifolium* orchid fragrance: (A) carvone, 25.58 min, $v_{(C=O)}$ = 1720 cm^{-1} and (B) orchid epoxycarvone, 26.27 min, $v_{(C=O)}$ = 1732 cm^{-1}.

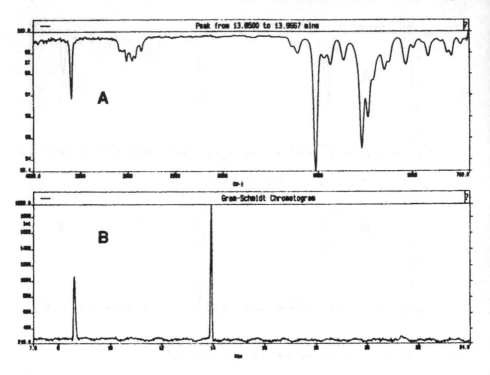

FIGURE 33. Supercritical fluid extract of Constant Comment Tea™ under relatively low pressure conditions, 125 atm, off-line with cryotrapping: (A) spectrum of eugenol peak at 13.91 min and (B) Gram-Schmidt chromatogram of 125 atm extract.

F. TG/FT-IR and TG/GC/FT-IR Applications

Aspartame is a popular artificial sweetener, which is recommended for use in soft drinks and other beverages. However, it is not recommended for use in cooked goods, such as cakes, pies, etc., because of its thermal lability. TG/FT-IR and TG/GC/FT-IR enable detailed investigation of the thermochemical properties of aspartame.[77] Figure 40A shows the thermal weight-loss curve of aspartame as measured by the TGA. The first derivative of the weight loss curve, shown in Figure 40B, reflects the relative rate of mass evolution during each of the weight loss events. The derivative of the weight loss curve (if inverted) closely matches the trace, shown in Figure 40C, of the Gram-Schmidt mass-evolution profile. The similarity results from the fact that the net concentration of evolved gases in the infrared cell depends on the rate of gas evolution.

Three thermally induced weight-loss events are in evidence during the main part of the run under a nitrogen atmosphere. A fourth weight loss was produced by switching to an air atmosphere at high temperature (800°C), which resulted in total oxidation of the sample residue, i.e., 100% weight loss by the sample. The high-temperature oxidation step is useful as a general procedure to test for nonvolatile additives or fillers such as silicon dioxide and titanium dioxide, among others. The high-temperature final oxidation can also be useful to minimize carry-over between samples.

The TG/FT-IR data show that the first weight loss corresponds to a nonstoichiometric loss of water. To test the hypothesis that simple drying is occurring, it is useful to attempt drying the native sample at several elevated temperatures and recovering a portion of the sample from

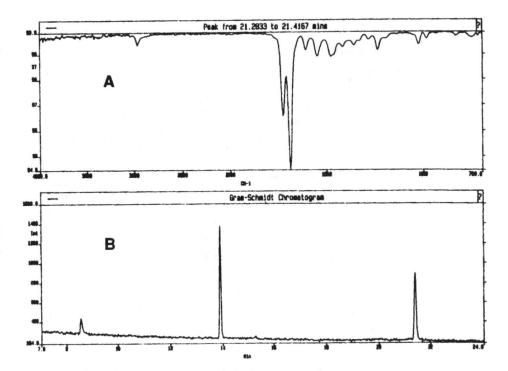

FIGURE 34. Supercritical fluid extract of Constant Comment Tea™ under relatively high pressure conditions, 500 atm, off-line with cryotrapping: (A) spectrum of caffeine peak at 21.35 min and (B) Gram-Schmidt chromatogram of 500 atm extract.

the TGA pan for evaluation by elemental analysis. The molecular formula for aspartame suggests the expected values shown in Line A of Table 6. The fact that the native sample is somewhat "wet" is indicated by the low carbon and nitrogen values shown in Line C of Table 6. Moreover, it is also clear that the native sample is not a true hydrate because the values are not low enough to fit those of an aspartame hydrate, shown in Line B of Table 6. Drying the sample to constant weight in the TGA pan at 120 and 150°C leads to elemental analyses which approach, but do not attain, the expected theoretical values, as shown in Lines D and E, respectively. Aliquots of the native and dried samples used for elemental analysis were also evaluated by FT-IR in potassium bromide (KBr) pellets, as shown in Figure 41. The differences in the spectra support the idea of a progressive drying of the sample, since the progressive changes in spectra A, B, and C are found in the broad OH stretch region around 3000 cm^{-1} and the OH bend region of water around 1600 cm^{-1}.

In the vicinity of 200°C, aspartame undergoes a second thermal weight-loss event. The vapor-phase infrared data show that the evolved material is mostly methanol, along with a small amount of residual water. Elemental analysis of the sample residue after methanol loss suggests a molecular formula consistent with that of aspartame less one molecule of methanol, as shown in Lines F and G of Table 6. The FT-IR spectra of this residue and the native sample of aspartame are contrasted in figure 42. Absorption bands appear to be fewer and weaker in the residue spectrum relative to that of the native aspartame specimens, which suggests that residue material may be a generally less polar, more symmetrical molecule.

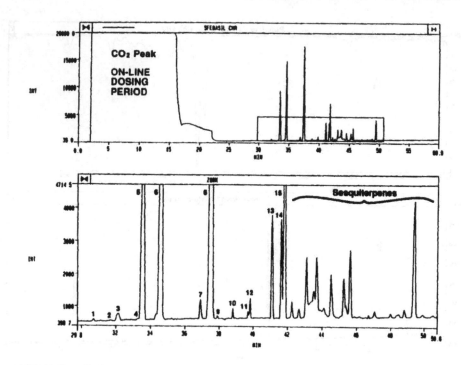

FIGURE 35. Chromatogram of sweet basil (*Ocymium basilicum*), by on-line SFE/GC/FT-IR with a thick film (5μm) column.

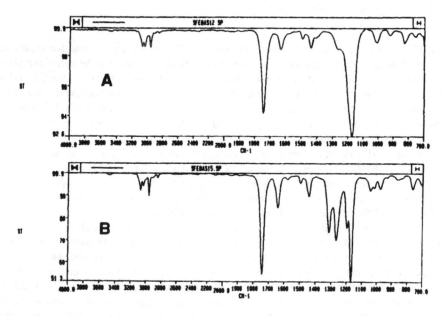

FIGURE 36. Spectra of cinnamate esters from sweet basil: (A) *cis*-methyl cinnamate and (B) *trans*-methyl cinnamate.

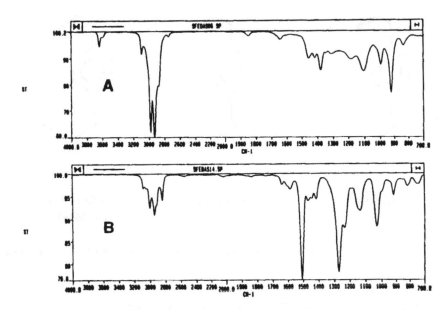

FIGURE 37. Spectra of two compounds previously unidentified in SFE extract of sweet basil: (A) linalool and (B) methyleugenol.

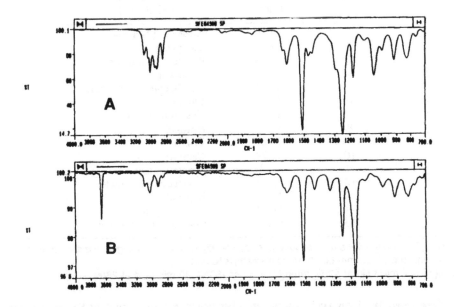

FIGURE 38. Spectra of monosubstituted propenylbenzene compounds identified in sweet basil: (A) estragole and (B) chavicol.

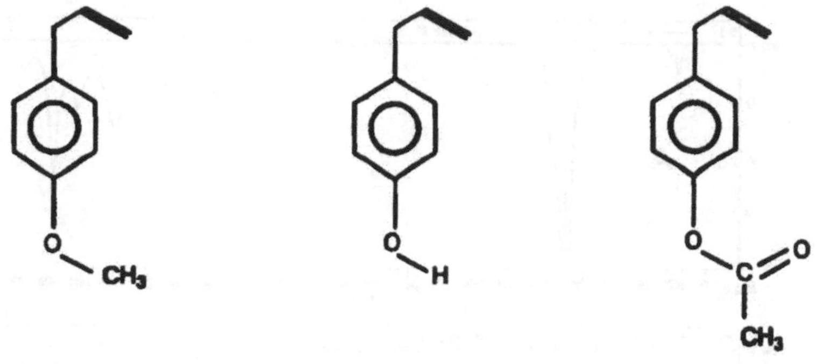

Estragole   **Chavicol**   **Chavicol Acetate**

FIGURE 39. Structures of monosubstituted propenylbenzene compounds identified or suspected in basil.

TABLE 5. On-Line SFE GC/FT-IR Analysis of Sweet Basil (*Ocymum basilicum*)

Peak #	Time (min.)	Identification (GC/FT-IR)
1	30.70	α-Pinene (2-pinene)
2	31.65	Camphene
3	32.13	β-Pinene (nopinene)
4	32.22	Limonene
5	33.57	1,8-Cineole (eucalyptol)
6	34.60	Linalool
7	36.90	Unsaturated ester
8	37.52	Estragole
9	38.27	Chavicol
10	38.79	Carvone
11	39.69	Bornyl acetate
12	39.82	*cis*-Methyl cinnamate
13	41.11	Eugenol
14	41.64	Methyleugenol
15	41.82	*trans*-Methyl cinnamate

Note: Remaining peaks are sesquiterpenes, typically $C_{15}H_{24}$, and oxygenated sesquiterpenes.
Column: 50 m, 0.32 mm i.d., 5 μm film (methylsilicone)
Program: 4 min hold at 40°C, ramp 8°C/min to 260°C, 30 min hold at 260°C.

The above data form a coherent picture when one takes into account the known structure of aspartame, which is the methyl ester of aspartylphenylalanine, as is shown in Figure 43A. The dipeptide methyl ester can orient to facilitate internal cyclization with loss of methanol to form a 2,4-diketopiperazine, as shown in Figure 43B. The piperazine is not an effective sweetener, and when aspartame undergoes this decomposition, it loses the value for which it was intended. Consequently, manufacturers strongly advise against its being used in cooked products such as breads, cakes, and pies, etc.

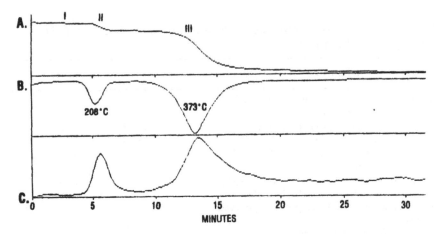

FIGURE 40. TG/FT-IR analysis of aspartame: (A) weight loss curve, (B) first derivative of weight loss curve, and (C) infrared (Gram-Schmidt) evolved gas profile.

Aspartame undergoes a secondary decomposition, as evidenced by the additional weight loss around 350°C, at which point, the piperazine undergoes what amounts to a general thermolytic decomposition to a variety of products. This third weight-loss event accounts for around 50% of the original native sample. Figure 44 shows the stacked plot of the TG/FT-IR analysis of aspartame, in which this secondary decomposition is the major feature. Materials of the secondary decomposition that can be identified immediately in the plot are ammonia, CO_2 and water. Other materials are detectable, but they are not readily identifiable in the mixture of evolved gases, which has a strongly disagreeable odor.

At this point, it is useful to take advantage of the fact that FT-IR is a nondestructive technique and that evolved gases can be recovered from the TG/FT-IR experiment and trapped for analysis by supplementary techniques. In this application, gases which exited the FT-IR cell were drawn by vacuum into a GC injector liner, which had been partially filled with Tenax-TA powder. When most of the gases from the secondary decomposition had evolved and been trapped, the liner was removed from the gas cell exit and vacuum source. The liner was then inserted into an unheated injector of a gas chromatograph. The oxygen was purged from the flow path with helium carrier gas, and the injector was heated ballistically to 290°C to desorb the trapped components onto the unheated capillary GC column. After the desorption step, the transferred materials were chromatographed with column temperature programming to separate the components. The resulting Gram-Schmidt chromatogram is shown in Figure 45. Peaks due to toluene and ethylbenzene can be recognized as decomposition products from the phenylalanine part of aspartame. Similarly, maleimide and succinimide can be identified as products derived from the aspartic acid portion of aspartame. A number of other spectra were isolated with good SNRs. However, they were not identifiable with the Sadtler vapor phase library. There remains the strong possibility that coelution of peaks may have degraded the quality of the spectral matches. This problem could be handled by further optimization of the chromatographic parameters.

G. Artifacts Commonly Encountered In GC/FT-IR

It is possible to observ GC peaks and IR spectra using GC/FT-IR systems even when no sample is introduced, and it is highly probable that all practitioners of GC/FT-IR will encoun-

TABLE 6. Elemental Analyses of Materials Obtained During TG/FT–IR Investigation of Aspartame

Identification	% C	% H	% N
I. Aspartame samples, native and dried			
A. Experimental results			
1. Native aspartame sample	54.93	6.46	9.15
2. Aspartame TGA dried to 120°C	56.38	6.13	9.33
3. Aspartame TGA dried to 150°C	56.63	6.18	9.34
B. Theoretical values			
1. Aspartame hydrate ($C_{14}H_{20}N_2O_6$)	53.83	8.96	8.96
2. Dry aspartame ($C_{14}H_{18}N_2O_5$)	57.14	6.16	9.52
II. Initial aspartame decomposition product, after methanol loss			
A. Experimental results			
1. Aspartame TGA residue	59.88	5.48	10.60
B. Theoretical values			
1. Cyclization product ($C_{13}H_{14}N_2O_4$), a 2,5-diketopiperazine derivative	59.54	5.38	10.68

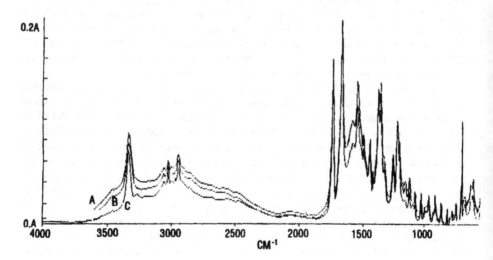

FIGURE 41. FT-IR spectra of aspartame under varied drying conditions: (A) native sample, (B) TGA dried at 120°C, and (C) TGA dried at 150°C.

ter some form of experimental artifact. There are four major types of experimental artifacts in GC/FT-IR. These arise from purge variation, column bleed, detector icing, and peak carry-over.

Variations in the purge of the optical path will obviously change the amount of water vapor and CO_2 in the beam. Fortunately, the absorption bands of CO_2 occur in regions which do not greatly interfere with the bands of interest for most molecules. However, water vapor fluctuations can be quite troublesome. The OH bending modes centered around 1500 to 1600 cm⁻¹

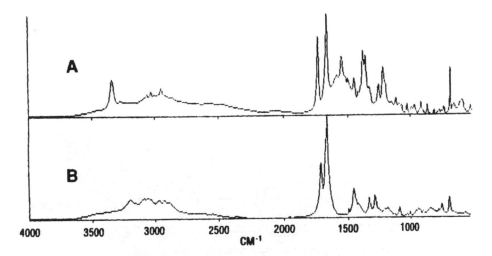

FIGURE 42. Effect of methanol loss on infrared spectrum of TGA sample residue: (A) native aspartame KBr pellet spectrum and (B) FT-IR spectrum of KBr pellet containing residue of TGA reacted aspartame sample after methanol loss.

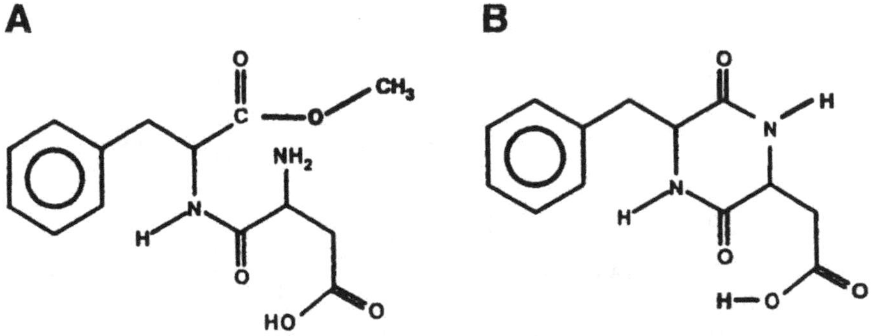

FIGURE 43. Structures of (A) aspartame and (B) postulated internal cyclization product.

span a large part of the fingerprint region of the mid infrared. The best approach is careful background subtraction of spectral slices before and/or after the peak.

Cross-linked, bonded stationary phases are relatively recent developments in capillary column technology. These new columns experience much less stationary-phase bleed. This feature is very positive for GC/FT-IR since the intrusion of silicone peaks into the data will be reduced. Nevertheless, it is possible to overheat the column to the point that bands due to stationary phase breakdown products can be observed eluting from the GC/FT-IR lightpipe as though they were sample peaks.

Figure 46 shows a spectrum obtained during a GC/FT-IR run in which all the features are due to artifacts. The bands in the low wavenumber region are due to silicone materials volatilized from the stationary phase, and the broad band around 3300 cm^{-1} is due to ice on the

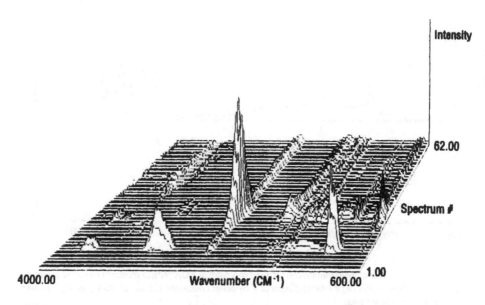

FIGURE 44.   Three-dimensional stacked plot of the TG/FT-IR analysis of aspartame.

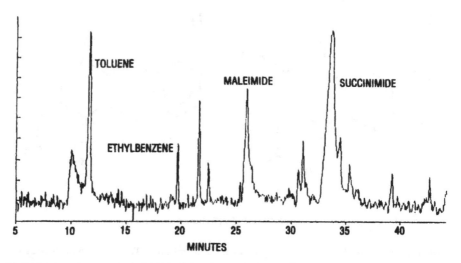

FIGURE 45.   Infrared chromatogram from GC/FT-IR analysis of materials evolved by aspartame secondary decomposition during TG/FT-IR experiment.

MCT detector element. MCT detectors are nominally evacuated and desiccated. However, the vacuum seal gradually leaks, and atmospheric water makes its way into the desiccant over a number of cycles of cooling and warming of the MCT. Gradually, the situation can develop that water stored in the desiccant is transferred to the MCT element when the detector is cooled from room temperature to operating temperature with liquid nitrogen. The water is said to cryopump to the detector element, where it resides as an ice crystal window and causes an OH stretching band of water-ice to grow over the period of the GC/FT-IR run. In cases of extreme

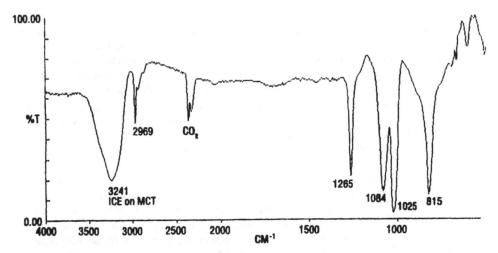

FIGURE 46.   Infrared spectrum of common artifacts in GC/FT-IR analyses.

water contamination of the MCT vacuum jacket, the water bending band in the 1600 cm^{-1} region may be observed. The problem can be alleviated by repumping the MCT detector at high vacuum (10^{-6} to 10^{-7} Torr) for an extended period (24 to 48 h). Most manufacturers of MCT detectors will perform this repumping service for a nominal fee.

It is always possible that a chromatographic run may be terminated before all components of the sample have eluted from the column. The result is that some chromatograms may contain peaks that were leftover from previous injections. This is most likely to occur when temperature programmed runs are prematurely terminated. However, it is probable that leftover peaks will be indicated by their significantly broader widths than those of peaks which belong to the current run. The remedy is usually to heat the column to condition it for a period when no sample is being introduced. Also, the addition of an extended high temperature hold-step to a temperature programmed run will minimize peak carry-over. With a spectroscopic detector, the effects of the broad carry-over peak can be minimized by treating it as part of the local background in a spectral subtraction step.

V. SUMMARY AND CONCLUSIONS

During the last decade, GC/FT-IR has evolved and developed into a formidable tool for molecular analysis, and a bifurcation of design modes has occurred. GC/FT-IR systems based on lightpipes provide on-the-fly GC/FT-IR, gas-phase IR spectra of eluted components. Systems based on eluate deposition on cooled surfaces offer the lowest detection limits, but these systems are relatively expensive and more complex to operate and maintain. Despite the differences in the design modes, both types of GC/FT-IR systems are capable of producing valuable insight into compositions of complicated samples.

The progress that has been made in GC/FT-IR was recently the subject of official acknowledgment in the form of publication of Method 8410 by the Environmental Protection Agency.[78] This method authorizes the use of GC/FT-IR as an alternative technique in the screening of unknown, possibly toxic samples for pollutants. A major strength of GC/FT-IR for environmental samples, as well as other complex samples, consists in its facility for isomer identification, particularly with aromatic compounds.

By comparison, GC/MS faces some limitations in its ability to differentiate structural isomers, but it is usually exceptionally strong in homolog identification and molecular weight determination. In general, the analyst is best served when information from both spectroscopies (MS and IR) is combined to improve the certainty of chromatographic peak identification. Because of the increasing importance of analytical certainty, both from the standpoints of regulatory compliance and economic efficiency, there may soon arise a significantly greater interest in multihyphenation, e.g., GC/IR/MS, of chromatographic detectors for a greater proportion of the work in molecular mixture analysis. When this occurs, the interest in enhancing the range of multihyphenated chromatospectrometric applications will probably expand into other chromatographies, notably SFE and HPLC.

Among sample preparation methodologies for chromatography, the most notable recent development is the rise in interest in SFE. Very recently, the EPA announced its intention to allow the use of SFE for preparation of samples for analysis of TPHs in soils and solid wastes by EPA Method 3560. Extracts may be evaluated by FT-IR using EPA Method 8440 or GC/FID by EPA Method 8015. Other environmental applications using SFE for sample preparation to be evaluated for possible future approval by EPA include those for PAHs, PCBs, certain pesticides, and several others. SFE is destined to become a standard chromatographic sample preparation technique and will replace organic solvent extraction in a great number of applications.

ACKNOWLEDGMENTS

The author gratefully acknowledges the assistance of the following people in the preparation of this manuscript:

Prof. Norris Williams and Dr. Mark Whitten of the Florida State Museum for the collaboration in the preparation of the orchid fragrances and discussion of the Euglossine Syndrome; Prof. Klaus Fischer of Louisiana State University for the many interesting and productive discussions; Dr. Alan Harmon of McCormick Inc. for his assistance with literature research and the useful discussions of molecular analysis strategy; Dr. Joe Tehrani of Isco Inc. for the original reference on Soxhlet extraction; Ms. Krista Swanson of Perkin-Elmer Elemental Analysis Department for performing the C, H, N analyses of the aspartame samples with the Perkin-Elmer Model 2400 Elemental Analyzer; Dr. John Hinshaw of Perkin-Elmer Gas Chromatography Product Department for his consultation and review of this manuscript; Dr. Brian McGrattan of Perkin-Elmer Infrared Product Department for his assistance in acquiring and processing the data from the SFE/GC/FT-IR experiments, using the Perkin-Elmer System 2000 FT-IR and PrepMaster SFE; and Charles Sadowski of the Perkin-Elmer GC/MS Product Department for the assistance in obtaining the supporting GC/MS data with the Perkin-Elmer Q-Mass 910.

REFERENCES

1. Mullens, J., Carleer, R., Reggers, G., Ruysen, M., Yperman, J., and Van Poucke, L.C., *Bull. Soc. Chim. Belg.*, 101(4), 266, 1992.
2. Jansen, J.A.J., Van der Maas, J.H., and De Boer, A.P., *Appl. Spectros.*, 46(1), 88, 1992.
3. Maurin, M.B., Dittert, L.W., and Hussain, A.A., *Thermochim. Acta*, 186(1), 97, 1991.
4. Solomon, P.R., Serio, M.A., Carangelo, R.M., Bassilakis, R., Yu, Z.Z., and Whelan, J., *J. Anal. Appl. Pyrolysis*, 19, 1, 1991.

5. Johnson, D.J. and Compton, D.A.C., *Am. Lab.*, 23(1), 37, 1991.
6. Ortego, J.D., Jackson, S., Yu, G.S., McWhinney, H., and Cocke, D.L., *J. Environ. Sci. Health, Part A*, A24(6), 589, 1989.
7. Khorami, J., Lemieux, A., Menard, H., and Nadeau, D., *ASTM Spec. Tech. Publ., 997 (Compos. Anal. Thermogravim.)*, 1988, 147-159.
8. Lephardt, J.O., *Appl. Spectros. Rev.*, 18(2), 265, 1982, 1983.
9. Brill, T. B. *Anal. Chem.*, 61(15), 897A, 1989.
10. White, R., *Chromatography/Fourier Transform Infrared Spectroscopy and Its Applications*, Marcel Dekker, New York, 1990, 328.
11. Griffiths, P.R. and deHaseth, J.A., *Fourier Transform Infrared Spectrometry*, Wiley-Interscience, New York, 1990, 656.
12. Nichikida, K., Housaki, T., Morimoto, M., and Kinoshita, T., *J. Chromatogr.*, 517, 209, 1990.
13. Fujimoto, C. and Jinno, K., *Anal. Chem.*, 64(8), 477A, 1992.
14. Taylor, L.T. and Calvey, E.M., Flow cell SFC/FT-IR, in *Hyphenated Techniques in Supercritical Fluid Chromatography and Extraction*, Jinno, K., Ed., Elsevier, Amsterdam, 1992, 65.
15. Griffiths, P.R., Norton, K.L., and Bonnano, A.S., SFC/FT-IR measurements involving elimination of the mobile phase, in *Hyphenated Techniques in Supercritical Fluid Chromatography and Extraction*, Jinno, K., Ed., Elsevier, Amsterdam, 1992, 83.
16. Jenkins, T.L., Kaplan, M., Simmonds, M.R., Davidson, G.D., Healy, M.A., and Poliakoff, M., *Analyst*, 116, 1305, 1991.
17. Healy, M.A., Jenkins, T.L., and Poliakoff, M., *Trends Anal. Chem.*, 10(3), 92, 1991.
18. Davidson, G. and Jenkins, T.J., *Spectros. Europe*, 32, 1992.
19. deHaseth, J.A. and Robertson, R.M., *Microchem. J.*, 40(1), 77, 1989.
20. Levy, J.M., Roselli, A.C., Storozynsky, E., Ravey, R., Dolata, L.A., and Ashraf-Khorassani, M., *LC.GC Magazine*, 10(5), 386, 1992.
21. Glauniger, G., Kovar, K.A., and Hoffmann, V., *Fresenius J. Anal. Chem.*, 338, 710, 1990.
22. Heres, W., *HRGC-FTIR: Capillary Gas Chromatography-Fourier Transform Infrared Spectrometry*, Huethig, A., Ed., Springer-Verlag, Heidelberg, 1987, 212.
23. Azarraga, L.V., *Appl. Spectros.*, 34, 224, 1980.
24. Griffiths, P.R., deHaseth, J.A., and Azarraga, L.V., *Anal. Chem.*, 55(13), 1361, 1983.
25. Hausdorff, H.H., *J. Chromatogr.*, 134, 131, 1977.
26. Freeman, S.K., in *Ancillary Techniques of Gas Chromatography*, Ettre, L.S. and McFadden, W.H., Eds., Wiley-Interscience, New York, 1969, 227.
27. Low, M.J.D. and Freeman, S.K., *Anal. Chem.*, 39, 194, 1967.
28. Erickson, M.D., *Appl. Spectros. Rev.*, 15(2), 261, 1979.
29. Bourne, S., Haefner, A.M., Norton, K.L., and Griffiths, P.R., *Anal. Chem.*, 62, 2448, 1990.
30. Sandra, P., *Sample Introduction in Capillary Gas Chromatography, Vol. 1*, Huethig, A., Ed., Springer-Verlag, Heidelberg, 1985, 265.
31. Grob, K., *Classical Split and Splitless Injection in Capillary Gas Chromatography*, Huethig, A., Ed., Springer-Verlag, Heidelberg, 1986, 324.
32. Analysis of volatiles, methods and applications, Proc. Int. Workshop, Wuerzburg, Germany, September 28-30, 1983, Schreier, P., Ed., Walter de Gruyter, Berlin, 1984, 460.
33. Sugisawa, H., Sample preparation: isolation and concentration, in *Flavor Research, Rencent Advances*, Teranishi, R., Flath, R.A., and Sugisawa, H., Eds., Marcel Dekker, New York, 1981, 11.
34. *Capillary Gas Chromatography in Essential Oil Analysis*, Bicchi, C. and Sandra, P., Eds., Springer-Verlag, Heidelberg, 1987, 435.
35. Soxhlet, F., *Dinglers Polytech. J.*, 232, 461, 1879.
36. *Analytical Supercritical Fluid Chromatography and Extraction*, Lee, M.L. and Markides, K.E., Eds., Chromatography Conferences, Inc., Provo, Utah, 1990, 325.

37. de la Tour, C., *Ann. Chim. Phys.*, 21, 127, 1822.
38. Andrews, T., *Philos. Trans. R.Soc.*, 159, 575, 1869.
39. *Supercritical Fluid Extraction, Principles and Practice*, McHugh, M.A. and Krukonis, V.J., Butterworths Publishers, Boston, 1986, 200.
40. Zurer, P.M., *Chemical and Engineering News*, June 22, 1992, 7-13.
41. Hannay, J.B. and Hogarth, J., *Proc. R. Soc. (London)*, 29, 324, 1879.
42. Krukonis, V.J., Supercritical fluid extraction in flavor applications, in *Characterization and Measurement of Flavor Compounds*, ACS Symp. Ser. No. 289, Bils, D.D. and Mussinan, C.J., Eds., American Chemical Society, Washington, D.C., 1985, 154.
43. Shaw, R.W., Brill, T.B., Clifford, A.A., Eckert, C., and Ulrich Frank, E., *Chemical and Engineering News*, December 23, 1991, 26-39.
44. Levy, J.M., Storozynsky, E., and Ashraf-Khorassani, M., Use of modifiers in supercritical fluid extraction, in *Supercritical Fluid Technology, Theoretical and Applied Approaches to Analytical Chemistry*, ACS Symp. Ser. No. 488, Bright, F.V. and McNally, M.E., Eds., American Chemical Society, Washington, D.C., 1992, 336.
45. Ashraf-Horassani, M., Taylor, L.T., and Zimmerman, P., *Anal. Chem.*, 62, 1177, 1990.
46. Hawthorne, S.B., Miller, D.J., and Krieger, M.S., *J. Chromatogr. Sci.*, 27, 347, 1989.
47. Levy, J.M., Ashraf-Khorassani, M., and Houk, R., *J. Chromatogr. Sci.*, 30(9), 1992.
48. Washall, J.W. and Wampler, T.P., *Spectroscopy*, 6(4), 36, 1991.
49. Cole, A. and Woolfenden, E., *LC.GC Magazine*, 10(2), 76, 1992.
50. Palmer, R.A., Childers, J.W., and Smith, M.J., *EPA Project Summary, EPA/600/S4-85/066*, 1986.
51. Dworzanski, J.P., Buchanan, R.M., Chapman, J.N., and Meuzalaar, H.L.C., *Preprints of Papers of American Chemical Society, Division of Fuel Chem.*, 36(2), 725, 1991.
52. Harrington-Fowler, L., *Am. Lab.*, August, 39, 1991.
53. Likens, S. and Nickerson, G., *Proc. Am. Soc. Brewing Chem.*, 5, 1964.
54. Godefroot, M., Sandra, P., and Verzele, M., *J. Chromatogr.*, 203, 325, 1981.
55. Godefroot, M., Stechele, M., Sandra, P., and Verzele, M., *J. High Resolution Chromatogr. Chromatogr. Commun.*, 5, 75, 1982.
56. deHaseth, J.A. and Isenhour, T.L., *Anal. Chem.*, 49, 1977, 1977.
57. *Applied Headspace Gas Chromatography*, Kolb, B., Ed., Heyden, London, 1980, 188.
58. Hachenberg, H. and Schmidt, A.P., *Gas Chromatographic Headspace Analysis*, Heyden, London, 1977, 125.
59. Schrynemeeckers, P.J., *Environ. Lab.*, October, 24, 1989.
60. Bellar, T.A. and Lichtenberg, J.J., *J. Am. Water Works Assoc.*, 66, 738, 1974.
61. Lawrence, A.H. and Elias, L., *Am. Lab.*, July, 88, 1989.
62. Manura, J.J. and Hartman, T.G., *LC-GC Magazine*, May, 46, 1992.
63. McClure, G.L., Presented in part at East. Anal. Symp., Abstract #133, New York, November 16 to 18, 1983.
64. McClure, G.L., Presented in part at the Arrowhead Conf. Soc. Appl. Spectrosc., Lake Arrowhead, CA, April 22, 1989.
65. McClure, G.L. and Roush, P.B., Presented in part at Pittsburgh Conference, Abstract # 469, Atlantic City, NJ, March 10 to 14, 1986.
66. McClure, G.L. and Coleman, P.B., Presented in part at Pittsburgh Conference, Abstract #869, New Orleans, LA, March 9 to 12 1992.
67. Masada, Y., *Analysis of Essential Oils by Gas Chromatography and Mass Spectrometry*, John Wiley & Sons, New York, 1976, 56.
68. McClure, G.L., Presented in part at FACSS, Abstract # 237, Philadelphia, PA, September 24 to 25, 1983.
69. Chaves dos Neves, H.J. and Vasconcelos, A.M.P., *Chromatographia*, March, (5,6), 233, 1989.

70. Coates, J.P., personal communication, 1982.
71. Prudhomme, P., *Chef Paul Prudhomme's Louisiana Kitchen*, William Morrow and Company, New York, 1984, 351.
72. McClure, G.L., Presented in part at Pittsburgh Conference, Abstract #756, New Orleans, LA, February 27 to March 1, 1985.
73. Williams, N.H., The biology of orchids and euglossine bees, in *Orchid Biology, Review and Perspectives, II.*, Arditti, J., Ed., Cornell University Press, Ithaca, NY, 1982, 119.
74. Silverstein, R.M., Rodin, J.O., Wood, D.L., and Browne, L.E., *Tetrahedron*, 22, 1929, 1966.
75. McClure, G.L., Willians, N.H., and Whitten, W.M., Presented in part at Pittsburgh Conference, Abstract #267, Atlantic City, NJ, March 3 to 9, 1984.
76. Lindquist, N.L., Battiste, M.A., Whitten, W.M., Williams, N.H., and Strekowski, L., *Phytochemistry*, 24(4), 863, 1985.
77. McClure, G.L., Cassel, R.B., and Patkin, A.J., Presented in part at Pittsburgh Conference, Abstract # 1326, New York, March 5 to 9, 1990.
78. EPA Method 8410, in *Test Methods for Evaluating Solid Waste, Physical/Chemical Methods, SW-846, 3rd ed., Proposed update 11*, Government Printing Office, Washington, D.C., June, 1990.
79. Hannay, J.B. and Hogarth, J., *Proc. R. Soc. London*, 30, 178, 1880.
80. Hannay, J.B. and Hogarth, J., *Proc. R. Soc. London*, 30, 484, 1880.

Chapter 8

QUANTITATIVE ANALYSIS — AVOIDING COMMON PITFALLS

Senja V. Compton and David A. C. Compton

CONTENTS

I. Introduction 219
II. Setting Down Criteria 219
III. Measurement Conditions Defined 221
    A. Standards Preparation 221
    B. Selection of Standards 222
    C. Overdetermination 222
    D. Precision and Significant Figures 224
    E. Composition of Reference Standards 224
    F. Verification 224
    G. Required Information 224
IV. Appropriate Sample Handling 225
    A. General Behavioral Considerations 225
        1. Solvent Effects 225
            a. Reactions 225
            b. Incompatibility 226
            c. Impurities 226
            d. Residue 226
            e. Evaporation 227
    B. Miscellaneous Considerations 227
        1. Number of Samplings 227
        2. Accessory Dependence 228
        3. Alkali Halide Salt Characteristics 228
        4. Chemical Compatibility 228
    C. Considerations for Solid Sample Examinations 228
        1. Polymers 228
            a. Preparation 229
            b. Dissolution 230
        2. Effects of Grinding 230
            a. Sample Alterations 230
            b. Uniformity 230
    D. Gases 230
        1. Theory 230
        2. Memory Effects 231
        3. Temperature and Pressure Effects 231
        4. Experimental Procedure 232
        5. Chemical Reactivity 232

0-8493-4203-1/93/$0.00+$.50
© 1993 by CRC Press Inc.

    E. Films 233
       1. Uniformity 233
       2. Preparation 233
       3. Pressure and Temperature Effects 233
       4. Cast Films 233
          a. Impurities 233
          b. Solvent Volatilization 234
       5. Rolled Films 234
       6. Opacity and Particulates 234
    F. Alkali Halide Disks 235
       1. Preparation 235
       2. Ion Exchange 235
    G. Liquid Cells 236
       1. Description 236
       2. Contamination during Filling 237
       3. Filling and Cleaning 237
    H. Mulls 238
       1. Preparation 238
       2. Sample Types 238
    I. Internal Reflectance Spectroscopy 239
       1. Description 239
       2. Reproducibility Issues 239
    J. Diffuse Reflectance Spectroscopy 239
       1. Description 239
       2. Sample Preparation 240
V. Recording the Spectrum 240
    A. Spectrophotometer Performance 240
       1. 100% (Transmittance) Line 240
       2. Single-Beam Spectrum Inspection 241
    B. Accessory Performance 241
    C. Data Collection Parameters 242
       1. Trading Rules 242
          a. Collection Time 242
          b. Resolution Selection 243
       2. FT-IR Terminology 243
          a. Interferogram and Centerburst 243
          b. Asymmetric/Symmetric Collection 243
          c. Undersampling 243
          d. Zero-Filling 243
          e. Apodization 244
    D. Detector Characteristics 244
    E. Filling the Infrared Beam 244
    F. Atmospheric Interferences 245
    G. Background Spectrum Selection 245
    H. Software Compatibility 245
VI. Treatment of Spectral Data 245
    A. Operator Bias 246
    B. Application Criteria 246
    C. Ordinate Axis Format 246
       1. Absorbance 246
       2. Kubelka-Munk 246
    D. Calibration Plots 247
       1. Deviation 247
       2. Piecewise Fit 247
    E. Baseline Correction 247

     F. Smoothing 248
     G. Resolution Enhancement 248
         1. Fourier Self-Deconvolution 248
         2. Maximum Entropy 248
         3. Derivative 249
     H. Spectral Subtraction 249
     I. Fringe Removal 249
     J. Interpolation 249
VII. Conclusions 250
Acknowledgments 250
References 250

I. INTRODUCTION

The major strengths of infrared spectrometry are applicability to both qualitative and quantitative measurements made upon substances that are not of metallic compositions. This allows for the determination of chemical functionality in a large number of covalently bonded species, including gases, liquids, polymers, and solids, both crystalline and amorphous. An overwhelming amount of literature has been devoted to the pros and cons of performing quantitative analyses by various mathematical treatments.[1-33] Although the promises made by the applications of these more advanced methods are very appealing, the resurgence of interest in infrared quantitative analysis has probably resulted mainly from the usage of computers to remove the tedium of the analytical processes.

Unfortunately, the actual sample treatment and laboratory technique necessary to acquire the initial data to be used in the mathematical approaches are very rarely discussed (usually behind closed doors) as they open a Pandora's box of systematic and random error sources. Too frequently, well-modeled data is used to illustrate the advantages and strengths by the various mathematical treatments. From this, it is assumed that good quality, reproducible infrared spectra just (magically) happen which conform to the models. As most experienced spectroscopists know, however, this is not the case. It requires constant diligence and "common sense" on the part of the analyst to avoid wasting time in the collection of nonpractical, useless data.

There exist a multitude of parameters which must be addressed when attempting to perform a quantitative measurement using infrared spectroscopy. Quantitative analyses encompass, but are not limited to, spectral subtraction, transmittance measurements of optical filters, discriminate analysis, library search comparisons, and pathlength determinations, in addition to the more familiar concentration reports. The categorization of quantitation into divisions for logical discussion is difficult to make and so we have devised the nomenclature "SMART" analysis. This mnemonic stands for: Setting down criteria, Measurement conditions defined, Appropriate sample handling, Recording the spectrum, and Treatment of spectral data.

The various mathematical approaches are described in detail elsewhere;[1-33] this chapter addresses the issue of obtaining spectra and calibration information for use by any of these methods. Therefore, the following sections will be devoted to investigating the issues which affect the capabilities and limitations of quantitative analysis by infrared spectrometry.

II. SETTING DOWN CRITERIA

The establishment of criteria to be met by a quantitative analysis procedure is often overlooked by the analyst. In the heated fervor to apply the most advanced mathematical treatment for quantitative analysis, the problem at hand is frequently made more difficult than necessary (i.e., blown out of proportion). The true nature of the problem to be solved needs to be defined (what), as well as the reason for the analysis (why). Some of the issues to be addressed follow. These are nontrivial considerations!

- Who is to perform the analysis?
- What is the skill of the analyst(s)?
- What is to be measured?
- Does the measurement need to be correlated to other data?
- When does the analysis need to be done?
- Where is the analysis to be performed?
- What is the environment where the spectrophotometer is to be placed?
- How quickly do the results need to be provided?
- What is the required format for the reporting of results?
- What are the accuracy and precision needed?
- Are there multiple suppliers for the ingredients?
- Are materials substituted for one another?
- Are components occasionally added to achieve a certain property?
- Are there multiple production methods?
- Do production batches vary with shift, weekday, seasonally, or vat-to-vat?
- What consumables are to be used in the analysis?
- What is the cost of method development (manpower, time, and money)?
- What is the final operation cost of the analysis?
- What are the expected concentration ranges to be covered?

Therefore, one needs to take pen in hand and set down realistic expectations, reasons, drawbacks, alternatives, time and monetary savings, expenditures, prospective variables, expected problems, and last, but not least, the benefits to be gained from the analysis. It is extremely advantageous to then have a colleague impartially review the evaluation (whether positive or negative) to insure that an important consideration item has not been overlooked.

In the production plant environment, numerous issues rear their ugly heads. Typically, the operator is not a trained spectroscopist and cannot be expected to know "good" data from "bad." A black box approach, where sample in ⇒ number out, is the expected mode of operation. Therefore, the analyst must question the necessity for the measurement, as it will undoubtedly be an unwelcome, additional job for the production personnel (unless a new hire is made) and will be regarded with suspicion (newfangled technology, the old way was better, *they* don't trust us, etc.) and even met with hostility.

After we ask "Is it really necessary?", we must then ask "How good does the method have to be?". Does the quantitative method have to be extremely rigorous, yielding precise, accurate amounts for a single component or for every component? Can it simply verify that the product is within acceptable limits? Also related to this same issue are environment and cost. High analytical precision and accuracy cannot be obtained if the environment where the analysis is to occur does not allow the spectrophotometer to perform well and will ultimately affect the quality of the reported analytical values. Additionally, do the results provided by analysis by infrared spectrometry defray the cost savings offered by less expensive, alternative methods which make similar measurements?

Speed and ease are oftentimes ignored in an analysis. The typical operator does not wish to exert much effort in acquiring or in preparing a sample for analysis. Therefore, any method must be extremely easy to perform (requiring minimal manipulations), while also insuring reproducibility. Many times analyses have to be *immediate* (a relative term) to avoid surcharges on waiting tankers, identification of incorrect blending, detection of degree of cure, etc. Even though infrared quantitative analysis may provide "better" answers, the timing may be too long per sample. Carefully evaluate the situation and do not make it more complicated than it really is.

As an aside, every analytical instrument has an expected lifetime (unfortunately, nothing works forever). Operation under harsh environmental conditions can shorten the lifetime of any infrared spectrophotometer. The analyst, therefore, should keep in mind the "downtime" involved in maintenance, repair, and replacement of the spectrophotometer. What are the backup plans for such a catastrophic occurrence? Can this method be transferred out of the development laboratory to another location for backup analyses? Also, how much is it going to cost for alternative backup analyses (in time and dollars) and what is the expected frequency of such behavior?

III. MEASUREMENT CONDITIONS DEFINED

So is quantitative analysis by infrared spectrometry a viable option? Now a formidable task is faced, one of defining how measurements are to be made, what ranges they are to encompass, and what conditions they are to be made under. Environmental variables such as temperature, pressure, and, oftentimes, humidity must remain stable so as not to introduce variability in data collection. However, their control is fairly straightforward. The spectroscopic region (far, mid, or near) for useful infrared measurements must be defined. Ultimately, the greatest task facing the analyst proves to be in the definition of appropriate standards for calibration of the analytical method.

A. Standards Preparation

The selection of the samples to be used for a calibration set requires a lot of forethought. A great deal of time and effort must be spent in defining the bounds of a quantitative method and generating the standards to adequately cover the expected range variations. This means that the extreme concentrations for each component must be included, as extrapolation outside of the calibrated concentration range is dubious at best.[30] If you are interested in only an expected narrow range of concentrations for your applications, the samples needed as calibration references need not include the entire range of possible concentrations (from 0 to 100% for each component). However, if you are specifically interested in accuracy at the extremes of a specification range, the concentrations covered by the reference samples must span a larger range than the expected specification range. This can be thought of by assigning the value c to the expected concentration for a component. The allowable range for the concentration may be expressed as c + d to c - d. For accuracy at the extremes of this range, it is strongly advised that the concentration range covered by the calibration standards be c + 1.5d to c - 1.5d.

The standards themselves must be composed of linearly independent concentration data. Linear dependence can be thought of as occurring in two differing fashions. The first case involves simple multiples (or dilutions) of one standard, so that any two components are in constant relative proportions to one another throughout the series. Given the case where three components (A, B, and C) are to be measured in a bulk matrix or in vapor phase, a situation may arise where the series of standards shown in Table 1 are prepared.

TABLE 1. Data Illustrating Linear
Dependence of Dilutions

| | Component wt% | | | Ratio |
	A	B	C	A:B:C
Standard 1	1	1	1	1:1:1
Standard 2	2	2	2	1:1:1
Standard 3	3	3	3	1:1:1

Clearly, this set of standards is doomed to failure unless each component has an isolated, unique band exhibiting no overlap with any other species, no interactions exist among the chemical species, and the samples are examined with a reproducible, known pathlength. In such a case, the integrated area or peak height associated with a component can be separately plotted vs. concentration to obtain its quantitative measurement (i.e., using separate quantitative methods to calculate the concentration of each component as an isolated entity, as in vapor phase analysis).

The second case of dependency occurs when the concentration of two or more components adds to a constant. This is illustrated by the set of data given in Table 2.

Comparing the ratios of any two components, no linear dependence of the data is seen. However, in each of the standard samples the sum of the concentrations of components A, B, and C always add to 12%. If an XYZ plot of this data is made (shown in Figure 1), the result is the definition of a plane. This does not prove to be a problem if this situation is encountered in every sample to be analyzed (e.g., such as the case where the components always sum to 100%), but what happens in this instance if the product only has the concentrations of A, B, and C summing to 10%? The situation is not described, and one would rely on extrapolation from this plane into unmapped, undefined space to obtain quantitative results. This is not a procedure which is highly recommended nor seriously suggested.

B. Selection of Standards

In the actual selection of mixtures to be used as standards, at least two components (in a system comprised of multiple species) should be at different levels in each blend. This will typically eliminate most occurrences of linearly dependent data sets. Additionally, all component levels should be represented the same number of times in the calibration blends. This prevents the spectral features of one component from dominating the spectral information used for quantitative measurements. Finally, the spacing of the standards should be made evenly across the expected range to allow for interpolation between data points in the calibration.

Oftentimes, the best source of standards (or blinds) are the "retains" from actual production runs. These are usually collected over long periods of time and incorporate many of the industrial idiosyncrasies which may be very difficult to reproduce under laboratory conditions but are necessary for a rigorous method development. However, a major assumption is made here: the material does not change in composition with time, so it can be regarded as a valid standard.

C. Overdetermination

The system under investigation should be overdetermined.[30,34] This action is reflected in the practice of recording the spectra of numerous standards and then performing a least-squares regression to calculate the absorptivity of the compound at a certain frequency. Typically, this

TABLE 2. Data Illustrating Linear Dependence of Concentration Constants

| | Component wt% | | | Sum |
	A	B	C	A + B + C
Standard 1	0	6	6	12
Standard 2	6	6	0	12
Standard 3	6	0	6	12
Standard 4	4	6	2	12
Standard 5	6	4	2	12
Standard 6	2	4	6	12
Standard 7	4	2	6	12
Standard 8	2	6	4	12

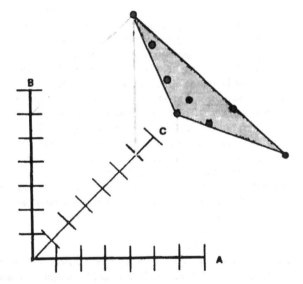

FIGURE 1. XYZ plot of the data given in Table 2 illustrating linear dependence.

dictates that more standard mixtures need to be examined than there are components in the system in order to determine error information. The use of more data than is strictly necessary reduces the effect of random errors. As a rule of thumb, two times the *minimum* number of standards proves to be adequate. The back-calculations resulting from this overdetermination of a system yield: an indication of the quality of the calibration matrix (average error); a detection of perturbation in a system, usually with one component being determined less accurately; a discovery of deviations from Beer's law at high concentrations; identification of a bad standard or poor measurement; and evidence for inappropriate spectral region selection for measurements in the quantitative method. However, in an exactly determined case where the minimum number of standards have been used, the back-calculated values will be exactly those that were input for each reference, reflecting a "perfect fit" of the data with no insight into the "goodness" of the quantitative method.

D. Precision and Significant Figures

Typically, the precision of the standard mixtures is generally better than the precision of the analytical system being developed. However, significant figures cannot be stressed enough. If standards are prepared with concentrations only good to two significant figures, the values reported for each of the components in the final analytical report are only good to two significant figures. Information cannot be obtained when no real data exists! The computer is perfectly capable of calculating values comprised of numerous digits, but they are not always meaningful. In cases where other analytical techniques are used to provide concentrations for the reference samples, the errors reflected in those methods are incorporated into the infrared analysis. In other words, the precision of the infrared quantitation cannot be better than the (instrumental) technique used to provide the concentrations used for the calibration standards.[35]

E. Composition of Reference Standards

In the laboratory preparation of standards, it is extremely important to use the actual materials used in the formulation of the final product rather than ultrapure grades of chemicals. This means that standards including alternative supplier sources of components must be prepared. The mutual interference of the spectra, not the number of components, is what is being determined in multicomponent analysis. The components cannot be determined singly, but all must be determined together. Take the case of a system in which the level of "solvent" is to be measured. Typically, in an industrial production facility, a grade of solvent is used. This solvent may be thought of as containing impurities, as it is not comprised of 100% of any one component, and, subsequently, there does not exist a "pure" spectrum for this solvent (which may differ in "impurity" level from batch-to-batch or from supplier-to-supplier). These "impurities" have specific infrared absorbances which contribute to the overall spectrum of a standard reference material. If their effects are not included in the calibration step, the analysis is doomed to failure as their absorbances cannot be compensated for in the measurements used to calculate concentrations. In this type of situation, many standards spanning the expected range must be included in the development of the quantitative method — an extremely tedious task.

F. Verification

Following the development of an analysis, it should be tested frequently to verify its validity. This can be done with "retain" material previously analyzed (provided its composition does not change with time) or synthetically generated blends. Blinds, consisting of both out-of-specification and in-specification products, should be analyzed. The testing should be done as part of a normal operating procedure, not as a special analysis, so that any user bias is eliminated.

G. Required Information

For the development of a rigorous method and a trustworthy analysis, in an industrial environment the analyst must be sure to address all measurement issues. Questions that must be addressed are as follows:

1. What is in the product?
   a. This should include all of the components, not just the ones to be quantified.
   b. The listing of all acceptable substitutes. These may have the same physical effect on the final product, but may be chemically different (e.g., the substitution of potassium hydroxide for sodium hydroxide to adjust pH level).

    c. The list of all vendors for ingredients and all acceptable component grades used in formulation.

    d. After collection of calibration spectra, it is extremely useful to do a factor analysis in order to verify that the information provided is consistent with the spectral data.

2. How well is the processed controlled? Is Monday morning production different from mid-week production? Is the winter-product different from the summer-product?

3. How good is the current working analytical procedure? What are the major problems and drawbacks with the existing analysis?

4. Is the method to incorporate real data, or should the data match (correlate to) some older, existing technique? This technique may have problems, but it may be the "standard" to which all products must conform.

5. What is the current practice for reporting the data? For instance, does 40/40/20 refer to 40% A, 40% B, and 20% C or is it the relative ratio of these components in a much more complex mixture?

6. How quickly does the analysis need to be performed, and how often?

IV. APPROPRIATE SAMPLE HANDLING

Sample manipulations required to record good quality, reproducible spectra for analysis must be very carefully defined and performed. It is extremely difficult to assign suitable headings to differentiate error sources, as simple categorization of solids, liquids, and gases incorporate numerous sample handling methods, each possessing its own inherent set of pitfalls. Thus, not only do problems exist with the sample form itself and its associated preparation technique, but also the sampling method used to acquire a spectrum, which may have quirks that need to be recognized. Some of the most commonly encountered situations when preparing samples for examination are discussed in the following sections, beginning with pitfalls common to a variety of situations.

A. General Behavioral Considerations

Several sample handling methods have a common denominator in introduction of error sources. Therefore, this section discusses behaviors which can be observed when samples are treated similar ways.

1. Solvent Effects

a. Reactions. The use of solvents in the preparation of a sample for examination can introduce problems in numerous fashions. First, the solvent may cause a reaction to take place.[26,36] This may take the form of intermolecular associations (e.g., hydrogen bonding, aggregation, preferential conformation of molecules, or unfolding) or chemical interactions (such as reacting directly with the sample under investigation to form a new species). The first case is generally more pronounced in the study of polar molecules. Typically, its effects are minimized by the examination of dilute solutions. The behavior can be detected by direct comparison (e.g., overlay plots, spectral subtraction) of two or more reference standard spectra and noting any shifts in band frequencies or appearance of new bands. The second case of chemical interaction is sometimes more difficult to discern. It is easy to note if the sample changes form, turning from a liquid into a solid after addition of a solvent. However, in most cases the change is not so obvious. A good test is to place a sample in the spectrophotometer and acquire spectra at set intervals for a period of time. Direct comparison of these spectra should then be made (overlay or 1:1 subtraction) to note the presence of any interloping species created by chemical reactions.

b. Incompatibility. In the examination of any sample dissolved in a solvent, the resultant solution should be inspected carefully to detect the presence of micelles, emulsions, suspensions, or precipitates. All such behaviors result in irreproducible data collections and, typically, preclude useful quantitative measurements from occurring. Additionally, when a sample is diluted to a predetermined concentration important for reproducibility in quantitation, care must be taken to eliminate volatilization of the solvent which would cause an artificial enhancement of solute levels. This means placing samples in sealed vessels for storage in a refrigerator prior to data collection, examining the sample immediately following preparation, and using an analysis method which allows the solution to be contained so that evaporation does not occur during the data collection step.[37] It is also important to insure that the solution under investigation does not have any entrapped gases (bubbles) as the infrared beam encountering the sample would not follow the same pathway in each sample and Beer's law would have two variables — pathlength and concentration.

c. Impurities. Solvents typically contain a number of impurities, some of which are purposely added for stability.[38] Therefore, problems resulting from their presence may rear their ugly heads. The most commonly encountered situation is that of alternate suppliers for the same chemical grade of solvent. Typically, a grade will be specified as a certain percentage of the one main chemical, with the remaining species listed as unknown. Any variations in the concentrations of these latter species will show corresponding differences in the spectrum for the solvent. Thus, for any quantitative method in which the solvent contributes to the overall spectrum used in the analysis, reference standards must be made with multiple sources of the solvent to compensate for the relevant spectral features due to the impurities. One commonly used solvent is water. Very often a buffer system comprises a portion of the aqueous medium for chemical stability of the material under investigation. These components must not be overlooked when developing a quantitative analysis method.

Not only can impurities wreak havoc with a quantitative method by having infrared absorbances, but they may chemically react with the system under investigation. Since impurities may be either inorganic or organic in nature, they may serve as a catalyst to encourage a reaction to occur or continue to occur; to form a complex with a specific component; to induce intermolecular interactions, such as hydrogen bonding; to adsorb to specific sites; or to create new chemical species in the system under examination. Again, the suggestions given for verification of such behavior of a sample with its main solvent should be followed here to detect any potential problems.

d. Residue. When volatile solvents are used for cleaning purposes, the analyst must make sure that any equipment (liquid cells, spatulas, mortar and pestles, etc.) must be completely dried after washing with the solvent so no carryover into a sample is made. However, incidences have occurred where residue remains following evaporation of a solvent (e.g., cleaning of liquid cells or internal reflection elements, casting polymer films, generating inorganic wafers). This residue is attributed to the presence of impurities or stabilizers with low vapor pressures. A quick method for verification of complete solvent volatility is to place some of the solvent onto an infrared transparent window and allow it to evaporate under the established experimental conditions used for analysis. An example of this is shown in Figure 2, where several drops of THF solvent were allowed to evaporate to "dryness" on a KBr window. The remaining residue is not THF as seen when compared to a reference spectrum of this solvent. A transmission spectrum through this window is then collected, recording spectral features (or lack thereof) of any species responsible for residue formation.

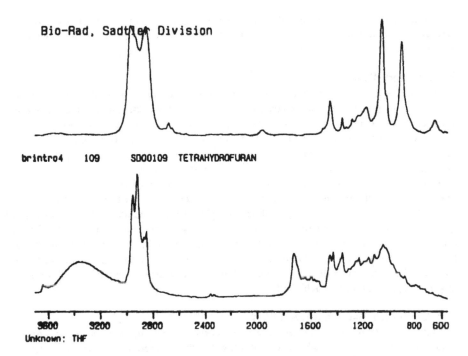

FIGURE 2.   Residue remaining on potassium bromide window after evaporation of a THF solvent compared to a reference spectrum of tetrahydrofuran. The residue is seen not to be residual THF but another species.

e. Evaporation.   The use of solvent evaporation by an analyst also may introduce two additional sources of error. The first situation is residual solvent from incomplete evaporation. This may occur by entrapment, adsorption to active sites, complex formation, or inappropriate evaporation conditions (e.g., too short of evaporation time period or insufficient heating temperature). This is very apparent in inorganic materials where the material may possess several different levels of water crystallization resulting from the presence of residual aqueous solvent. The second case involves the introduction of moisture (water) into a sample by too rapid evaporation of the solvent. As evaporation is a cooling process, quick volatilization of a solvent with a low boiling point (e.g., ether, acetone, Freon, and methylene chloride) may cool the surface sufficiently enough that atmospheric moisture will condense onto the sample.[39] Spectral features due to water are very strong, and they create significant interferences when attempting to make quantitative measurements. Additionally, the presence of water may cause a chemical reaction to occur in moisture-sensitive samples.

B. Miscellaneous Considerations

1. Number of Samplings

For every reference standard and each sample analyzed, a minimum of two complete samplings should be made (duplication). This procedure minimizes random errors and verifies reproducibility in measurements. Additionally, with only small amounts of sample needed for most measurements, the analyst makes the naive assumption that the tiny portion quantified is representative of the bulk. Multiple examinations verify the homogeneity of the material and minimize the chances of incorrect analysis.

2. Accessory Dependence

A method should never be developed which is solely dependent upon a single holder, crystal, or cell. This item refers to an analysis based on a certain pair of windows, bottle of halide powder, a particular diffuse reflectance cup, a single internal reflection element (IRE), a lucky pipette, etc. It is only a matter of time before that equipment is lost, damaged, or broken, and the analytical method dies an ignoble death. This is not to be confused with using a fixed pathlength liquid cell or long pathlength gas cell, where the sample thickness is of importance.

Cells used to contain a sample during examination (gas cells, liquid cells, environmental chambers, etc.) all have the characteristic ability to acquire a "memory". This means that spectral features due to previously analyzed samples are retained and appear in subsequent spectra. Typically, this problem is remedied by flushing the container with an inert material and followed by the sample to be analyzed or taking frequent backgrounds through the cell. However, in instances where this problem may be severe (i.e., long path gas cells or liquid internal reflection accessories), a spectrum of the sample must be acquired in a flowthrough mode for equilibrium to be achieved.

3. Alkali Halide Salt Characteristics

Halide salts (powders, windows) are hygroscopic. Therefore, they should be stored in desiccators or warm ovens which do not allow for the absorption of water. Additionally, reference spectra of the materials should be recorded (ratioed to open beam) for detection of any impurities (Figure 3). The analyst should not store these materials near organic materials that have substantial vapor pressure, as they will exhibit surface adsorption of the volatile species. Calcination, in air, using a high temperature oven, can eliminate the presence of most organic impurities in halide salts without adversely affecting crystallinity. Rather than purchasing powdered halide salts, which typically have more impurities than crystals used as windows comprised of the same salt, it is sometimes advantageous to grind discarded halide salt plates to obtain purer material.

4. Chemical Compatibility

The last generic observation to be made regarding samples in general is chemical compatibility with all materials with which they come in contact. Nothing can take the place of good common laboratory practices and general chemistry principles when examining samples. Just because the analyst wishes to make a quantitative measurement upon a sample, basic chemical reactions do not become invalid. This includes such events as chemical reactivity with halide compounds used in making pressed disks or comprising IREs and cell windows; water incompatibility with most ionic salts used for making pressed disks and cell windows; partitioning coefficients in extraction procedures; drying of a sample while residing in the inert atmosphere of the spectrophotometer; and reaction of chelating agents from metal complexes with materials comprising an accessory.

C. Considerations for Solid Sample Examinations

1. Polymers

Throughout the years, numerous publications have concerned themselves solely with the infrared analysis of polymers and associated additives; in 1964, a compilation of established

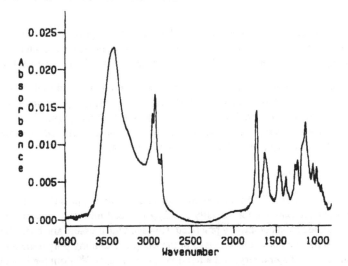

FIGURE 3. The spectrum collected for a sample of potassium bromide powder as a pressed pellet showing contamination by organic species.

quantitative analysis methods was published.[40] Unfortunately, the authors are unfamiliar with an update to this extremely useful review, and the reader must either perform an extensive literature search to discover a particular application or develop an analysis based upon the following guidelines.

a. Preparation. Standard reference samples for polymers are oftentimes the most difficult to obtain. This is because of the slight variations which may be present in their formulations. For example, the analyst cannot assume that a method yielding quantitative information regarding the vinyl chloride and vinyl acetate concentrations in a polymer can be applied to every formulation of this blend. The method used to manufacture the polymer becomes an important factor, and perhaps the limiting factor, in the determination of which polymers may be analyzed by a particular method.[41] This means that a polymer generated from a copolymerization process will not exhibit identical features to that corresponding to a block polymer,[42] or to that comprised of individual homopolymers due to different bond formations (e.g., cross linking) and the functional groups residing in different environments (i.e., resulting in associations such as hydrogen bonding).[43] This can be represented by a polymer composed of monomer units A and B. Copolymerization will create random strings of ABABBAABAA while a block polymer will consist of recurring sequences, AAABBBAAABBB. Blending of individual homopolymers will not have any chemical bonding between the two units but will just be physical mixtures of AAAA and BBBB.

The reaction mechanism used to generate a polymer causes crosslinking at specific sites and, ultimately, determines the polymer's crystallinity[43] and stereochemistry. Various catalysts[35,41] may be used (both organic and inorganic) to control the reaction pathway resulting in different polymer structures from the same reactive ingredients.[41] Not only do these catalysts cause different molecular structures reflected in dissimilar infrared spectra, but they also have spectral features themselves, including light scattering from particles. Thus, compositions of polymers prepared from different catalysts realistically cannot be analyzed by the same calibration.[41] Another issue related to polymer formulation is the presence of additives. Depending upon the final use of a polymer, or its inherent instability, various compounds, such as UV inhibitors, antioxidants, flame retardants, plasticizers, and antiozonants, may be incorporated at trace levels.[44] Again, these species all have infrared spectral absorbances which may perturb the polymer matrix and must be taken into account when quantitative analysis is performed (e.g., limiting the reported accuracy, negating use of the same calibration).

b. Dissolution.   The preferred method of polymer examination is dissolution in a solvent, as this overcomes crystallinity effects.[43] However, it is not always possible to find a suitable solvent, nor can inorganic species (typically used as fillers and pigments) be expected to dissolve in a solvent. Therefore, the polymer form often must be altered to allow for examination. This alteration typically takes the form of grinding.

2. Effects of Grinding

a. Sample Alterations.   When the morphology of a sample requires that it acquire the form of a fine powder for examination, usually grinding in a ball mill or in a mortar with a pestle are the two methods most commonly practiced. However, the pressure and heat generated from the process can induce numerous changes in the sample, including polymerization, degradation, oxidation, loss of symmetry, and alteration of crystallinity.[3,36] Both the induction of, and destruction of, polymorphism (ability of a chemical compound to crystallize in several forms which are structurally distinct)[45] are sources of error providing potential pitfalls for the unwary analyst. Structural damage caused by the grinding process may be minimized or eliminated by performing a "moist" grind or using ultrasonic radiation. In the case of many inorganic materials, the addition of isopropyl alcohol to the material during the milling process minimizes the damage.[45] This must be followed by thorough evaporation of the solvent prior to examination. The possible reaction of the alcohol with/in the sample and contamination by solvent impurities must be investigated.

Performing the grinding in the presence of a halide salt is not advised for several reasons; thus, the ground powder is typically added to the sample after grinding. The first reason is that differences in hardness between the sample and the salt will cause nonuniformity of particle size between the two materials, with the "softer" material being ground finer or simply deformed. More importantly, ion exchange (e.g., an organic acid becoming an acid salt)[42] and isomorphous substitution of cations in a latticework may occur causing examination of the "wrong" material.

b. Uniformity.   A major problem, which arises when a finely powdered sample is needed for the acquisition of a spectrum, is making the particle size small, uniform, and reproducible. Light-scattering takes o   r from absorption with increasing particle size and may become the predominant effect reflected in an infrared spectrum. Therefore, size should be small, less than the wavelength of radiation used for its examination. The likelihood of reproducibility in particle size is increased by always grinding the same amount of sample for the same length of time. An aid to assuring sample uniformity is using wire mesh sieves for isolation of particle size fractions and using only one of the fractions for analysis. However, in this latter case, extreme caution must be exercised so that the fractions are representative of the whole sample and are not comprised exclusively of one component.

D. Gases

1. Theory

The infrared absorption spectra of many gases at low pressure consist of a large number of extremely sharp, narrow bands corresponding to transitions between individual vibration-rotation energy levels. Since there are less absorbing molecules in a given volume of gas than in a condensed phase sample, a greater sample thickness is required to record its infrared spectrum (from ten centimeters to several hundred meters in length). The minimum detectable concentration level depends upon the specific gas being measured.[46]

Long path gas-phase measurements permit quantitative determination of trace organic molecules down to concentrations well below 1 ppm.[46] Multicomponent spectra of small molecules exhibiting prominent vibration-rotation structures can be evaluated quantitatively by selection of characteristic vibration-rotation lines without extensive mathematical procedures. However, the quantitative determination of a component's contribution to a multicomponent spectrum with broad, heavily overlapping bands requires least-squares fitting or principal component analysis.

In infrared spectroscopy, the spectrum of the cell without a sample (I_o) and the spectrum of the cell containing the sample (I) must be recorded. As shown in Equation 1, for quantitative analyses the absorbance spectrum, $A_i(\tilde{v})$:

$$A_i(\tilde{v}) = \log[I_o(\tilde{v})/I(\tilde{v})] = a_i(\tilde{v}) \cdot b \cdot c_i \tag{1}$$

is calculated with $a_i(\tilde{v})$ = absorptivity of component i at wavenumber $\tilde{v}$, b = sample path length, c_i = concentration (partial pressure) of component i, and $(\tilde{v})$ = wavenumber in cm^{-1}.

The spectrum of a multicomponent mixture A_u which is the sum of n individual spectra A_i is given in Equation 2:

$$A_u(\tilde{v}) = \sum_{i=1}^{n} A_i(\tilde{v}) \tag{2}$$

Equations 1 and 2 are known as the Beer-Bouguer-Lambert law (commonly referred to as Beer's law). Beer's law strictly holds only for isolated molecules, interacting with the infrared radiation. This requirement is met ideally when gases or vapors at low partial pressures are diluted in a nonabsorbing matrix, e.g., parts per million (ppm) in nitrogen. Deviations from the ideal behavior can arise from optical interferences, nonlinearities at high absorbance values, or different total pressure or temperature[38] of the sample and the reference spectra, rather than from intermolecular interactions,[43,46] as in liquid phase spectra.

2. Memory Effects

The results from trace-component analysis in long-path gas cells can be adversely affected due to selective adsorption or desorption of materials from the cell walls. The cell "memory" should be erased by emptying (evacuating) and filling it several times before a spectrum is recorded.[3,38,46] It is often convenient and cheap to use very dry, high-purity nitrogen gas to flush a cell. In some instances, it may even become necessary to keep fresh sample continuously flowing through the cell to obtain an accurate representation of its composition.

3. Temperature and Pressure Effects

It is extremely important to carefully control both temperature and pressure when performing any quantitative analysis upon gas phase species. Absorption band intensities, widths, and areas are dependent upon both of these parameters. In fact, the extinction coefficient (absorptivity) of a component in terms of its partial pressure may be a function of the total pressure of the sample,[37,46] as explained below.

The infrared absorption spectrum of a gas at low pressure consists of narrow bands corresponding to transitions between individual vibration-rotation energy levels. The distribution of molecules in the various rotational (and vibrational) energy levels follows the Boltzmann distribution, which is a function of temperature. As the temperature increases, the population of the molecules in the higher-energy states increases, and the contours of the vibration-rotation band change accordingly. In addition, more frequent collisions induce broadening of the absorption bands, which does not necessarily correlate linearly with concentration. Additionally, if rotational isomers are possible, a temperature change may result in a different distribution of the isomers with corresponding variations in the spectra.

As the total pressure of a vapor phase sample is increased, the number of molecular collisions per unit time increases. The molecule cannot rotate freely during these collisions, and if its rotational energy changes by absorption of radiation, one or both of the energy levels involved are displaced. The actual result in measurement terms is that the average absorption band is widened with an apparent increase in its intensity. The phenomenon is known as pressure broadening, and its effect must be taken into account during calibration by recording all standards at the same total pressure. Usually this is accomplished by pressurizing all gas samples to 760 Torr (millimeters of mercury) with an inert gas such as nitrogen.[3,27,37] In addition to minimizing intensity variations from pressure broadening, this procedure gives better sensitivity for weak absorbers and precludes any intensity changes resulting from accidental air leakage into the cell. The analyst must also be aware that different absorption bands of the same gas may respond differently to pressure broadening (i.e., its effect is not simply a factor applied across an entire spectrum).

4. Experimental Procedure

A typical experimental sequence used to collect vapor-phase spectral data would be as follows. First, the desired pathlength cell is selected. The cell is evacuated using an appropriate vacuum pump, which does not allow for accidental introduction of pump oil vapor into the system. The cell is filled to 760 Torr with high-purity nitrogen gas using a manifold system. The single-beam background reference spectrum is collected through the cell by the spectrophotometer. The cell is evacuated and filled with the sample. It is evacuated again and refilled with the sample to be studied, in addition to any high-purity nitrogen gas necessary to achieve the total pressure of 760 Torr, making sure that the absorbance of the species to be measured will be within acceptable limits for quantitation. The sample is allowed to achieve thermal equilibrium. A spectrum of the gas sample is collected, using the previously collected single-beam spectrum as the background for ratioing. The cell is evacuated and flushed with high-purity nitrogen gas to be made ready for the next sample.

5. Chemical Reactivity

As with all samples, the analyst should be familiar with the species present in the samples to be analyzed and their compatibilities with the materials composing the gas cell. Hydrogen fluoride and tungsten hexafluoride gases are extremely corrosive materials requiring the use of specially constructed cells for their analysis (attacking most glass and halide salt surfaces). Many organics will attack polymer O-rings used to seal some gas cells, causing leaks or new species to appear from the chemical reactions. Additionally, the use of grease for ground glass joints (e.g., manifold connections, stopcocks) should be minimized as the grease itself has a vapor pressure and may appear in the sample spectrum, or it may absorb some gaseous species providing a "memory" effect from their subsequent degassing. If it must be used, a heavy grease with low volatility should be selected. However, contamination from grease is difficult to remove by simple flushing of the system, as it coats all surfaces and has low volatility. It generally requires dismantling of the accessory and washing with a suitable solvent which does not cause condensation nor leave a residue.

E. Films

1. Uniformity

A sample in the physical form of a film can simply be examined by standard transmission techniques, and is typically easy to make and store. However, one of the most difficult problems in such transmission measurements is that of uniformity, both in sample thickness and homogeneity. Free-standing films may have their thicknesses measured directly through use of a micrometer. It is best to obtain an average thickness based upon several measurements taken across the film. Samples are expected (realistically) to exhibit some variation in thickness due to preparation practices, but those possessing significant nonuniformity, or nonconformity to the reference standards or other samples being prepared as references, should be discarded. Additionally, several spectra for each sample should be recorded with the sample being rotated to a different orientation in the beam. The quantitative results then are reported as an average from the measurements, thus limiting the error introduced by nonuniformity in sample pathlength. Typically, a retaining ring or a spacer (a washer may be used if more expensive equipment cannot be purchased) in a mold can be used to achieve reproducible uniform films.

2. Preparation

Films are usually prepared on or between Teflon-coated sheets or aluminum foil to aid in their removal for mounting into suitable holders. The film should be taut and flat, free of crevices and holes, and mounted without stretching (which would alter film thickness and possibly introduce orientation due to stress). Oftentimes, fringing is observed in a spectrum of a film, which is based on its thickness. A matted film surface, which allows a "clean" spectrum to be recorded (eliminates the shiny parallel sides allowing internal reflections), can be obtained by placing a fine grain carborundum paper in the mold used to prepare the film on the side of the Teflon sheet or aluminum foil opposite from the sample.[27] Occasionally, the positioning of the film against a halide salt crystal may eliminate the multiple reflection interference effect, causing the fringe pattern to disappear.

3. Pressure and Temperature Effects

Pressed films, usually prepared between the platens of a hydraulic press, can exhibit spectral features due to the conditions used during their preparation. The literature should be consulted for more extensive details concerning the effects of the conditions upon the quality of the film prepared.[47,48] When samples require elevated temperatures to allow the deformation necessary for making a film, it must be carefully controlled so as not to introduce irreproducible crystallinity effects or chemical changes.[41] The pressure used to press the material thin enough for examination by transmission must be reproducible to insure that the same physical and chemical state of the sample is achieved. Spacers may be used to ensure that all films are produced with the same thickness. The methods for the preparation of hot-pressed films are more thoroughly described elsewhere.[27] Most importantly, the length of time that the sample is exposed to heat and pressure must be carefully regulated.

4. Cast Films

a. Impurities. Films resulting from the dissolution of a material in a solvent, followed by evaporation of the same solvent (cast films), have a multitude of potential error sources. First and foremost of the problems is the introduction of solvent impurities into the material to be quantified. Butylated hydroxytoluene (BHT) is a well-known inhibitor in tetrahydrofuran (THF), an excellent organic solvent. However, since BHT has a relatively high boiling point, it is rarely removed during simple evaporation of the THF at room temperature. The spectral absorbances due to this interloping species will vary randomly in any standards or samples prepared from dissolution/evaporation unless steps are taken to: insure that the quantitative method compensates for the presence of the BHT spectral features, remove the BHT from the material prior to analysis, or make the concentration of BHT identical in every standard and sample, negating its spectral contributions.

b. Solvent Volatilization.   During the evaporation of a solvent for the generation of a cast film for quantitative analysis, a variety of other events may occur. If the solvent volatilizes quickly (e.g., ether, acetone, and methylene chloride), condensation of atmospheric water vapor may occur due to the cooling process effected by the evaporation process.[39] This will then introduce interfering spectral absorbances due to the included moisture in the sample or the formation of a poor, uneven film surface. Additional problems include evaporation errors from the loss of volatile components in the sample and the occurrence of polymorphism and formation of aggregated domains based upon evaporation rate. These effects are generally accentuated with the application of heat to speed up the solvent loss. Elevated temperatures may also cause the degradation (e.g., oxidative decomposition) of many compounds. The inclusion of entrapped bubbles of air or other gaseous species from the evaporation of a solvent can also create an inhomogeneous film. Last, but not least, are the problems of solvent retention in a cast film by incomplete evaporation[36] (not waiting long enough, adsorption to specific components, complex formations, etc.), requiring heating in an vacuum oven for removal, or actual chemical reaction of the sample with the solvent used for dissolution.

5. Rolled Films

One last way of preparing films is the rolling of powders into sheets. At times the rolling surfaces are heated to make the materials more malleable. However, the analyst must be aware that this method offers several disadvantages: (1) carryover contamination from previous samples from incomplete cleaning, (2) pressure and temperature are difficult to reproducibly control and may cause reactions to occur in the sample, (3) the heating may cause evaporation of any volatile materials, and (4) the rolling process can introduce orientation into the final film which may not be repeatable. In the event that a method is developed using production-rolled film exhibiting orientation, all samples should be marked with the roll direction. The analysis must involve positioning of the film in the spectrophotometer in the same direction since the infrared beam will almost certainly show preferential polarization in one vector.

6. Opacity and Particulates

There are two final situations that may arise which will cause problems when performing quantitative analysis on films by a transmission method of analysis. The first involves a system in which species are present that may cause opacity. Generally, this means that energy throughput will be low resulting in poor signal-to-noise ration (SNR) or the sample will have spectral regions which exhibit total infrared absorbance. The second event involves a sample which possesses particulate matter. Scattering of the infrared radiation will occur, based upon particle size, and cause various nonlinear anomalies in the recorded spectra.[45] Additionally, these particulate species are not uniformly distributed throughout a film, generally residing as aggregated domains and making quantitative analysis extremely difficult.

F. Alkali Halide Disks

1. Preparation

The same problems which exist for films also are applicable to the recording of samples prepared as halide disks. In this method of analysis, the sample is introduced as a finely ground powder into a similar mesh halide salt diluent (e.g., NaCl, KBr, KCl, etc.) at a low concentration (typically 2 to 10% by weight). The powders are mixed, placed into a die, and pressed into a disk under pressure (e.g., hydraulic or handheld press). The die should be rotated by hand on the mixture prior to pressing to insure a good distribution of the powder across the area of the pellet. The best quality pellets are normally obtained using a hydraulic press at 10 tons pressure while evacuating the air from the die with a vacuum pump. The resultant disk should be translucent and exhibit uniform coloring. Many of the concerns regarding the homogeneity of a film apply for a sample examined by this method including thickness, wedging, particulate distribution, air voids, pressure effects, etc.[2] However, even more potential pitfalls may arise.

For quantitative measurements to be made, the powders (sample and diluent) must be carefully weighed prior to mixing so that repeatability in sample concentration can be achieved. A weight of 350 mg of alkali halide powder (e.g., KBr) mixed with 1 to 5 mg of sample is appropriate for the preparation of a 13-mm disk, which will yield absorbances in a linear range for quantitative measurements.[27,49] All of the mixture should then be placed into the die for pressing. Alternatively, a predetermined weight of a mixture can be placed into the die.[2,50] A blank pellet is made in a similar manner (same weight) for the same batch of alkali powder.[27] This is used as the background reference material to obtain a ratioed spectrum.

Occasionally, in quantitative analyses of powdered materials, the use of an internal standard for normalization is preferred. Sodium azide, with a strong peak at 2140 cm[-1], is useful for many organic samples.[47] Other materials found to be useful for internal standards are lead thiocyanate, calcium carbonate, and hexabromobenzene.[24,47]

A problem that can arise when examining a pressed pellet is when the beam size at the sample is smaller than the area of the pellet. Since it is extremely difficult to prepare a pellet that exhibits complete homogeneity and has an equal density across, a small beam focus may record the spectrum for a nonrepresentative region of the pellet. In such instances, it is highly recommended that a pellet size that is smaller than the standard 13-mm pellet diameter be used in order to appropriately match the beam size and acquire the spectrum of the whole pellet.

2. Ion Exchange

Since a halide salt is used as a diluent, caution must be taken to insure that ion exchange with the sample does not occur.[2,3,27] This is a potential danger whenever polar molecules, or inorganic molecules possessing lattice symmetry or cation sites, are examined. Hydration, both internal and external adsorption, provides a convenient medium through which ions may exchange. Therefore, moisture sensitive materials should be prepared for analysis in a dry box or under an inert atmosphere.[3] The halide diluent should be kept dry (dried in a suitable oven without any organic vapors present) and stored in a desiccator until used. Mixture of the sample with the diluent should not occur until just prior to examination. Procedures in which the two materials are ground together in a mill or with mortar and pestle are considered bad practice.[3,49] Rather, they should be powdered separately to the same mesh size and then intimately mixed, without the introduction of pressure. Many inorganic materials require that they be ground while moist (usually isopropyl alcohol is used) to avoid structural damage, followed by drying to remove the wetting agent.[45] Alternatively, ultrasonic radiation may be used to create the particulate size necessary for examination of the sample. Needless to say, all labware used in the preparation of moisture sensitive materials should be extremely dry.

G. Liquid Cells

1. Description

As the name implies, the sample must be in a liquid form for analysis by this method. Liquid cells come in various forms — sealed, demountable, variable pathlength, microcavity, and variable temperature.[37]

The macro-liquid cells consist of a spacer which determines pathlength, two windows, two plugs, and an assembly form which holds the windows and spacer in position. The assembly form typically has two Luer Lok fittings which allow a syringe to be attached for sample loadings. The top window has two small openings which are aligned with the Luer Lok fittings for the introduction and removal of the sample. The plugs fit into the Luer Lok fittings in order to prevent the sample from evaporating or leaking out during an analysis.

The sealed liquid cells are of a fixed pathlength. This means the analyst must decide on which thickness cell is required for a useful spectrum to be obtained from a sample (i.e., the desired infrared absorbances are between 0.2 and 0.7 absorbance units). The spacer cannot be changed, and, therefore, the pathlength of the cell cannot be changed. Likewise, the cell cannot be disassembled for cleaning purposes. It is advisable to periodically measure the cell thickness since some window materials slowly dissolve with usage.

The demountable liquid cell can be used in a fixed pathlength mode as well. The advantages of this cell are that it can be disassembled for cleaning and that the spacer can be changed to provide flexibility in pathlength choice. One major disadvantage is that once the cell is reassembled following cleaning or spacer replacement, the pathlength must be redetermined. The easiest method for determining or verifying the pathlength of a cell (or thin film) is to use the interference fringes superimposed on the baseline of its recorded spectrum. This is accomplished by collecting a spectrum of the empty cell (ratioed against the empty beam of the instrument) which is comprised of fringes from constructive and destructive interference of radiation reflected from the internal surfaces of the cell windows. The distance can be calculated from Equation 3:[3]

$$\text{pathlength} = \frac{n}{2} \cdot \frac{10}{(f_1 - f_2)} \tag{3}$$

where pathlength is the cell thickness in millimeters, f_1 and f_2 are the frequencies (cm^{-1}) of two maxima from the interference fringes observed in the spectrum of the empty cell, and n is the number of maxima occurring between f_1 and f_2.

Liquid micro-cavity cells permit the collection of solution spectra from samples within the range of 10 to 100 μg without a beam condenser. In most cases the cells are disposed of following analysis, as they are extremely difficult to clean.

Variable-temperature liquid cells generally are demountable liquid cells. They can be cooled or heated, depending upon the particular design. The main usage of this type of cell is for the study of temperature sensitive materials in which the temperature must remain constant; otherwise, phase changes due to temperature are measured. The analyst must allow sufficient time prior to data acquisition so that thermal equilibrium of the sample is reached, otherwise spectral artifacts will be introduced based upon changing temperature.

The best results for quantitative measurements are obtained if the cell is not moved after the introduction of the first sample, in which case sampling should be a flow-through method.[27] The reason for this is that sample-cell positioning is difficult to reproduce accurately. When this flow-through method is not possible, the cell should fit snugly into its holder to protect

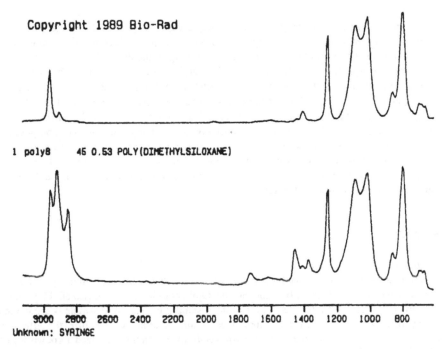

Copyright 1989 Bio-Rad

1 poly8   45 0.53 POLY(DIMETHYLSILOXANE)

Unknown: SYRINGE

FIGURE 4. The spectrum recorded for a silicone lubricant extracted from a disposable syringe by using *n*-hexane as a solvent.

against any motion. Additionally, the cell should always be inserted into the holder in the same direction each time.

2. Contamination during Filling

The standard liquid cell is normally filled using a syringe or pipette. Care must be exercised to prevent contamination of the sample as most plastic disposable syringes are internally coated with a silicone lubricant (refer to Figure 4), easily extracted into most solvents. Additionally, the plasticizers present in some rubber plunger tips can be selectively extracted by organic solvents resulting in contamination. Therefore, the use of glass syringes is highly recommended.

3. Filling and Cleaning

Typically, sample introduction is made from the bottom fitting so that the air is forced upwards through the top opening in the cell. For cells thinner than 0.1 mm, the two-syringe technique is recommended.[49] A syringe containing the sample (or cleaning solvent) is placed in one port. A second, empty syringe is attached to the second port. The plunger of the empty syringe is withdrawn to produce a vacuum in the cell, causing the sample to flow into the space. This procedure eliminates the high pressures produced when filling thin cells by forcing the sample through the cell when using a single syringe. The thin cells generally retain their specified pathlength longer and do not develop pockets between the spacer and windows, which cause carryover effects from one filling to the next. Additionally, the analyst must make sure that no air bubbles are present between the windows as they will distort the spectrum and change the sample pathlength.

Cleaning is typically accomplished by pulling a vacuum through the cell (e.g., vacuum pump, syringe) to evacuate the sample, followed by copious flushing of the cell with a solvent to remove any residue or trace evidence of sample, and drying by vacuum or passage of dry air through the cell. The air must be "dry" in order to prevent the fogging of hygroscopic crystal windows (most halide salts) by moisture. Since this procedure is a cooling process, the dimensions of the cell will change. Therefore, time must be allocated for the cell to thermally equilibrate prior to examination of the next sample; otherwise, band intensity changes may be recorded. Additionally, since many windows are composed of alkali halides, they are subject to attack by water and water vapor, which cause pitting or "fogging." Fogging may be avoided by keeping the cells in an atmosphere of low relative humidity (i.e., desiccator) or by ensuring that they are stored at a temperature sufficiently above ambient to prevent water adsorption.[37]

H. Mulls

1. Preparation
Solid samples which are insoluble in the usual solvents can be prepared using an oil mull of the ground powder. This may be a mineral oil (Nujol), chlorinated/fluorinated oil (Fluorolube), or hydrocarbon grease (in the case of far-infrared sampling). The solid sample is ground into a very fine powder. The particles should be small, in order to minimize the effects of diffraction, and of uniform size. To a known weight of this sample is added a known weight of finely ground material to serve as an internal standard. This material should not react with the sample and should possess isolated spectral features which can be used for ratioing in the quantitative analysis. These two powders are then intimately mixed. A small amount of one of the aforementioned oils is added to the powder, with stirring, until a thick paste consistency is reached. The oil is added one drop at a time and the mixture blended thoroughly after each addition. Too much oil will cause the sample to be too fluid and, also, create an extremely strong spectrum which may mask the features due to the sample under analysis. The mixture is smeared onto a salt plate as a thin film and another plate is placed on top, with a slight twisting motion to remove any air spaces and create a uniform film thickness. It is then mounted in a holder and placed in the beam path for data acquisition.[37,49]

2. Sample Types
This method is used for any solid material which can be ground into a powder form. Polymers can be frozen with liquid nitrogen or dry ice for the grinding process if they are soft and malleable at room temperature. Oftentimes a file or rasp is useful in providing fine dust from a large object. Normally, however, a ball mill or mortar with pestle are the best grinding equipment.

Samples which do not give good spectra by this technique are those containing particulates (mainly metals) which cause spectral distortions due to mirror-like reflections from their surfaces or graphite/carbon black particulates which act as severe baseline distorting agents and absorb most of the available infrared energy.

The background to be used for ratioing purposes are the windows prior to the application of the mull. The oil should not be used for ratioing as it will not have the same pathlength as in the sample, and the oil bands will appear as residuals in the ratioed spectrum. For removal of the oil bands which are in the linear range, spectral subtraction may be used. This requires that a spectrum of the oil between the windows be acquired. However, the analyst must be made aware that nonlinearities may be introduced by bias in performing spectral subtractions.

Two of the more commonly used accessories employed to acquire spectra are internal reflection attachments (IRAs) and diffuse reflectance units. Therefore, brief descriptions of their use follows.

I. Internal Reflectance Spectroscopy

1. Description
Internal reflectance spectroscopy (IRS) is a technique where a sample is placed in contact against a special crystal, with a high index of reflection, termed an internal reflectance element (IRE). The infrared beam is focused onto the beveled edge of an IRE, reflected through the crystal, and directed to the detector in the spectrophotometer. At each point of reflection where the sample contacts the IRE, the radiation penetrates a short distance into the sample and is absorbed.[51] The resultant spectrum closely resembles that of a transmission spectrum for the same sample. A more extensive discussion of the IRS technique is presented in Chapter 3 of this book.

IRS is usually employed when a sample cannot be conveniently studied by traditional transmission techniques.[52] It is also the preferred technique when a nondestructive method of analysis is desired.

2. Reproducibility Issues
Carefully designed IRS experiments can yield adherence to linear relationship of absorbance vs. concentration and, thus, permit quantitative measurements. However, it is essential that data collection is reproducible. Therefore, experimental parameters must be extremely well-defined and controlled. The angle of incidence and number of reflections can be assumed to be nonvariables for each IRA, provided that the crystal is not changed nor the accessory misaligned. The main variable encountered in quantitative measurements is uniform, repeatable contact of the sample against the IRE. This includes reproducibility in IRE area coverage and quality of contact. Constant film thickness and the use of torque wrenches[51] may be used to duplicate sample contact efficiency in solid samples. Alternatively, a band-ratioing technique may be employed to perform quantitative measurements, as it is somewhat independent of pressure so long as the bands used are in the same spectral region[3] (typically no more than 100 to 200 cm^{-1} apart). The analyst must also assure that air bubbles/pockets are not present against the IRE, introducing irreproducible sample contact. The reader is referred to Chapter 3 in this book for further discussions on the pitfalls of analysis by the IRS technique.

J. Diffuse Reflectance Spectroscopy

1. Description
Diffuse Reflectance Infrared Fourier Transform Spectrometry (DRIFTS) is a reflection/absorption technique where the samples to be analyzed are either "neat" (pure) or prepared as dispersions in nonabsorbing matrices such as halide salts or diamond powder.[53] Neat samples should only be used when measurements are to be made on weakly absorbing spectral regions. Reflection does not occur at exactly the same frequency as absorption for strong bands, which causes shifts and distortions (Reststrahlen). The Kubelka-Munk format of data presentation is favored by some spectroscopists. This format is discussed further in Section VI.C.2. The reflection contribution, comparable to stray light, can be reduced by decreasing the particle size, or more tightly packing the material.[54] These procedures cause more interactions of the incoming radiation with the sample.

The effect of different geometries of diffuse reflectance attachments has been studied in detail.[55] It has been shown that the off-axis accessory design exhibits a linear absorbance response over wider range of usable sample concentrations than the straight-through design.

2. Sample Preparation

Typically, the sample is ground into a fine powder for examination. Therefore, most samples are solids which can be altered to form small particulates. All the criteria discussed previously concerning the manipulation of a sample to acquire a fine powder apply here as well, including the physical mixing of the diluent with the sample. Some matrix materials, such as diamond and polyethylene powders, do not require any further processing and are used "as is" from the manufacturer.

A mixture consisting of between 1 to 10 % by weight of sample[52] in an inert diluent is made by thorough mixing of the two powdered fractions. The diffuse cup (holder) is loaded by spooning the powder into the bowl until the cup is overfilled. The sample is packed by dropping the cup several time from a height of one-eighth to one-fourth inch. The knife edge of a spatula is run over the surface in a single motion, even with the edges of the cup lip, to obtain a flat, reproducible surface. Alternatively, the cup can be slightly overfilled and inserted into a hand press (used for pressing halide disks). Pressure is applied to uniformly pack the powder and create a flat surface. After filling, the cup is loaded into the diffuse reflectance accessory and a spectrum is collected. A background spectrum of the matrix material is used for ratioing purposes. It should be acquired from the same batch (lot and grinding session) of matrix material used to dilute the sample, using the same cup loading procedure as that for sample examination.

At times, a sample may not lend itself to grinding because of location, size, or texture. A piece of carborundum (silicon carbide) paper can be lightly rubbed over the surface of the material to be examined. A piece, the same diameter of the diffuse reflectance accessory cup, can be cut from the area of the paper which came in contact with the sample. The circular cutout is place onto a cup loaded with powder in order that its surface is even with the top of the cup. A spectrum is then collected, using a clean piece of carborundum paper and loaded in the same manner as the background for ratioing purposes.

Additional details on diffuse reflectance spectroscopy can be found in Chapter 4 of this book.

V. RECORDING THE SPECTRUM

A. Spectrophotometer Performance

Also important to the discussion of quantitative infrared methods of analysis are the instrumental conditions needed for good quantitative analysis. High analytical precision and accuracy cannot be obtained if the spectrophotometer is not performing properly. Gross misalignment of spectrophotometer optics will introduce major sources of nonlinearity and irreproducibility.[3] The analyst should keep a log book of instrument performance to monitor long-term operating behavior. A standard practice for routine monitoring of instrument performance has recently been issued by the ASTM.[56]

1. 100% (Transmittance) Line

Suggested test measures include the regular collection of a 100% line under predefined conditions (i.e., co-addition time, resolution, apodization, aperture size, etc.). Typically this will consist of two single-beam spectra collected sequentially and ratioed to obtained a % Transmittance (T) spectrum. The recorded spectrum should be flat, featureless, and positioned at 100% T. The sources for any spectral artifacts should be isolated and removed (e.g., electrical spikes and vibrations). A flat spectrum demonstrating an offset from 100% T is typical of temperature instability in the spectrophotometer's source, in a thermally cooled detector which hasn't reached thermal equilibrium, or in fluctuations of the surrounding environment. Sloping baselines are usually due to misalignment of the spectrophotometer optics. This data should be collected prior to introduction of an accessory or sample. The analyst should follow the established procedure for realigning the spectrophotometer as required by the recorded instrument performance. Additionally, the noise level (RMS and/or peak-to-peak) should be measured in a predefined region, typically between 2200 to 2000 cm^{-1} in the mid-infrared as there exists little spectral interference in this region and the detector response is greatest here. A rise in noise level will indicate problems with spectrophotometer performance, which may be due to misalignment or degradation of source or detector. In any event, good quality data cannot be collected when this behavior occurs and maintenance of the system is required.

2. Single-Beam Spectrum Inspection

Next, one of the single beam spectra collected to generate the 100% line data should be examined carefully. The maximum of the blackbody curve should remain constant as it is a measure of the total amount of energy passing through the spectrophotometer based upon the temperature of the source. If the value should drop suddenly, it is usually indicative of the source beginning to "burn out" and requires immediate replacement, although spectrophotometer alignment should first be verified. Additionally, the "icing" of a liquid-nitrogen cooled detector (e.g., mercury-cadmium-telluride or MCT), due to loss of vacuum, can be observed by examining the single beam spectrum for a triangular-shaped absorbance band near 3250 cm^{-1}. If an increase in the intensity of the absorbance in this area is noted, the dewar needs to be reevacuated, typically by shipment back to the vendor where the appropriate equipment exists for the procedure. Finally, the area of the single beam following the detector/optics cutoff should be examined carefully. This is the spectral region which will typically reflect detector nonlinearity (from saturation) as the detector should record the behavior corresponding to no infrared energy being transmitted. This should not be confused with A/D overflows. In a system where the detector is responding linearly, after the cutoff is reached, the single beam energy drops to zero response and remains flat at zero (refer to Figure 5). In cases where the detector is exhibiting nonlinearity, the cutoff will never reach zero response either remaining positive (seeing energy when none is present) or descending below zero response (recording negative energy) or will exhibit various strayings from the zero level.[56,57] This behavior is usually more severe in high sensitivity detectors (e.g., MCTs) and may require that the infrared beam be extensively attenuated (with metal mesh screens, optical filters, and apertures) prior to obtaining a linear response. All metal mesh screens used for attenuation of the infrared beam should be handled as optical elements, carefully shielded from dust, and never touched, as contamination will manifest itself as spurious infrared absorbances in any spectra collected during their use.[23]

B. Accessory Performance

These same observations should be made after the accessory for sample examination has been introduced into the spectrophotometer. In this case, alignment of the accessory, rather than the spectrophotometer, is performed. Additionally, if attenuation of the beam is necessary

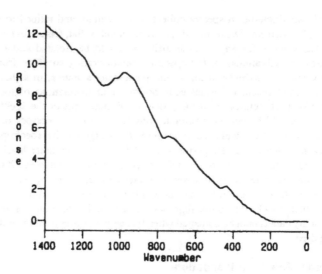

FIGURE 5. Expansion of the cut-off region from a single beam spectrum illustrating good detector linearity.

to achieve detector linearity, both background and sample spectra must be collected with the same configurations of metal mesh screens, optical filters, or apertures to maintain a reproducible beam path.

C. Data Collection Parameters

Several data collection parameters need standardization for a rigorous quantitative analysis method to be developed. This means that the analyst must collect the data under identical conditions under which the calibration was performed. These parameters include collection time, resolution, zero-filling factor, undersampling ratio, interferogram symmetry, and apodization function. In many low-cost instruments, selections of these parameters are fixed and are, therefore, not accessible.

1. Trading Rules

a. Collection Time. The signal-to-noise ratio (SNR) exhibited by a spectrum is related to ordinate reproducibility and affects the precision of an absorbance measurement. It improves by a factor of the square root of the collection time (approximated by the number of scans co-added).[32,54,58] Although the noise level reflected in an infrared spectrum may be reduced by signal-averaging, a fourfold increase in measurement time will decrease the noise level exhibited in a spectrum only by a factor of two. An additional point to consider when attempting to minimize the noise in a ratioed spectrum is that both the sample and background single beam spectra used for its calculation each contain a measure of noise. Typically, the single-beam spectrum which has the least sensitivity or has been acquired for the shortest measurement time will define the noise level in the final ratioed spectrum. That is to say, if the magnitude of the sample's signal is twofold weaker than that for the background signal, the sample collection time should be extended by a factor of four before equaling the sensitivity of that exhibited by the background.

b. Resolution Selection.   Trade-offs are in order when selecting the conditions to collect spectral data. First, high enough spectral resolution, which is related to the accuracy of an absorbance measurement with respect to the true spectrum of the material, should be used to take advantage of the spectral features. The instrument resolution used to record a spectrum should, theoretically, be a factor of five narrower than the narrowest spectral feature measured to produce a spectrum from which quantitative measurements can be made.[2] However, at higher resolution the spectral noise becomes greater, data collection times become longer, and computational time for manipulations increases dramatically. That is to say, per scan, a 4 cm^{-1} resolution spectrum will be twice as noisy as a spectrum measured at 8 cm^{-1} resolution.[32] In terms of measurement time, this means that the 4 cm^{-1} resolution spectrum will take 8 times as long to collect in order to obtain the same SNR. Thus, time constraints and practical considerations require a compromise in resolution selection and typically a factor of two to three narrower is adequate.

2. FT-IR Terminology
a. Interferogram and Centerburst.   To better understand the effect of parameter selections, a basic understanding of terms used to describe the data collection from a Fourier-transform infrared (FT-IR) spectrometer is needed. Explanation of FT-IR operation is given elsewhere[32] so only a brief description is given here. In an FT-IR spectrometer, the signal "seen" by the detector is termed an interferogram. The region of maximum signal in the interferogram is called the centerburst and corresponds to the point when the moving and fixed mirrors in the interferometer are equidistant from the beamsplitter. The interferogram is transformed mathematically by the Fourier transform to produce a single beam spectrum. The Fourier transform takes the data recorded as a function of mirror movement in centimeters (distance) and produces data as a function of frequency in cm^{-1} (wavenumber). The single beam spectrum shows the instrument energy profile which is affected by characteristics of the source, the beamsplitter, and the sensitivity of the detector at different wavelengths. Two single-beam spectra are ratioed to obtain an absorbance or transmittance spectrum. This means that an FT-IR spectrometer is inherently a single beam spectrophotometer, and, thus, it is imperative to record a background spectrum for each sample examined.

b. Asymmetric/Symmetric Collection.   An interferogram is essentially symmetrical about the centerburst. Therefore, in most instances it is only necessary to measure one side of the interferogram to obtain the spectral information, and so the default conditions for many instruments is to do an asymmetric (single-sided) data collect. However, experience has shown that the line shape of intense, narrow bands is described most accurately when a symmetric, double-sided data collect of the interferogram is performed. This allows for the most accurate phase correction of these unusual interferograms (e.g., optical filters) by providing a more precise definition of band shape, position, and, more importantly, intensity.

c. Undersampling.   For standard mid-infrared operation, the electronics need only sample the detector voltage at every second (or third) zero crossing of the laser detector.[32] This is called undersampling the interferogram. Advantages of undersampling are that the resulting data files are decreased in size and that mathematical operations are performed faster.

d. Zero-filling.   A 4 cm^{-1} resolution asymmetric interferogram collected with an undersampling ratio, which samples the detector voltage at every second zero crossing of the laser detector, contains 4096 (i.e., 4K) data points. The Fourier transform takes an interferogram data file as input and returns two spectral data files as output. These two files are the real and imaginary part of the single-beam spectrum and, to retain mathematical consistency, each output file contains 2K data points. The 2K data points in the output file, therefore, give a data point spacing every 4 cm^{-1} (i.e., one data point for every unit of resolution). At this data point spacing, sharp spectral bands are not well-defined, and may appear triangular or flat topped. In order to remedy this problem, it becomes necessary to calculate more spectral points in the output file. This is accomplished by increasing the number of points (input) in the interferogram by simply adding strings of zeros to the end of the interferogram. This technique is known as zero filling, and is, effectively, a means of forcing the computer to interpolate the data.[1,32] The typical default level of zero filling in the mid-infrared region is one. In this case, the interferogram is doubled in length before the Fourier transform, so that twice as many data points are computed. This would correspond to 8K data points in the interferogram, given in the above example, and would result in a point every 2 cm^{-1} in the spectrum. A zero filling-factor of four or eight is recommended for high resolution gas phase work to adequately define the narrow bands recorded in the spectrum, or for solids and liquids where the band width is close to the available instrument resolution.

e. Apodization. When collecting spectral data, the interferogram is only sampled over a finite distance and effectively truncated to zero at the end of the file. The information present at a larger distance from the centerburst (greater retardation) is lost. This discontinuity in the data file results in a distortion of the shape of narrow lines because the information for these lines was truncated at the limit of resolution. This is usually only noticeable for gas phase spectra. The resultant distortion of the band shape appears as negative side lobes at the base of an absorption peak in the transformed spectrum. The lobes are more pronounced for peaks with a half bandwidth close to the resolution used for examination. Apodization is the process used to remove these side lobes (or feet) from the base of the peaks, and application of the mathematical functions affect the removal, but at the expense of some decrease in resolution.[32] It represents a compromise between the half width of the instrument line-shape function and its side lobes. Typically, boxcar apodization produces the best resolution as it simply multiplies the sampled interferogram by one, but it does leave the side lobes at the peak bases.[58] It is the preferred apodization function when analyzing high-resolution gas-phase spectra with rotational-vibrational fine structure. Generally, triangular, Norton-Beer, or Happ-Genzel apodization functions are used in the examination of condensed phase samples where bands widths are broader.

D. Detector Characteristics

The standard mid-infrared detector is a room temperature DTGS (deuterated triglycine sulfate) detector. Liquid-nitrogen cooled detectors provide greater analytical sensitivity, but they are more expensive to operate, and it is more difficult to achieve detector linearity with them, a necessary condition for performing quantitative analysis measurements.[59]

E. Filling the Infrared Beam

The whole beam area at the sample position should be filled to insure maximum sensitivity over the whole absorbance range.[37] If this is not possible, a mask should be used so that the part of the beam which does not pass through the sample is not permitted to reach the detector. The effect on the spectrum acquired when the sample does not fill the beam is distortion by the intensities of absorption peaks being artificially depressed. Thus, it follows that there should be no "holes" in the sample.[54]

F. Atmospheric Interferences

Atmospheric carbon dioxide and water vapor have significant infrared absorption bands which can adversely affect any quantitative analysis method which use measurements in these corresponding spectral regions. Therefore, sealed and desiccated units should be inspected on a routine basis to make sure that the desiccant has not become saturated. Those methods using a purged spectrophotometer system (nitrogen gas or dry air) must establish a reasonable time (which must be made part of the whole quantitative method) for the removal of these species prior to data acquisition. It is not so much that these species must be removed completely from the instrument, but rather that they reach a reproducible concentration level so as not to adversely affect the results reported by quantitative measurements. Ideally, an analysis would avoid these regions precluding any possible interferences by these species.

G. Background Spectrum Selection

The single beam background for computing a ratioed spectrum should be collected through the accessory or holder used for examining the sample. This is to assure that (1) the infrared beam is attenuated in the same manner as when the sample spectrum is recorded, (2) any spectral features due to the accessory will be ratioed out in computation of an absorbance spectrum, and (3) the infrared beam path with be identical between reference and sample examinations. Frequent background collections should occur in those sampling methods which are prone to carryover effects.

H. Software Compatibility

The final topic concerns an established quantitative analysis. First, known standard reference materials (comprised of current batch ingredients) should be examined as blinds on a routine basis to verify the validity of the reported results. This process will also flag any changes in formulation of which the analyst may not be aware. Second, when a new release (upgrade) of software is loaded on a computer system, the quantitative results reported from the upgrade should be compared to those obtained from the older release using the same set of spectra.[60] Additionally, a spectrum of a well-characterized sample should be collected under identical conditions using both versions of software. These spectra should then be compared to confirm reproducibility in the data collection routines. Slight modification in algorithms or bugs in the software release may cause a working method to become ineffective. In this case, the analyst can either reload the old release of software (depriving himself/herself of any improvements and new capabilities offered in other programs) or develop a new quantitative method based upon the current software release.

VI. TREATMENT OF SPECTRAL DATA

A number of data manipulation methods have become available to the analyst due to the powerful computer required to operate an FT-IR spectrometer. There is an alarming increase in the tendency to wantonly alter spectroscopic data before performing quantitative measurements. The main reason for this is that the novice analyst often attempts to make the spectrum "look nice" in the mistaken belief that a good appearance will improve the quantitative results. The methods of data treatment used in an analysis need to be as carefully defined and understood as those used for sample handling.

A. Operator Bias

Many analysts do not realize that employing data enhancement techniques will often introduce errors in the final result by application of an irreproducible bias. One type of bias is operator-dependent bias, which may be difficult to uncover. Usually, the effect is only discovered several months into implementation of an analysis, when a new analyst suddenly cannot get the method to work. Upon investigation, it may be found that the originator of the method had performed selected, undocumented manipulations (e.g., baseline corrections or smoothing) upon the spectra before making quantitative measurements, based solely upon personal observation of certain spectral features. Even a more serious situation may arise from the use of arbitrary bias, by the introduction of random errors in reported results.

B. Application Criteria

There are three rules for the application of data manipulation when performing quantitative analysis: (1) insure that all spectra are in a linear format (e.g., absorbance), (2) use only data treatment that is absolutely necessary to make the required quantitative measurements, and (3) insure that any manipulations used are performed identically on all spectra (calibration reference and unknowns).

C. Ordinate Axis Format

1. Absorbance

All spectra to be used for quantitative measurements need to be examined in a format where the ordinate axis is linear with sample concentration (provided adherence to Beer's law is followed). This means that if the thickness of a sample is doubled (increasing the level of a component by a factor of two), then the recorded band intensities should be doubled. The less rigorous quantitative analysis routines require that the ordinate be linear with concentration, but the more advanced routines (e.g., principal component regression and partial least squares) can cope with data which exhibit nonlinearities arising from chemical interactions in the sample.

The most commonly used format for the ordinate axis is Absorbance. It is particularly important that the common % Transmittance format not be used, since this is a logarithmic scale with respect to concentration. Other formats that may be used for quantitative measurements are Kubelka-Munk for diffuse reflectance and photoacoustic units for PAS, although the potential for PAS quantitation is usually very limited due to severe detector saturation problems.

2. Kubelka-Munk

The use of Kubelka-Munk units for quantitation in diffuse reflectance requires some comments. We have obtained quantitative results from diffuse reflectance experiments using both Absorbance and Kubelka-Munk units. Both formats seem to work equally well over reasonable quantitation ranges of mixtures at low concentrations in a suitable diluent (i.e., a few percent by weight in KBr). Neither format is mathematically rigorous for diffuse reflectance data, the Kubelka-Munk format actually being an empirical correction as seen in Equation 4.[52,54]

$$f(R_\infty) = \frac{(1 - R_\infty)^2}{2R_\infty} = \frac{k}{s} = \frac{2.303ac}{s} \tag{4}$$

where k = absorption coefficient, s = scattering coefficient, a = absorptivity, c = concentration, and R_∞ = reflectance. The expression for Kubelka-Munk may introduce spectral anomalies when values greater than 100% Reflectance are measured. The spectral consequence is "folding" of bands above 100% Reflectance values. This commonly occurs in two fashions. The first case is when the sample exhibits a greater reflectivity than the diluent material, which is also used as the background reference material. Second, reproducible cup-filling is difficult to achieve, with any small change in the focal height of the surface affecting the energy throughput and, thus, the positioning of the baseline. Various algorithms have been developed to alleviate this problem (e.g., as performing a baseline offset on or applying a scaling factor to the percent Reflectance data to bring it below 100% before performing the Kubelka Munk correction). However, the application of a baseline correction to a logarithmic format will only be linear in Absorbance at values very close to 100%.

D. Calibration Plots

1. Deviation

An exact straight line plot of Absorbance vs. concentration for a real system is less common than many analysts would like to believe, except for some isolated cases involving dilute mixtures of gases or powders. The lack of any deviation in such a plot would indicate a complete absence of any intermolecular interactions among the species in the mixture, as well as being devoid of any instrumental or experimental effects.

2. Piecewise Fit

Many mixtures will give calibration plots that are straight over a selected range, or at least adequate enough for analytical purposes. However, it may become necessary to use separate calibration plots for different concentration ranges.[3,39] One potential problem may be encountered in the examination of solid mixtures, such as copolymers. Such a mixture can exist in two differing phases, based upon concentration. Therefore, the recorded infrared band intensities will be nonlinear over the whole concentration range. However, a plot of the band intensities vs. concentration may show two reasonably linear regions which may be used for individual analyses, with a sharp intercept at the phase diagram boundary. Therefore, the analyst needs to be aware of the behavior of the sample system in order to establish rugged quantitative methods.

The more commonly used, and abused, data treatment methods will now be addressed in detail, with a view to emphasizing the merits and problems associated with each.

E. Baseline Correction

Ideally, baseline correction should never be applied directly to spectral data as it is an "inspired guess".[58] Rather, the measurement of absorbance bands ought to be made relative to baseline definitions within a quantitative analysis package. In isolated cases, where the spectrum may possess an elevated, or tilted, baseline, it may be removed using a standard one- or two-point tilt baseline correction without adversely affecting quantitative measurements, especially when calculating peak area or height relative to a local baseline definition. However, all spectral files (reference standards and unknowns) must be treated identically.[27]

The analyst should be aware of some of the consequences from the application of baseline corrections. First, an elevated baseline, such as an absorbance of 0.5, may well push strongly absorbing spectral bands into a nonlinear range and also cause the peak maxima to exhibit a higher level of noise. Second, the baseline itself may actually contain spectral information which sophisticated software can correlate to the concentration of a component, such as carbon black. Third, if the baseline exhibits curvature which needs more than a two-point tilt correction, it probably means that the experimental conditions are ill defined. Fourth, it is essential that no baseline correction is performed on a spectrum in the percent Transmittance format prior to conversion to an absorbance format, as nonlinear factors are introduced by the logarithmic function.

In the situation where it becomes necessary to remove a curved baseline from a series of spectra (e.g., which may occur with a sample on a very thin substrate), the use of a high-pass filter in Fourier-space is superior to a manual baseline correction which is influenced strongly by analyst bias. A Fourier-space filter operates on all spectra in a similar fashion, which may not yield a "perfect" baseline removal, but it does avoid adding irreproducibility.

F. Smoothing

The use of spectral smoothing prior to quantitation is an extremely poor analytical practice. Smoothing degrades spectral resolution, and the use of a moving smoothing function, (e.g., Savitzky-Golay smoothing) also affects band shape and peak position.[60] If a spectrum is too noisy for good quantitative measurements, then the analyst has two options: (1) record the spectrum at lower resolution or (2) co-add data for a longer time period.[60] The spectroscopic trading rules show that by halving the resolution, the SNR will improve by a factor of two per scan. Additionally, since SNR is proportional to the square root of the collection time, for a given resolution, a factor of four in collection time is needed for only an improvement of two in signal. Therefore, the first alternative of decreasing resolution is typically more attractive for improvement in SNR, since time is always at a premium and the application of a smoothing function would degrade the resolution anyway.

If smoothing does become a necessity, a Fourier-space smoothing function (high-pass filter) is preferable, as this has been shown to affect the band shape less than a moving smooth.

G. Resolution Enhancement

Several different methods of performing resolution enhancement exist for cases in which overlapping spectral bands are broader than the instrument resolution. If a quantitation method is unable to extract useful measurements from severely overlapping spectral regions, then resolution enhancement may be used to assist in better characterization of the features. Again, the same function must be applied to all spectra in an identical fashion. The commonly used functions are Fourier self-deconvolution,[34] maximum entropy calculations, and derivation.[53]

1. Fourier Self-Deconvolution

The Fourier deconvolution function reduces the spectral band widths, but does not change the band area,[34] which means that band intensity increases. Additionally, the spectral noise increases rapidly with the degree of band narrowing, requiring only spectra exhibiting high SNRs to be used.[34,60] Only a conservative level of resolution enhancement should be applied to data for quantitative analysis and the same values applied to all spectra (reference standards and unknowns).[34]

2. Maximum Entropy

The maximum entropy method function is similar to Fourier deconvolution, but controls the noise level better. It may, however, give rise to extraneous bands, and, therefore, it needs to be used with caution.

3. Derivative

The derivative calculation has traditionally been used for resolution enhancement by UV/VIS/NIR spectroscopists, where the absorption bands are generally much broader than the instrument resolution. There exist several reasons that derivation is not recommended in conjunction with quantitation of FT-IR data (which have much narrower bands). First, the intensity of a derivative peak depends on the width (and data point spacing) of the original band. Any slight changes in band width lead to large alterations of the derivative intensity. This effect is marginal on very wide bands,[60] (e.g., in the near-IR), but very marked on the narrower mid-IR bands. Next, the second derivative shows side lobes that point in the same direction as the main derivative peak, which makes for a very complex spectrum. Third, derivation greatly increases spectral noise,[1,60] and, thus, necessitates the application of additional smoothing functions (more manipulations and possible sources of bias).

H. Spectral Subtraction

There are two manners in which spectral subtraction may prove to be useful for quantitative analysis. First, it can be used as a preparative step for spectra, where a sample spectrum requires removal of interfering absorbance bands due to a solvent, water vapor, or a substrate. However, if the solvent or substrate exhibits interaction with the sample, then spectral subtraction is no longer valid and will lead to residual features that are due to this interaction.

Spectral subtraction can, itself, be used as a quantitative tool under carefully controlled circumstances. In general, this means that the pathlength of the sample must be constant. This can be achieved by using a fixed pathlength liquid or gas cell, liquid IRA, diffuse reflectance (if careful weighing of the sample and diluent is performed and particle size remains uniform), or pressed films made with spacers of known thickness. Under these conditions the spectral features due to a "pure" component (recorded from a known amount of sample) are removed by spectral subtraction from the spectrum of a mixture. The spectral subtraction factor reported for complete removal of the "pure" component from the mixture can be used to calculate the concentration of that component.[34] In some instances where very small amounts of solute exist, then a 1:1 subtraction for removal of the solvent or matrix material may be performed to leave the residual bands of the solute for measurement.[34]

I. Fringe Removal

The presence of spectral fringes, which arise from multiple internal reflections inside a cell or film, can be a major problem when implementing a quantitative method. The fringes do allow for calculation of the sample thickness (refer to Section IV.G.1), allowing for normalization of the spectra of different thicknesses. Generally, it is necessary to remove these fringes from all spectra before good quantitative measurements can be made.

If the fringe is a high frequency wave, then the interferogram can be inspected for the presence of a secondary (and perhaps also tertiary) centerburst. Removal (software) of this extraneous centerburst(s) will remove the wave from the computed spectrum. If the fringe is a low frequency wave, then it may be better to treat this as a baseline correction problem, and use a Fourier high-pass filter (as noted above) for its removal.

J. Interpolation

There exists a number of procedures for interpolating points to enhance the resolution of a spectrum artificially, providing a better band shape for manipulations. The simplest method involves the use of fitting a polynomial to known points. However, the best procedure is to add zeros to the end of an interferogram prior to computing the Fourier transform.[1] If the spectrum has already been computed, then the zero-filling can be applied using Fourier manipulation software.

VII. CONCLUSIONS

There exists a great potential in performing quantitative analysis of both inorganic and organic materials by infrared spectroscopy. Good quantitative methods by infrared spectroscopy are possible, as recorded by the multitude of published methodologies in the literature. However, the analyst must recognize that the mathematical approaches for quantitative analysis comprise only a part of the solution for a working analytical method. The abilities to completely and thoroughly define the system to be analyzed, selecting appropriate standards, and then to obtain spectra suitable for analysis are often the limiting factors in real world applications. A summary of the twenty rules to apply for verification of a quantitative method's feasiblity are presented in Table 3.

The spectroscopist needs to have knowledge of the potential and limitations of other analytical techniques, so as to avoid blindly using infrared spectroscopy where another technique would be more applicable. It is also the responsibility of the analyst to adequately describe the instrumentation and performance required to duplicate the repeatability and accuracy of the developed method in terms that may be understandable to others who are to use the methodology and the reported results. This latter step is the one aspect of the analytical process which still needs substantial improvement — the problem of preparing the sample and presenting it to the instrument (i.e., the analysis is only as good as the sample which has been introduced into the spectrophotometer). However, if the analyst makes a true hunt for gremlins, keeps a touch of paranoia, and holds a deep abiding trust in Murphy's law when developing a quantitative method,[58] the greater the likelihood that a rigorous, trustworthy analysis will result.

ACKNOWLEDGMENTS

The authors would like to thank Dr. Richard A. Crocombe (Bio-Rad, Cambridge, MA) for all of his assistance in the preparation of this manuscript.

Table 3. The Twenty Golden Rules for Good Quantitative Analysis

Rule 1:   The standards must be composed of linearly independent concentration data, not simple multiples or dilutions of one standard.

Rule 2:   The standards must be composed of linearly independent concentration data which do not allow the concentrations of two or more components to add to a constant other than 100%.

Rule 3:   In the selection of mixtures to be used as standards, at least 2 components (in a system comprised of multiple species) should be at different levels in each blend.

Rule 4:   All components should be represented the same number of times in mixtures comprising the calibration set.

Rule 5:   The concentrations of species in the calibration standards should be evenly spaced over the expected range.

Rule 6:   The system under investigation should be overdetermined.

Rule 7:   Precision of the infrared quantitation cannot be better than the (instrumental) technique used to provide the concentrations used in the calibration standards.

Rule 8:   The actual materials used in the formulation of the final product, and not ultra-pure grades of chemicals, should comprise the composition of calibration standards.

Rule 9:   Carefully examine the physical chemistry of a solution prior to using it in a quantitative method.

Rule 10: For every reference standard and each sample analyzed, a minimum of two complete samplings should be made.

Rule 11: A method should never be developed which is solely dependent upon a single holder, crystal, or cell.

Rule 12: Ensure that the analysis method does not affect the sample.

Rule 13: Sample history may be an important factor.

Rule 14: Data for both calibration references and unknowns must be collected under identical spectrometer operating conditions.

Rule 15: The single beam background for computing a ratioed spectrum should be collected through the accessory or holder used for examining the sample.

Rule 16: The calibration should be checked frequently with blinds exhibiting characteristics for both in and out of specification production material.

Rule 17: Use only those mathematical treatments that are absolutely necessary to make the required quantitative measurements.

Rule 18: Ensure that all spectra are in a linear format (absorbance, Kubelka-Munk, etc.) for quantitative measurements or spectral manipulations.

Rule 19: All spectral manipulations used must be performed identically on all spectra (calibration references and unknowns).

Rule 20: Review the developed method to see if it makes spectroscopic sense (i.e., the analytical measurements can be correlated to logical infrared absorbances).

REFERENCES

1. Gillette, P. C., Lando, J. B., and Koenig, J. L., A survey of infrared spectral data processing techniques, in *Fourier Transform Infrared Spectroscopy*, Vol. 4, Ferraro, J. R., Basile, L. J., Eds, Academic Press, Orlando, 1985, chap. 1.
2. McClure, G.L., Quantitative analysis from the infrared spectrum, in *Laboratory Methods in Vibrational Spectroscopy*, Willis, H. A., van der Maas, J. H., and Miller, R.G.J., Eds, John Wiley & Sons, New York, 1987, chap. 7.
3. Smith, A. L., *Applied Infrared Spectroscopy: Fundamentals, Techniques and Analytical Problem Solving*, John Wiley & Sons, New York, 1979.
4. Brown, C.W., *JTEVA*, 12(2), 86, 1984.
5. Maris, M.A., Brown, C.W., and Lavery, D.S., *Anal. Chem.*, 55, 1694, 1983.
6. Kisner, H.J., Brown, C.W., and Kavarnos, G.J., *Anal. Chem.*, 55, 1703, 1983.
7. Brown, C.W., Lynch, P.F., Obremski, R.J., and Lavery, D.S., *Anal. Chem.*, 54, 1472, 1982.
8. Kisner, H.J., Brown, C.W., and Kavarnos, G.J., *Anal. Chem.*, 54, 1479, 1982.
9. Antoon, M.K., Koenig, J.H., Koenig, J.L., *Appl. Spectrosc.*, 31(6), 518, 1977.
10. Koenig, J.L. and Kormos, D., *Appl. Spectrosc.*, 33(4), 349, 1979.
11. Haaland, D.M. and Easterling, R.G., *Appl. Spectrosc.*, 36(6) 665, 1982.
12. Haaland, D.M. and Easterling, R.G., *Appl. Spectrosc.*, 34(5) 539, 1980.
13. Gendreau, R.M. and Griffiths, P.R., *Anal. Chem.*, 48(13) 1910, 1976.
14. Lin, C.L., Niple, E., Shaw, J.H., and Calvert, J.G., *Appl. Spectrosc.*, 33(5), 487, 1979.
15. Lin, C.L., Niple, E., Shaw, J.H., and Calvert, J.G., *Appl. Spectrosc.*, 33(5), 481, 1979.
16. Bauman, R.P., *Absorption Spectroscopy*, John Wiley & Sons, New York, 1962, chap. 9.
17. McCue, M. and Malinowski, E.R., *Anal. Chim. Acta*, 133, 125, 1981.
18. Leggett, D.J., *Anal. Chem.*, 49(2), 276, 1977.
19. Sternberg, J.C., Stillo, H.S., and Schwendeman, R.H., *Anal. Chem.*, 32(1), 84, 1960.
20. Barnett, H.A. and Bartoli, A., *Anal. Chem.*, 32(9), 1153, 1960.
21. Sasaki, K., Kawata, S., and Minami, S., *Appl. Op.*, 23(12), 1955, 1984.
22. Jochum, C., Jochum, P., and Kowalski, B.R., *Anal. Chem.*, 53, 85, 1981.
23. Perry, J.A., *Appl. Spectrosc. Rev.*, 3(2), 229, 1970.
24. Potts, W.J., *Chemical Infrared Spectroscopy*, Vol. I, John Wiley & Sons, New York, 1963.
25. Tufts, L.E. and Davis, A., Infrared quantitative analysis: general comments on instrumental conditions for quantitative analysis, in *Progress in Infrared Spectroscopy*, Vol. I; Szymanski, H.A., Ed., Plenum Press, New York, 1962, 151.
26. *Infrared Spectroscopy: Its Use in the Coatings Industry*, Federation of Societies for Paint Technology, Philadelphia, 1969, chap. 5.
27. Practices for General Techniques of Infrared Quantitative Analysis, *ASTM Practice E 168–88*, ASTM: Philadelphia, 1991.
28. Osten, D.W. and Kowalski, B.R., Multicomponent calibration and quantitation methods, in *Computerized Quantitative Infrared Analysis*, McClure, G.L., Ed., ASTM, Philadelphia, 1987, 6.
29. Haaland, D.M., Methods to include Beer's law nonlinearities in quantitative spectral analysis, in *Computerized Quantitative Infrared Analysis*, McClure, G.L., Ed., ASTM, Philadelphia, 1987, 78.
30. Crocombe, R.A., Olson, M.L., and Hill, S.L., Quantitative Fourier transform infrared methods for real complex samples, in *Computerized Quantitative Infrared Analysis*, McClure, G.L., Ed., ASTM, Philadelphia, 1987, 95.
31. Malinowski, E.R., Applications of target factor analysis to quantitative absorption spectroscopy, in *Computerized Quantitative Infrared Analysis*, McClure, G.L., Ed., ASTM, Philadelphia, 1987, 155.
32. Griffiths, P.R. and de Haseth, J.A., *Fourier Transform Infrared Spectrometry*, John Wiley & Sons, New York, 1986.

33. Cahn, F. and Compton, S., *Appl. Spectrosc.*, 42(5), 865, 1988.
34. Compton, D.A.C., Young, J.R., Kollar, R.G., Mooney, J.R., and Grasselli, J.G., Some applications of computer-assisted quantitative infrared spectroscopy, in *Computerized Quantitative Infrared Analysis*, McClure, G.L., Ed., ASTM, Philadelphia, 1987, 36.
35. Battiste, D.R., Butler, J.P., Cross, J.B., and McDaniel, M.P., *Anal. Chem.*, 53, 2232, 1981.
36. Conley, R.T., *Infrared Spectroscopy*, Alyn and Bacon, Boston, 1972, chap 8.
37. Price, W.J., Sample handling techniques, in *Laboratory Methods in Infrared Spectroscopy*, Miller, R.G.J. and Stace, B. C., Eds., Heyden, London, 1972, chap. 8.
38. Practices for General Techniques for Qualitative Infrared Analysis, *ASTM Practice E 1252–88*, ASTM, Philadelphia, 1991.
39. Haslam, J., Willis, H.A., and Squirrell, D.C.M., *Identification and Analysis of Plastics*, Heyden, London, 1965.
40. Smith, A.L. and Kiley, L.R., *Appl. Spectrosc.*, 18(2), 38, 1964.
41. Tosi, C. and Ciampelli, F., Applications of infrared spectroscopy to ethylene-propylene copolymers, in *Advances in Polymer Science*, Vol. 2, Springer-Verlag, Berlin, 87, 1973.
42. Hampton, R.R., *Rubber Chem. Technol.* 45, 546, 1972.
43. Zichy, V.J., Quantitative infrared analysis of polymeric materials, in *Laboratory Methods in Infrared Spectroscopy*, Miller, R.G.J. and Stace, B.C., Eds., Heyden, London, 1972, chap 5.
44. *Encyclopedia of Chemical Technology*, 2nd ed., John Wiley & Sons, New York, 1970.
45. Russell, J.D., Infrared spectroscopy of inorganic compounds, in *Laboratory Methods in Vibrational Spectroscopy*, Willis, H. A., van der Maas, J. H., Miller, R.G.J., Eds., John Wiley & Sons, New York, 1987, chap. 18.
46. Hanst, P.L. and Hanst, S.T., *Gas Analysis Manual for Analytical Chemists*, Infrared Analysis, Anaheim, 1990.
47. Higgins, G.M.C. and Miller, R.G.J., Insoluble polymers and rubbers, in *Laboratory Methods in Infrared Spectroscopy*, Miller, R.G.J. and Stace, B.C., Eds., Heyden, London, 1972, chap. 16.
48. Willis, H.A., Preparation of polymer samples for IR examination, in *Laboratory Methods in Vibrational Spectroscopy*, Willis, H.A., van der Maas, J.H., and Miller, R.G.J., Eds., John Wiley & Sons, New York, 1987, chap. 9.
49. Hannah, R.W., Simple sampling, in *Laboratory Methods in Vibrational Spectroscopy*, Willis, H.A., van der Maas, J.H., and Miller, R.G.J., Eds., John Wiley & Sons, New York, 1987, chap. 8.
50. Brown, J.M. and Elliott, J.J., The quantitative analysis of complex, multicomponent mixtures by FT-IR: the analysis of minerals and of interacting organic blends, in *Chemical, Biological and Industrial Applications of Infrared Spectroscopy*, Durig, J.R., Ed., John Wiley & Sons, New York, 1985, 111.
51. Wilks, P.A., A practical approach to internal reflection spectroscopy, in *Laboratory Methods in Infrared Spectroscopy*, Miller, R.G.J. and Stace, B.C., Eds., Heyden, London, 1972, chap. 14.
52. Westra, J.G. and van Woerkem, P.C.M., Reflection spectroscopy, in *Laboratory Methods in Vibrational Spectroscopy*, Willis, H.A., van der Maas, J.H., and Miller, R.G.J., Eds., John Wiley & Sons, New York, 1987, chap. 10.
53. Willis, H.A., Chalmers, J.M., Mackenzie, M.W., and Barnes, D.J., Novel quantitative polymer analysis through computerized infrared spectroscopy, in *Computerized Quantitative Infrared Analysis*, McClure, G.L., Ed., ASTM, Philadelphia, 1987, 58.
54. *Optical Spectroscopy: Sampling Techniques Manual*, Harrick Scientific Corporation, Ossining, 1987.
55. Hembree, D.M. and Smyrl, H.R., *Appl. Spectrosc.*, 43(2), 267, 1989.
56. Practice for describing and measuring performance of Fourier transform infrared spectrometers: level zero and level one, *ASTM Practice E 1421-91*, ASTM, Philadelphia, 1991.
57. Chase, D.B., *Appl. Spectrosc.*, 38(4), 491, 1984.

58. Hirschfeld, T., Quantitative FT-IR: a detailed look at the problems involved, in *Fourier Transform Infrared Spectroscopy*, Vol. 2, Ferraro, J.R. and Basile, L.J., Eds, Academic Press, New York, 1979, chap. 6.
59. Griffiths, P.R., Mid-infrared Fourier transform spectrometry, in *Laboratory Methods in Vibrational Spectroscopy*, Willis, H.A., van der Maas, J.H., and Miller, R.G.J., Eds., John Wiley & Sons, New York, 1987, chap. 6.
60. Hirschfeld, T., Computerized infrared: the need for caution, in *Computerized Quantitative Infrared Analysis*, McClure, G.L., Ed., ASTM, Philadelphia, 1987, 169.

Chapter 9

ERRORS IN SPECTRAL INTERPRETATION CAUSED BY SAMPLE-PREPARATION ARTIFACTS AND FT-IR HARDWARE PROBLEMS

Robert C. Williams

CONTENTS

I. Introduction 256
II. Examples 257
   A. Silicone Contamination 257
      1. Microscope Slides 257
      2. Silicone-Treated Wipes 258
      3. Silicone Lubricant 258
   B. Orientation/Crystallinity Effects 259
      1. Cooled Melts 259
      2. Crystallinity in Polyurethanes 261
      3. Pressure-Induced Effects 261
   C. Other Sample-Preparation Artifacts 264
      1. Cast Films 264
      2. Pinholes/Too-Thick Samples 266
      3. Heterogeneous Samples 267
      4. Improper Reference Standards 267
   D. Software-Induced Artifacts 267
   E. FT-IR Hardware-Induced Artifacts 268
      1. Out-of-Tune Beamsplitter 268
         a. Cause of Detuned Condition 270
         b. Importance of Diagnostic Maintenance 270
      2. Fogged Beamsplitter 270
         a. Cause of Fogging 270
         b. Importance of Purge 270
      3. Incorrect Electronic Filtering 271
III. Conclusion 273
References 273

0-8493-4203-1/93/$0.00+$.50
© 1993 by CRC Press Inc.

I. INTRODUCTION

Modern Fourier-transform and microprocessor-controlled infrared spectrometers make it easy to obtain an infrared spectrum — perhaps too easy! Previous-generation infrared spectrometers were often big, expensive, slow, and temperamental; successful operation of these systems frequently required a judicious mixture of science, engineering, and art. Interpretation of the spectra obtained was generally based on first-principal functional group analysis and previous experience; computer-searchable libraries were the nascent dreams of some visionary spectroscopists. The number of infrared laboratories was limited by the cost and complexity of the instrumentation, while the number of spectroscopists was limited by the relatively long apprenticeship required to become proficient in sample preparation, instrument operation, and spectral interpretation.

In contrast to the situation in the "bad old days", the actual dollar cost of modern instrumentation is generally less than the previous generation; in inflation-adjusted dollars, the current instruments are only a fraction of the cost of their predecessors. Moreover, the compact designs of these instruments require much less bench- or floor-space than their earlier counterparts. In addition, spectral acquisition times are much shorter. Spectra with 2 cm^{-1} resolution, which used to require 20 to 30 min to record on an older dispersive instrument, can now be acquired in well under a minute on a Fourier transform system and only slightly longer on a ratio-recording dispersive instrument. A 0.3-cm^{-1} resolution spectrum , which literally took this author all night to acquire over a very narrow frequency range 20 years ago, can now be acquired in the same 20 min it used to take to acquire a 2- or 4-cm^{-1} spectrum; good 0.02-cm^{-1} resolution spectra can be acquired in less than 1 h. Modern instruments are far more reliable, require much less routine maintenance, and are infinitely easier to operate than their predecessors; after only the briefest of training, virtually anyone with minimal technical skills can obtain a spectrum. Moreover, once a spectrum has been acquired, computerized library searching and other chemometric techniques will often facilitate interpretation, and in some instances, quantification of several components in a complex mixture. These innovations are of such magnitude that they can truly be called revolutionary.

As a result of these advances, there are now several times as many laboratories with infrared capabilities as there were a generation ago. At the same time, the ease of instrument operation and the availability of computer-assisted interpretation and identification have effectively reduced the apprenticeship time required to become a practicing, if not an accomplished, spectroscopist.

Unfortunately, there are several serious drawbacks associated with these revolutionary innovations. Many operators no longer have the requisite sample-preparation skills necessary to obtain photometrically correct spectra; to some extent, the instrument companies have exacerbated this problem, by implying that the new instruments are so good they can "compensate for bad sample preparation". Moreover, many recently trained spectroscopists lack the experience necessary to recognize spectral artifacts caused by inadequate sample preparation. When this deficiency is coupled with an uncritical dependence on computerized library searches, the resultant errors can range from embarrassing to disastrous. In addition, many computer-assisted systems use vendor-written proprietary and undocumented, "black-box" data-enhancement/manipulation/quantitative analysis procedures; both these and user-written programs can introduce subtle and artificial changes in the resultant spectra.

A variety of sources discuss various aspects of general sample preparation.[1-4] These discussions are augmented and updated in this volume in Chapter 2, Chapter 10, and Chapter 11. (Analysis-specific sample-preparation techniques are outlined in the chapters which discuss these particular analysis methods.) Reference 3 also gives a brief description of some

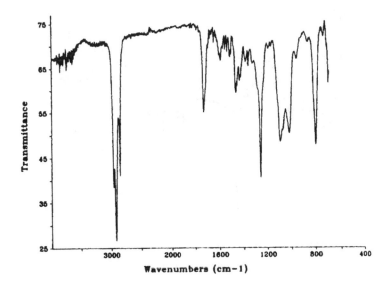

FIGURE 1.   Transmission infrared spectrum of a PVC gel obtained using an Analect XAU Microscope mounted on a Nicolet 7199 FT-IR. The spectrum exhibits silicone contamination from a glass microscope slide on which the sample was prepared.

commonly observed spurious infrared absorption bands. In the present chapter, specific examples are presented to illustrate some of the spectral distortion and/or misinterpretation problems cited in this introduction. These examples are culled from the actual work of commercial and industrial laboratories with which the author has dealt.

II. EXAMPLES

A. Silicone Contamination

The spectrum of poly-dimethylsiloxane is one of the first spectra that many industrial spectroscopists learn to recognize. Unfortunately, even when a silicone spectrum is identified, it is not always recognized as a contaminant rather than a legitimate component of the analyte sample. Silicone contamination is ubiquitous. Whenever silicone is identified, the possibility of contamination should *always* be considered. Several examples illustrate this.

1. Microscope Slides

Figure 1 is a spectrum taken through an infrared microscope of a "fisheye" removed from a plasticized polyvinyl chloride (PVC) film. Although the spectrum does contain bands typical of plasticized PVC, it also indicates the presence of a very significant level of silicone. The fisheye was cut out with a scalpel and placed by the analyst, an experienced optical micros-copist, on the surface of a glass microscope slide, prior to transfer to a potassium bromide (KBr) crystal. (For a detailed discussion of sample-preparation techniques used with microsamples, see Chapter 6.) The glass slide used had been taken from a new box; informa-tion on the box indicated that all the slides within had been "pre-cleaned" by the manufacturer. When a spectrum was acquired of the methanol wash of one of these slides, silicone was

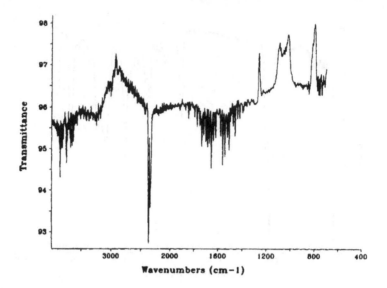

FIGURE 2. ATR spectrum obtained after removal of a weathered PVC compound sample. The negative bands indicate that the supposedly clean ATR crystal run for the background spectrum was contaminated with silicone.

clearly evident in the wash spectrum. Apparently, a silicone lubricant was used on or in the equipment that cut, stacked, and boxed the glass slides. The silicone lubricant does not pose a significant problem in optical microscopy and, indeed, is virtually undetectable by that technique, but it poses a serious problem for samples prepared for infrared spectral analysis. The lesson is important. Benign contaminants which do not interfere with other analyses can seriously interfere with infrared analyses. Vendor-cleaned or certified materials should be recleaned and/or tested to verify that they are not introducing contamination!

2. Silicone-Treated Wipes

In a second case, a series of ATR spectra were taken of weathered PVC samples. (For more detail regarding the preparation of ATR samples, see Chapter 3.) The formulation was not known, but it was believed that silicone might constitute a part of the formulation. Several spectra appeared to confirm this, and all appeared to be going well until the spectrum in Figure 2, which shows *negative* silicone bands, was obtained. Investigation determined that the supplier of wipes used to clean the ATR plates had been changed and that the new wipes, which were designed for polishing eyeglass lenses, were impregnated with silicone. When the analysis was repeated using properly cleaned crystals, no silicone was detected in any of the spectra.

3. Silicone Lubricant

As a final example of silicone contamination, the acetone wash of a material deposited in the transport tube during a TGA-IR experiment is shown in Figure 3. This particular TGA-IR interface was homemade and used an unheated transfer line; condensation within the transfer tube is normally not a problem when the transfer tube is heated to above 200°C. (See also Chapter 7.) In this case, because the formulation under analysis was known to contain silicone, the detection of silicone in the TGA-IR spectrum appeared reasonable. However, further investigation revealed that a supposed fluorocarbon grease used on the ball joint which

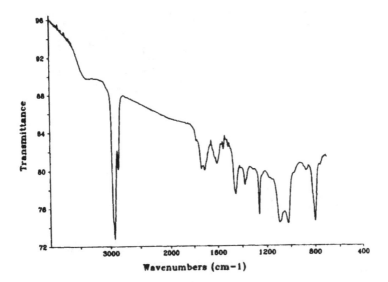

FIGURE 3. Transmission spectrum of the condensate observed in the transfer tube after a TGA/FT-IR experiment.

connects the TGA output line to the IR cell input line was instead silicone. When the experiment was rerun using a fluorocarbon grease, no silicone was detected.

While good experimental technique and proper sample handling protocols, such as those discussed in Chapters 2, 10, and 11, will help minimize the instances of contamination of the sample while it is under analysis, they can do nothing to prevent contamination prior to receipt of the sample. Because silicone contamination is so ubiquitous, the possibility of contamination should always be thoroughly investigated any time silicone is identified, even (or perhaps, especially) when its presence appears reasonable.

B. Orientation/Crystallinity Effects

Many materials are capable of forming crystalline, or at least ordered, domains. The speed and ease of formation of the ordered form will depend on the material involved. However, a spectroscopist must always be aware of the possibility that an analyte material may have an ordered form, whose spectrum can be significantly different from that of the amorphous form.

1. Cooled Melts

Figure 4A is a spectrum of a rubber additive which was submitted for routine identification; Figure 4B is a spectrum of an authentic sample of the suspected antioxidant which was submitted at the same time. On first glimpse, the spectra contain enough features in common that the unknown appears to be of the same general class as the reference; however, the spectra are sufficiently different that they do not appear to be an exact match. An attempt was made to identify the unknown more completely. Manual and computerized searches of both internal and commercial reference collections failed to locate a better match than the submitted reference compound. To determine whether the unknown might be a mixture which contained the reference compound, a melting point was obtained. The melting point had a 1.5°C range, which indicated that the unknown was a relatively pure material; moreover, the melting point was an excellent match for the reported melting point of the reference compound!

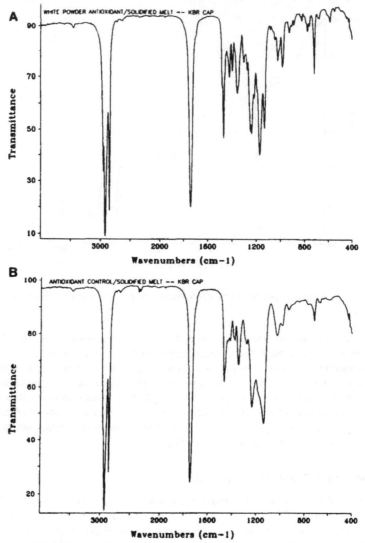

FIGURE 4. (A) Spectrum of a white powder rubber additive, prepared as film solidified from the molten state. (B) Spectrum of an authentic sample, believed to be the same material as (A).

Note that the spectrum headers in Figures 4A and 4B indicate that both samples were prepared as hot melts on KBr. (In this laboratory, sample preparation information is *always* included in the spectrum header, because the analyst who interprets the spectrum may not be the person who prepared the sample and acquired its spectrum. With this information, the spectroscopist is much less likely to misinterpret a sample-preparation artifact. For example, in some cases, a weak hydroxyl band may have great significance; on the other hand, the significance is highly questionable if the sample was prepared as a KBr pellet.) In this instance, the person who prepared the samples was a chemist with over 25 years general experience, but with a somewhat limited experience in infrared sample preparation. He had prepared each sample by melting a small amount on a KBr crystal under a heat lamp; when the sample melted, a second crystal was placed on top of the molten sample, to form a capillary film between the two crystals. The crystal sandwich was then removed from under the heat lamp and allowed to cool to solidity; a spectrum of the solidified capillary film was acquired sometime after the molten material had cooled to solidity.

Both the unknown and the reference samples were believed to have been prepared under the same conditions. However, given the method of sample preparation, as well as all the other information available, the analyst began to suspect a crystallinity effect. A second spectrum of the unknown was acquired *immediately* after it had solidified from the melt; the resultant spectrum was identical to Figure 4B.

The bands of the unknown in Figure 4A are significantly sharper than those in Figure 4B. The difference between the two spectra is that the former spectrum represents an ordered, crystalline form of the sample, while the latter spectrum represents an amorphous form. Either the unknown was inadvertently allowed to cool for a longer time before its spectrum was acquired, or it was of a higher purity than the reference material and, therefore, crystallized more rapidly under the same sample-preparation conditions.

2. Crystallinity in Polyurethanes

In a second example, the polyurethane whose spectrum is shown in Figure 5A was rejected by a customer, because the relative band intensities in the 1800 to 1500 wavenumber region did not match the customer's previously run reference spectrum shown in Figure 5B. In this laboratory, a quick check of the literature indicated a known propensity of polyurethanes to crystallize rapidly at room temperature.[5, 6] A series of experiments confirmed that spectra similar to Figure 5A were obtained *immediately* after a sample was prepared, and spectra similar to Figure 5B were obtained after the sample was allowed to anneal at room temperature for a while. It is also significant to note that essentially the same behavior was observed both for films prepared by hot-pressing and for films cast onto KBr from solution in tetrahydrofuran. The failure of the customer's spectroscopist to recognize a well-documented polyurethane crystallinity effect resulted in the rejection of a perfectly good shipment of material!

3. Pressure-Induced Effects

A third example of an orientation effect is shown in Figure 6A, which is the spectrum of a PVC resin, run as neat particles between the diamond faces of a miniature, high-pressure, diamond-anvil optical cell purchased from High Pressure Diamond Optics, Inc. (Tucson, AZ). When the same resin sample was run as a hot-pressed film which was quenched within a second or two of pressing, the spectrum in Figure 6B was obtained. (The Nicolet 7199 spectrometer used in this particular experiment has a very highly collimated sample beam. When the spectra of thin films are obtained on the Analect XAU Infrared Microscope which is mounted on the 7199, fringing is usually a problem. For this reason, both surfaces of the film, whose spectrum is shown in Figure 6B, were manually roughened with 1000-mesh crocus cloth to eliminate fringes. An equivalent spectrum was obtained for the unroughened film, in a different spectrometer which has a less-highly collimated sample beam.) Clearly, although Figures 6A and 6B are generally similar, there is a small but significant difference in the apparent resolution, which is most evident in the 1430 to 1420 cm^{-1} doublet structures. (In PVC spectra, this 1430 to 1420 cm^{-1} doublet is sometimes taken to be one of the indicators of polymer microstructure, i.e., tacticity and/or crystallinity.[7-9])

When this phenomenon was first observed, the possibility was considered that the film structure shown in Figure 6B represents amorphous PVC, where any order present in the resin had become randomized during the pressing procedure; however, attempts to anneal this film completely failed to introduce any measurable changes in the doublet ratio. In addition, when

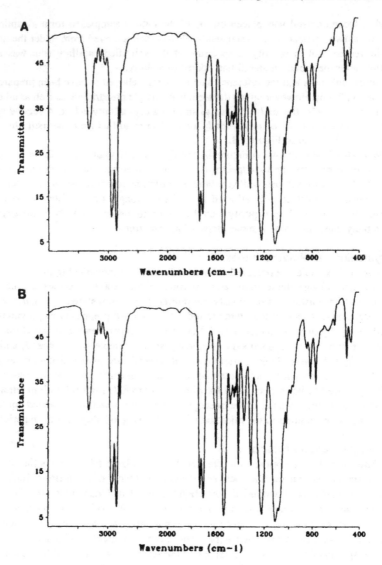

FIGURE 5. Spectrum of a polyether urethane. (A) Prepared as a pressed film; spectrum acquired immediately after pressing and quenching. (B) After the film was allowed to anneal at room temperature.

a portion of the same film used to generate Figure 6B was mounted in the diamond cell, the spectrum shown in Figure 6C was obtained. In this spectrum, the 1430 to 1420 cm^{-1} region is very much like that observed in the diamond-cell spectrum of the neat resin particles shown Figure 6A, except that the doublet is even less well-resolved. It was also observed that intermediate pressures on the diamond cell produced spectra which were intermediate between Figures 6B and 6C, and that the process is reversible.

It now appears that the PVC experiences a pressure-induced change in the high-pressure domain between the two diamond-anvil faces. It is not clear whether this represents a localized partial orientation of the PVC or a different physical change such as a densification which may

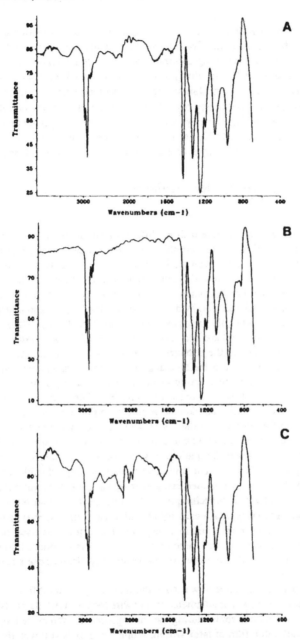

FIGURE 6. (A) Transmission spectrum acquired on an Analect XAU Infrared Microscope (mounted on a Nicolet 7199 FT-IR) of PVC resin in a miniature, high-pressure, diamond-anvil, optical cell. (B) Transmission spectrum of a PVC film prepared by hot-pressing the resin. (C) Spectrum of (B) mounted in a miniature diamond cell.

effect the polymer's refractive index, etc. Although this artifact does not preclude qualitative identification of the materials involved, it would certainly play havoc with any attempt to obtain quantitative results (including spectral subtraction) which rely on any of the bands that are pressure dependent. Although this phenomenon has not been studied in detail in this laboratory, if this is indeed an orientation effect, one would expect to see analogous behavior with terephthalate polyesters and any other material which can become oriented under stress; the same cautions regarding quantitative analysis would then be expected to hold for this class of materials. If the effect observed is not the result of orientation, but of some more universal physical change, then *all* quantitative results obtained from diamond cell spectra are suspect.

C. Other Sample-Preparation Artifacts

1. Cast Films

Figures 7A, 7B, and 7C represent three different attempts to cast a film from the same ethylene/vinyl acetate (EVA) latex sample onto a germanium crystal. Figure 7A is clearly different from the other two, but even the latter two figures differ from each other by almost 10% (relative) in their ratio of the C–H stretch band near 2920 cm^{-1} vs. the acetate band near 1020 cm^{-1}. Note also that the two bands which are ratioed together are of approximately the same intensity. Worse results are obtained when two bands of significantly different intensities are used; the 2920:1235 band ratio varies by over 15% between Figures 7B and 7C! The rather significant differences observed are attributed to inadequate film preparation. These films are either (1) so thick that many of the stronger bands lie outside the linear range of the detector; (2) are nonuniform in thickness in the area analyzed by the infrared beam; and/or (3) do not cover the full beam width of the infrared beam (which is really just an extreme example of Case 2). The films were cast by applying a small droplet of latex with a disposable micropipet to a germanium crystal (actually, a broken ATR plate); the side of the micropipet was then used as a blade to spread the latex across the crystal and remove excess material, after which the crystal-and-film were dried either under a heat lamp or on a warm hotplate. When this particular sample preparation technique was examined closely, however, repeated attempts to use this technique to cast films containing quantitatively reproducible band ratios were unsuccessful. The best results, obtained using extreme care, yielded a ratio reproducibility in the range of ±5% (relative). This technique has been in use in this and other laboratories for a great many years, primarily as a means for obtaining qualitative spectra; when quantitative infrared analysis became more common, the same technique was used to prepare these samples, too. As in the case of the diamond cell, the artifacts introduced do not necessarily constitute a problem for qualitative analysis; however, they clearly pose a severe obstacle to good quantitative analysis.

The film preparation technique now used in this laboratory is a modification of a procedure suggested by Clara Craver. A 1-in. diameter copper wire loop is inserted into the jar containing the latex; the loop is kept horizontal during the entire procedure. When the loop is withdrawn, it normally contains a thin film of latex within. While the latex is still wet, the loop is placed over a 25-mm diameter KBr disk, effectively depositing the film on the disk. A second disk is then placed over the film, and a small amount of pressure is applied to the top disk. So long as uniform pressure is placed upon the top disk, the resultant capillary film between the KBr disk sandwich appears to be reasonably uniform. The disk-film-disk sandwich assembly is then dried (either on a hotplate or under an infrared heat lamp), and a spectrum is acquired. Using this technique, this laboratory has achieved an average replicate reproducibility of the EVA band ratios of better than 1%, relative.

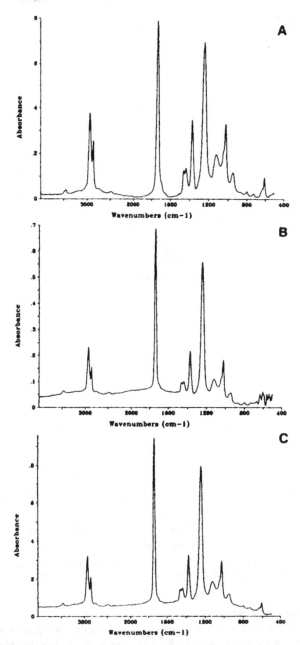

FIGURE 7. Transmission spectrum of an EVA latex, prepared as three film casts (A, B, and C) from the latex onto germanium.

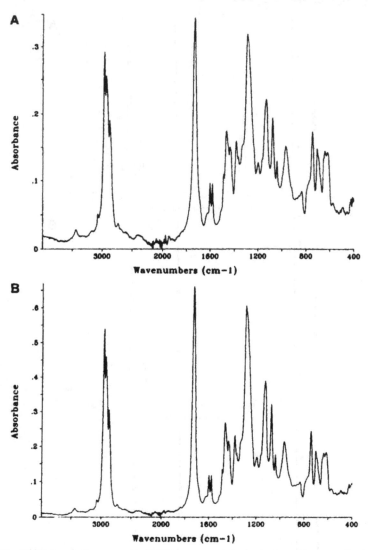

FIGURE 8. (A) Transmission spectrum of a PVC compound obtained using a diamond cell, without any beam condenser, on the Nicolet 710 FT-IR. (B) Spectrum of a second piece of the same PVC compound; acquisition conditions as in (A).

2. Pinholes/Too-Thick Samples

A problem similar to that observed for cast films is illustrated by Figures 8A and 8B, which are the diamond cell spectra of an extruded PVC compound. Originally, Figure 8A was to serve as a control, to which spectra of various gel samples removed from the strip would be compared. After this spectrum had been acquired, the analyst noted that the ester carbonyl overtone band near 3440 cm^{-1} is significantly stronger than is normally observed in similar compounds, especially when the apparent total spectrum thickness only corresponds to 0.35 absorbance units. When another portion of the same control sample was rerun, the spectrum in Figure 8B was obtained. This particular artifact has not been replicated, and its source has never been completely identified; however, nonuniform sample thickness and/or voids in the film are the primary suspects.

3. Heterogeneous Samples

A third example of a sample-preparation artifact occurred when a commercial testing laboratory rejected a previously qualified PVC polymer compound, because the spectrum did not match that of their reference on file. Investigation revealed that the samples at the commercial testing laboratory were prepared by dissolving them in hot orthodichlorobenzene (ODB) and then casting a film from the hot ODB slurry onto a salt crystal. The difficulty of casting a uniform film from a homogeneous solution has already been discussed. Compounding this difficulty is the fact that many polymer compounds contain components (i.e., fillers and inorganic pigments, various stabilizers, and in some cases, high molecular-weight polymeric modifiers) which do not dissolve completely in ODB. At best, these undissolved components constitute a heterogeneous slurry in the ODB solution. The likelihood of reproducibly casting from this slurry a uniform film, which maintains the exact ratio of ODB-insoluble to ODB-soluble components present in the original compound, is vanishingly small. In inexperienced hands, this fractionation may be so bad that the cast film sample is not even acceptable for qualitative interpretation of the resultant spectrum. In experienced hands, this technique is probably suitable for qualitative analysis; however, this technique should never be used for a *quantitative* comparison.

4. Improper Reference Standards

The inherent flaw in the film cast from the heterogeneous ODB solution would probably have been sufficient to explain many of the discrepancies observed between the two spectra. However, in this instance, there was an even more serious problem. The reference material was a natural color (i.e., one with relatively little filler/pigment) but the rejected sample was a beige compound, which contained significant amounts of both calcium carbonate and titanium dioxide. Peaks representing the latter filler/pigment materials were identified as contaminants by the testing lab. The beige compound had been rejected because it was compared to the wrong reference spectrum! Unfortunately, this occurs much more commonly than is generally realized. It is extremely important to verify that a reference standard to which a sample will be compared is authentic, uncontaminated, and corresponds *exactly* to the sample being analyzed.

D. Software-Induced Artifacts

In addition to two research-grade instruments, whose access is restricted to experience spectroscopists, the author's laboratory also supports an "open" infrared laboratory unit, which is accessible to everyone at the Technical Center. The "open" laboratory is basically a small, benchtop Fourier transform-infrared (FT-IR), which is located in the main infrared laboratory and is maintained by IR-laboratory personnel, but which is available for use by any of the 500 plus people at the Technical Center. The system is used primarily for qualitative comparison of hot-pressed polymer films and solution-cast latex films against the reference spectra of known products. IR-lab staff are available for consultation on more difficult problems; however, most of the sample preparation and the more basic spectral comparisons are performed by the open-lab users. Because a menu-based system was not available when this unit was installed, a set of macro routines was written which allow these relatively unsophisticated users to operate the spectrometer without mastering the powerful, but complex, system-command language.

Figure 9 is the spectrum obtained by one of these open-laboratory users of a gel in an extruded polyurethane strip. Most experienced spectroscopists will recognize that the sample is much too thick, as many of the bands appear "bottomed out". However, when this observation was made, the user replied that the sample could not be too thick, because the strongest

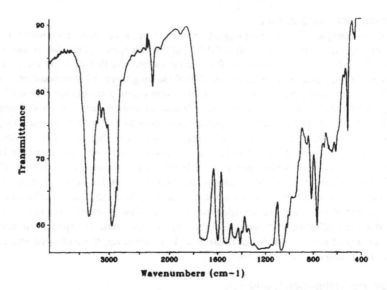

FIGURE 9. Transmission spectrum of a gel found in a polyurethane part. The sample is much too thick, but this attribute is obscured by a combination of pinholes/incomplete coverage of the analysis aperture and software which normalized the spectrum to 98% T.

band was only 56% Transmittance (T)! There were probably several factors contributing to the problem observed in this spectrum. The sample was difficult to press and the pressed film did not appear to be uniform in thickness; visual observation of the sample indicated a few small pinholes, which certainly constituted part of the problem. Moreover, the pressed film was quite small; although it was masked with an aperture, the sample probably did not cover the entire aperture. Finally, there was another hidden and more insidious problem. The acquisition macro used on this FT-IR system automatically normalized all spectra so that the spectrum maximum was 98% T. When similar samples were examined under comparable conditions, but using system command language which bypassed the autonormalization routine, the spectra typically exhibited baselines under 20% T. The autonormalization software hid the evidence of this problem from this novice user who, in turn, did not have sufficient interpretive experience to recognize the result of an inadequate sample preparation! The autonormalization procedure has subsequently been deleted from the open-lab macro package. However, this example also illustrates the potential havoc which can be caused by undocumented, vendor-proprietary software, which may not have been fully tested under the varied operating conditions which innovative and/or inexperienced users may encounter.

E. FT-IR Hardware-Induced Artifacts

1. Out-of-Tune Beamsplitter

Figure 10A is the spectrum of a chlorinated-PVC (CPVC) pipe compound intended for use with potable hot water. For this particular product, macros have been written which measure the ratios of various peaks in the spectrum vs. the 1255 cm^{-1} peak of the CPVC base polymer; the resultant numerical data are compared against a stored list of ratios from a reference control to validate production material and to test suspected complaint samples. This particular sample appeared to be badly out of spec, and a replicate analysis appeared to confirm this result. Note,

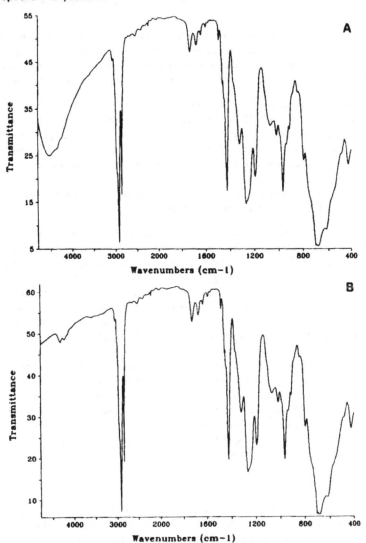

FIGURE 10. (A) Transmission spectrum of a CPVC pipe compound, which shows unexpected absorption in the region 4800 to 3000 cm⁻¹. This extraneous absorption is one of the indicators of an out-of-tune beamsplitter. (B) Spectrum of the same compound after realigning the beamsplitter.

however, the baseline in the 4800 to 3000 cm⁻¹ region, which did not match, even qualitatively, previously run samples of this product line. None of the compound constituents, nor any of the known set of possible contaminants, would account for this unusual baseline. In addition, there is also more noise present in the 4800 to 3000 cm⁻¹ region than there should be.

At that point, the FT-IR used to obtain these spectra was examined more closely. This examination revealed an interferogram whose zero path difference (ZPD) centerburst maximum was significantly lower than normal. Moreover, after the interferogram was transformed, the relative intensity of the high-frequency (i.e., 4800 to 3000 cm⁻¹) portion of the resultant single beam spectrum was much lower than normal. Finally, when a 100% line was acquired, the overall noise level was worse than normal, and the high-frequency noise level was particularly bad. These results identify an interferometer which is badly out of tune. After the system was retuned, the spectrum in Figure 10B, which exhibits a normal baseline, was obtained; the quantitative macros now indicated that this particular compound was in-spec.

a. Cause of Detuned Condition. An investigation was initiated to determine why this particular instrument was so badly out of tune. The investigation revealed that the laboratory had suffered a power outage early on the morning on the day the original spectrum was acquired. When the power was restored and the FT-IR was brought back into normal operation, the system was allowed to equilibrate for over one-half hour before any spectra were acquired, but performance diagnostics were not run. Apparently, the thermal shock or some other mechanism associated with the interruption of power and bringing the bench back into operation was sufficient to detune the beamsplitter.

b. Importance of Diagnostic Maintenance. This particular example illustrates several important points. First, routine performance diagnostics recommended by the instrument manufacturer or the ASTM Standard E1421-91, "Practice For Describing And Measuring Performance of Fourier Transform Infrared Spectrometers: Level Zero And Level One", should be run on a regular and routine basis.[10] Second, any time a bench experiences any sort of abnormality which results in a shutdown (power, water, air-bearing gas, major changes in environmental temperature, etc.), these performance specifications should be rerun. Note, however, that the operator must allow sufficient time (a minimum of 30 min, and perhaps more, depending on the manufacturer's recommendation) for the instrument to stabilize prior to running these performance tests. If this procedure had been followed, the out-of-tune instrument would have been detected and retuned before the sample was ever run. Third (and perhaps, most important), the operator should be trained to observe and note both the empty-beam ZPD center-burst maximum value and the resultant Fourier-transformed single-beam spectrum and to recognize when either of these are abnormal; any abnormality at all is a warning sign which should not be ignored, because it normally indicates a significant degradation in the instrument's performance.

2. Fogged Beamsplitter

A second example of an improperly maintained instrument is illustrated in Figure 11. This spectrum was obtained on a mid-1980s vintage benchtop instrument located at a manufacturing facility. Note the very high noise level in the 4000 to 3000 cm^{-1} region. This is *not* uncompensated water vapor; it is noise. The source of this problem was traced to a partially fogged beamsplitter.

a. Cause of Fogging. Investigation revealed that the FT-IR was situated in a fairly humid environment but, because the instrument purge came from a bank of nitrogen cylinders, the system was only purged when the instrument was in operation. By the time this situation was discovered, the instrument was still functional, but its performance was degraded, particularly above 2800 cm^{-1}. This problem could have been detected sooner, and perhaps before instrument performance had degraded so badly, if the user had run the performance standards suggested above on a routine basis.

b. Importance of Purge. The above example illustrates an important issue which is often overlooked by first-time FT-IR customers. Virtually all of the commercially available mid-infrared FT-IRs use rocksalt optics of some kind. Although some of the newer systems now

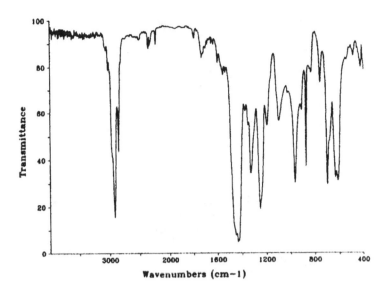

FIGURE 11. Transmission spectrum exhibiting severe baseline noise in the region 4000 to 3000 cm⁻¹. A fogged beamsplitter was identified as the source of this baseline noise.

use totally enclosed optics and others use proprietary coatings (usually KRS-5) to render the salt optics less sensitive to attack by moisture, any rocksalt optic is potentially subject to attack in a high-humidity environment. In a system without sealed optics, a source of clean, dry air or nitrogen is mandatory; even in a sealed system, a purged sample chamber will significantly enhance the ZPD center-burst maximum over what is obtained in room air. Yet many first-time customers forget to include an air dryer when preparing their capital requests. This omission can be expensive; replacement costs for a fogged beamsplitter typically run $3000 to $5000!

3. Incorrect Electronic Filtering

Figure 12A is the single-beam spectrum obtained on an older research-grade FT-IR, whose operator reported a pronounced decrease in the size of the ZPD center-burst maximum, while Figure 12B is the single-beam spectrum typically obtained when the instrument is operating properly. This particular instrument has a series of computer-selectable high- and low-pass filters, which are applied to the detector signal (i.e., the interferogram); the specific filters chosen generally depend on the scan speed and desired frequency range. On this spectrometer, the circuit which selects the specific high- and low-pass filters depends on a series of optical isolators which appear to have a finite lifetime; when one of these isolators goes bad, the net effect is to select the bit associated with that isolator, even when the operator has not attempted to select it. In Figure 12A, one of the isolators had, indeed, gone bad, with the result that a low-pass filter was selected which begins to cutoff at approximately 1400 cm⁻¹. The infrared signal just above 1400 cm⁻¹ is attenuated only slightly; however, there is increasingly greater attenuation at higher frequencies. This can be seen more clearly in Figure 12C, which is the ratio plot of the filter setting resulting from the "stuck" bit vs. the desired filter setting. (Note that the spectrum is plotted with a scale change at 2000 cm⁻¹, which appears to cause a change in the slope of the 100% line.) This particular problem was caught as soon as it commenced, because the operator observed that the empty-beam interferogram was unusual.

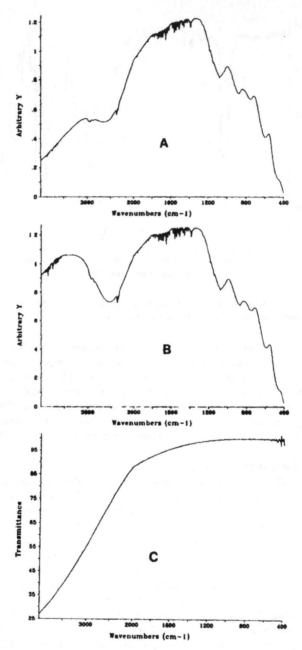

FIGURE 12. (A) Single-beam spectrum of an empty sample compartment. The energy at 4000 cm^{-1} is significantly lower than is normal on this instrument. (B) Spectrum acquired when the instrument is operating properly. (C) Ratio of (A) to (B). This spectrum identifies an improperly set low-pass filter, which begins to attenuate the infrared signal at approximately 1400 cm^{-1}.

III. CONCLUSION

This chapter has discussed some of the common problems in sample preparation and the interpretation of data containing sample-preparation or other spectral artifacts which the author has personally encountered during almost 20 years of industrial spectroscopy. There are a good many other artifacts which could be discussed, if space permitted. However, the examples which have been given, and which constitute only the proverbial tip of the iceberg, should convince the reader of the need for more comprehensive training of both the operators of modern FT-IR spectrometers and of those who interpret the spectra, as well as the need for more diligence looking for artifact effects in spectra. It has been said that a little knowledge is a dangerous thing; the examples presented here illustrate that too little understanding of the basic physical principals of the FT-IR instrument and the associated sample-handling techniques can lead to disaster. Virtually all of the examples presented from the author's laboratory were caught by experienced spectroscopists at the time that the spectra were acquired (i.e., *before* incorrect results were reported); however, several examples have also been presented from other laboratories, whose spectroscopists were apparently not sufficiently skilled to recognize these problems before a report was made. The dollar costs of these erroneous analyses, which sometimes resulted in the rejection and decertification of in-spec material are significant, and the potential liability associated with these erroneous results is high. The FT-IR instruments of today are marvelous systems, which are capable both of doing many powerful things and of generally making a spectroscopist's life easier. But the spectroscopist must always realize that the FT-IR and its associated computer enhancement are only tools; the ultimate responsibility for the quality of the spectra obtained and for the correct interpretation of those spectra still lies with the spectroscopist!

REFERENCES

1. Kendall, D.N., In *Applied Infrared Spectroscopy*, Kendall, D.N., Ed., Reinhold, New York, 1966, chap. 4.
2. Stine, K.E., *Modern Practices in Infrared Spectrrroscopy*, Beckman Instruments, Fullerton, CA, 1970, chap. III-IX.
3. Miller, R.G.J. and Stace, B.C., Eds., *Laboratory Methods in Infrared Spectroscopy*, Heyden, New York, 1972.
4. Smith, A. Lee., *Applied Infrared Spectroscopy*, John Wiley & Sons, New York, 1979, chap. 4.
5. Haslam, J, Willis, H.A., and Squirrel, D.C.M., *Identification and Analysis of Plastics*, 2nd ed., Heyden, Philadelphia, 1972, 453.
6. Cornish, P.J., Identification and analysis of polyurethane rubbers by infrared spectroscopy, *Anal. Chem.*, 31(8), 1298, 1959.
7. Krimm, S., Berens, A.R., Folt, V.L., and Shipman, J.J., *Chem. Ind.*, 1512, 1958.
8. Kawasaki, A., Furukawa, J., Tsuruta, T., and Shiotani, S., An infra-red sutdy of the crystallization of poly(vinyl chloride), *Polymer*, 2(2) 143, 1961.
9. Hummel, D.O., *Infrared Atlas of Polymers, Resins and Additives — An Atlas*, Vol. I, Part 1, John Wiley & Sons, New York, 1971, 147, 149.
10. ASTM Standard E1421-91, Practice for Describing and Measuring Performance of Fourier Transform Infrared Spectrometers: Level Zero and Level One.

Chapter 10

SAMPLING TECHNIQUES FOR INFRARED ANALYSIS IN THE ENVIRONMENTAL AND FORENSIC LABORATORIES*

Kathryn S. Kalasinsky and Victor F. Kalasinsky

CONTENTS

I. Introduction 276
II. Transmission 276
    A. Techniques 276
    B. Examples 276
III. Reflectance 279
    A. Techniques 279
    B. Examples 279
IV. Microspectroscopy 281
    A. Techniques 281
    B. Examples 281
V. Chromatographic Effluents 282
    A. Techniques 282
    B. Examples 283
VI. Conclusions 285
References 288

---

*The opinions expressed herein are the private views of the authors and are not to be construed as official or as representing the views of the Department of the Army or the Department of Defense.

0-8493-4203-1/93/$0.00+$.50
© 1993 by CRC Press Inc.

I. INTRODUCTION

In the environmental and forensic laboratories, typically the most difficult and crucial factor in solving a problem is determining the appropriate analytical technique. Secondarily, the sampling method used in the analytical technique of choice can help answer the question or preclude any definitive results. This decision-making process has received little attention in the analytical literature but is the crux of all analytical applications. The advantage of vibrational techniques relative to many other analytical procedures is the absolute identification that is possible through the selectivity of molecular fingerprinting, if the reference spectrum is available. The evolution of Fourier transform instrumentation has allowed the sensitivities to reach such levels that the accessories and methodologies can accommodate a vast range of sampling.

Hyphenated techniques can simplify many problems and streamline the selection process by doing multiple analyses with one sampling. Oftentimes, sample size can dictate some analytical methodology decisions. Infrared (IR) and Raman microscopy techniques have aided the field in this area by not only reducing the necessary sample size but by also reducing sample preparation and the need for separation from the matrix.

Sampling techniques must not only consider the matrix in which the sample resides, but also the phase in which the sample exists (gas, liquid, or solid). In general, infrared analytical sampling can be divided into four main areas: transmission, reflectance, microspectroscopy, and chromatographic effluents.

II. TRANSMISSION

A. Techniques

Transmission is the time-honored traditional method for infrared spectroscopy. Liquids were first analyzed in transmission cells with variable pathlengths attainable through various spacers between salt plates. Capillary or thin films sandwiched between two salt plates are also a common method of obtaining infrared spectra of viscous liquids. The salt pellet, nujol mull (thin film), and cast film are all established methods of transmission infrared spectroscopy that still have wide utility for solid materials. Gases can be analyzed in transmission mode by using single-pass, fixed-pathlength cells or by enhancing the signal with multi-pass cells, with either a fixed or variable pathlength.

B. Examples

The "Analytical Approach" column of *Analytical Chemistry* has featured many problem-solving strategies that have employed infrared transmission along with other analytical techniques in a decision-making process that is mapped out for the particular samples of interest.[1-3] A prime example of the decision processes which employ infrared transmission can be seen in a recently reported[4] sample of a shoe box which was submitted to an analytical services laboratory. A sealed container held portions of a "wet" shoe box which had caused shoe store employees to become ill. Upon receipt of the sample, the chemist noted a strong, unidentifiable odor emanating from the shoe box as it was transferred to a sample bulb for lab work. The sample was freeze-thawed on a vacuum line to remove all the non-condensable gases, then frozen in liquid nitrogen and allowed to warm up slowly. The first volatile gases emitted from the sample were trapped in a single-pass 15-cm pathlength infrared gas cell, and the transmission

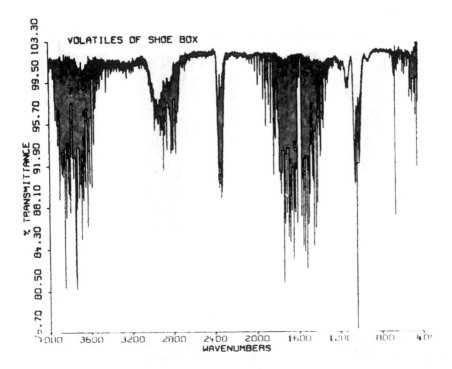

FIGURE 1. Infrared spectrum of volatile gases from shoe box sample. (From Kalasinsky, K.S., *Trends Anal. Chem.*, 9(3), 83, 1990. With permission.)

spectrum was obtained (see Figure 1). Atmospheric water and carbon dioxide were evident as well as an identifiable methanol band at approximately 1036 cm⁻¹. The C–H stretch region displayed the expected methanol peaks along with some other unidentified bands. The sample in the sample bulb was then allowed to warm up to room temperature, and the equilibrium mixture of gases was trapped in the infrared cell and the infrared transmission spectrum obtained (see Figure 2). This spectrum shows the atmospheric gases as before, but the methanol peak is now greatly diminished in size. The remainder of the peaks which were in the C–H stretch region were again unidentified. It was then noted that the infrared cell windows had begun to deteriorate. The sample was pumped out and the spectrum of the remaining material obtained (see Figure 3). These data revealed a peak identified as formaldehyde at approximately 1742 cm⁻¹ which was previously disguised by the atmospheric water absorptions. (The atmospheric absorptions appeared in this spectrum as negative peaks.) All the rest of the C–H stretches were then accounted for from the formaldehyde frequencies. Methanol and formaldehyde were existing in an equilibrium mixture in the shoe box sample. Final inquiry into the sample origin uncovered the shipping information which noted that biological samples had been shipped with the shoes by a public carrier, and one of the biological sample containers had broken, leaking formaldehyde on the shoe boxes.

Quantitative information, in addition to qualitative results, is one of the advantages of using infrared absorption data. Much has been published in the literature on this topic, and a review can be found elsewhere.[4] An example of the use of this feature in a sampling technique determination is represented in Figure 4. These spectra are air samples in a 20-m

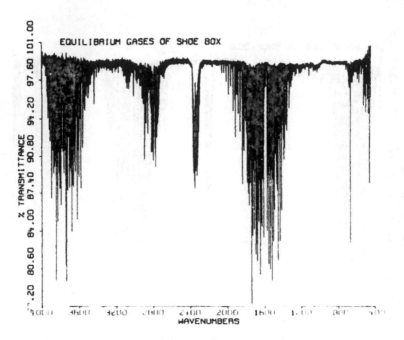

FIGURE 2. Infrared spectrum of equilibrium mixture of gases over liquid from shoe box sample. (From Kalasinsky, K.S., *Trends Anal. Chem.*, 9(3), 83, 1990. With permission.)

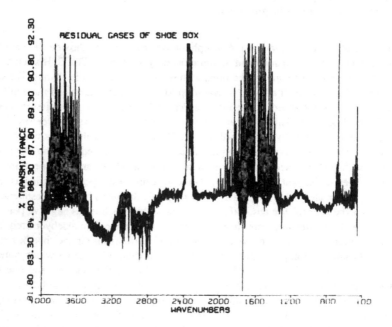

FIGURE 3. Infrared spectrum of residual gases from shoe box sample. (From Kalasinsky, K.S., *Trends Anal. Chem.*, 9(3), 83, 1990. With permission.)

multi-pass cell, the top spectrum being contaminated air and the lower spectrum normal air. The contaminated air sample came from the office of a manufacturing plant where the employees were losing consciousness at their desks. The difference in the two spectra can be seen at approximately 2145 cm^{-1} in a region where carbon monoxide (CO) absorbs. An expansion of the CO region with both spectra overlaid is shown in Figure 5. It can be seen from this representation that the contaminated air has about ten times the normal concentration of CO. The air samples from which these spectra were obtained were collected approximately 4 h after the syncopal episodes began with the employees and the building had been opened up and aired out with fans. The CO levels were estimated to be much greater at the time of the events than those during sample collection. The analytical conclusions helped lead the inspection team to the source of the problem, which was a crack in the factory gas heater system.

III. REFLECTANCE

A. Techniques

Infrared reflectance spectroscopy covers a very broad area of applications. These techniques include internal reflectance (ATR), diffuse reflectance, and specular reflectance and can be used in a number of different configurations for most liquid and solid samples. In most cases, industrial analytical laboratories have been using reflectance infrared spectroscopy as one of their prime tools for quite some time, but for proprietary reasons the published applications remain scant. Thus, reflectance work has not received the attention proportionate to its use. Several other chapters in this book are dedicated to this topic; thus, the discussion here will be brief.

B. Examples

Applications of reflectance work and the sampling decision process can be found in the literature,[5-7] and an example of this type of analysis can be found in a study of vinyl coatings, as shown in the ATR spectra of Figures 6 and 7. In these two cases, a manufacturer of vinyl material for office furniture had complaints from customers that the material did not hold up to wear and soil. The manufacturer contended in both cases that cleaning products made of isopropyl alcohol had been used on the material which had removed the protective coatings. (A warning stating such was issued with the product when sold.) For the case in Figure 6, spectra of the following vinyl materials were obtained: standard lot sample from the manufacturer, standard lot sample after treatment with isopropanol, portion of the customer's sample from area under furniture where there was no access for cleaning, and soiled portion of the customer's sample from an area to which cleaning personnel had access (cleaned with soap and water prior to analysis as recommended by the manufacturer). It can be seen that the spectra are all similar, with the exception of the standard lot sample from the manufacturer, indicating that the protective coating was never placed on the material used in the customer's furniture coverings.

For the case in Figure 7, spectra of the following materials were obtained: standard lot sample from manufacturer, dark portion retaining color of customer's sample, and light portion of faded area of customer's sample. These results show similar spectra, with the exception of the faded portion of sample, indicating that the protective coating was on the original customer's sample but had been removed by cleaning products.

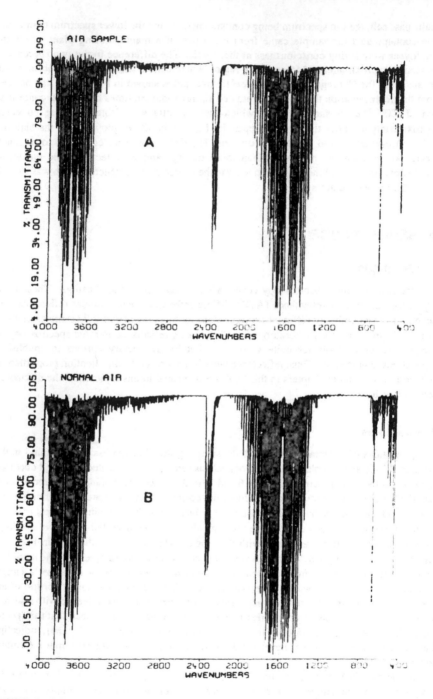

FIGURE 4. Infrared spectra of contaminated (A) and normal air sample (B). (From Kalasinsky, K.S., *Trends Anal.Chem.*, 9(3), 83, 1990. With permission.)

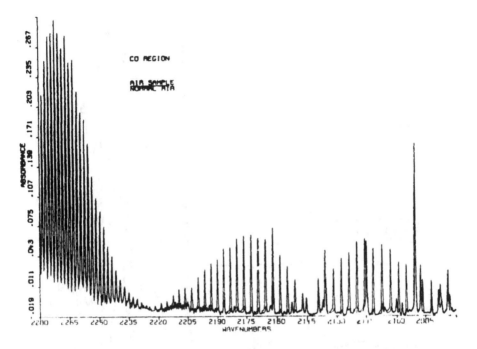

FIGURE 5. Infrared spectra of carbon monoxide region of contaminated and normal air sample overlaid. (From Kalasinsky, K.S., *Trends Anal. Chem.*, 9(3), 83, 1990. With permission.)

These two previous examples were identical in nature, but infrared reflectance spectroscopy was able to determine unequivocally that two different situations were present. The issues involved dealt with large sums of money in replacement costs, and litigation was resolved out of court.

IV. MICROSPECTROSCOPY

A. Techniques

Microspectroscopy has recently had a resurgence in the vibrational analysis field because of the advantages of Fourier transform infrared (FT-IR) and has gained popularity in the environmental and forensic areas due to the minute amount of sample needed for analysis and the minimal sample preparation that is required.[8,9] The analytical beam separates the sample from the matrix by virtue of the size of the detection area. Applications include fibers, polymer particulates, single crystals, and spectral mapping.[10-13]

B. Examples

Identification of foreign materials in biological samples has been greatly assisted by the development of infrared microscopy.[14] Microparticles of polymers such as PTFE (polytetrafluoroethylene), nylon, polyethylene, polyurethane, polysulfone, and silicon-based polymers have been determined to be the cause of some foreign-body reactions. These substances have originated from heart grafts, ocular valve implants, prosthetic implants, and

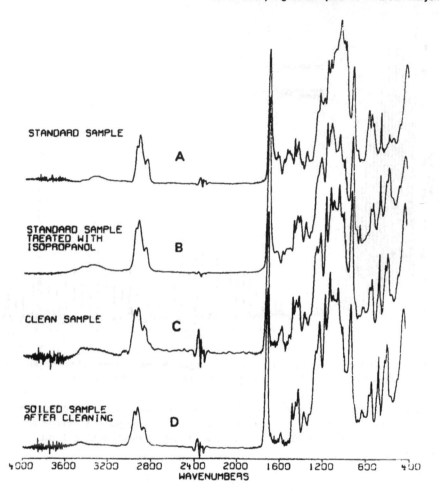

FIGURE 6. Attenuated total reflectance spectra of vinyl upholstery material without protective coating: standard sample (A); standard sample treated with isopropanol (B); clean sample (C); and soiled sample after cleaning (D). (From Kalasinsky, K.S., *Trends Anal. Chem.*, 9(3), 83, 1990. With permission.)

skin biopsies. An example of nylon microparticles from an ocular valve implant imbedded in tissue can be seen in the photomicrograph of Figure 8. The infrared spectra obtained from infrared microscopy are shown in Figure 9. The top trace (A) is a spectrum of the nylon and surrounding protein in the tissue. The middle trace (B) is the reference material of nylon 6/6, and the bottom trace (C) is the protein in a tissue section away from the imbedded particles. Data manipulation can ascertain that the foreign particles compare with the reference materials.

V. CHROMATOGRAPHIC EFFLUENTS

A. Techniques

Chromatographic interfaces with infrared spectrometers have probably served more to popularize the technique of vibrational spectroscopy than any other application. A review of

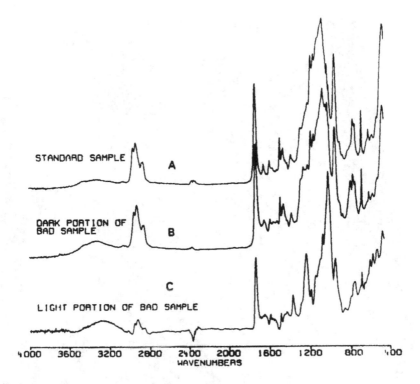

FIGURE 7. Attenuated total reflectance spectra of vinyl upholstery material with protective coating: standard sample (A); dark portion of bad sample (B); and light portion of bad sample (C). (From Kalasinsky, K.S., *Trends Anal. Chem.*, 9(3), 83, 1990. With permission.)

the development and application of the various chromatographic interfaces can be found elsewhere.[15] The variety of infrared chromatographic interfaces include gas chromatography (GC), high-performance liquid chromatography (HPLC), thin-layer chromatography (TLC), and supercritical fluid chromatography (SFC). These interfaces have served to separate complex mixtures, identify compounds with differentiation of isomers, and make quantitative measurements, with the detection limits in some developing areas reaching picogram levels. Head-space GC/FT-IR is one technique that has received little attention, but has great utility in the analytical field for odors and volatile gases which are evolved from sample matrices.[16]

B. Examples

One application that has evolved due to the recent developments in interfaces which have increased detection limits to levels comparable to those of mass spectrometry (MS) is forensic drug analysis.[17] With comparable detection limits, the tandem technique of GC/IR/MS has also found more utility not only in forensic drug analysis but also in environmental fields.[18] The studies of drug metabolites which exist in very low concentration in complex matrices of body fluids present unique sampling difficulties.[19, 20] Monitoring drug metabolites from new drugs is essential. It was found that one drug that was approved for pilots as an antihistamine was metabolized to a drug that is prescribed as a sedative. Shown in Figure 10 is the infrared spectrum of that drug metabolite obtained by cryogenic deposition GC/FT-IR. Studies are underway to determine metabolite blood concentration as well as physiochemical interactions.

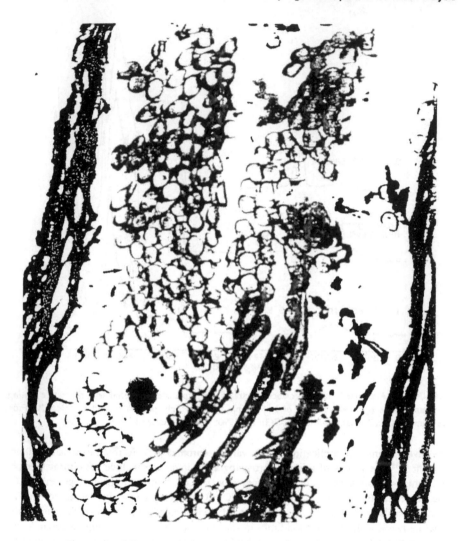

FIGURE 8. Photomicrograph of a tissue section with nylon particles. (Magnification 10X.) (From Centeno, J.A., Kalasinsky, V.F., Johnson, F.B., Vinh, T.N., and O'Leary, T.J., *Lab. Invest.*, 66, 123, 1992. With permission.)

Studies of infrared method development for drugs of abuse are currently being pursued by several laboratories for routine drug testing.

Vapor phase GC/FT-IR techniques have been successfully developed for absolute concentrations of low nanogram amounts which are equivalent to 200 ng/ml urine. Cryogenic deposition GC/FT-IR techniques are being developed which indicate low picogram amounts of material are detectable. Figure 11 shows some of the preliminary data obtained in this study for amphetamines. The top trace is 3 ng of amphetamine obtained in real time on a Digilab Tracer system. The middle trace is 600 pg of the material obtained in real time. The bottom trace is the same 600 pg of material that was deposited signal-averaged for 200 scans. The spectral quality of the signal-averaged spectra is comparable to that of the higher level reference spectrum, with the exception of the ice band between 3000 and 3500 cm^{-1}. The ice

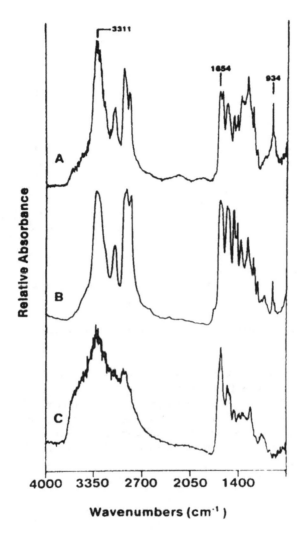

Wavenumbers (cm⁻¹)

FIGURE 9. Infrared spectra of nylon and surrounding protein in a tissue section (A); Reference material of nylon 6/6 (B); Protein in a tissue section away from nylon particles (C). (From Centeno, J.A., Kalasinsky, V.F., Johnson, F.B., Vinh, T.N., and O'Leary, T.J., *Lab. Invest.*, 66, 123, 1992. With permission.)

band is an indication of the attention that must be paid to the mechanical set-up of a system that can detect lower limits of material. Fine-tuning of a system like this one for body fluid metabolites will possibly provide new levels of knowledge about human processes.

VI. CONCLUSIONS

Numerous examples of chromatographic sampling, solid sampling, and gaseous sampling techniques have been presented along with a discussion of the selection of an appropriate method for a given problem. There are still other methods which have not been discussed here

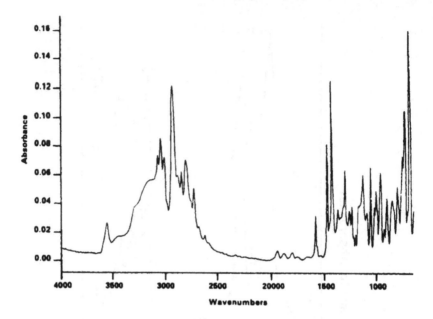

FIGURE 10.   Infrared spectrum of known sedative produced as drug metabolite from marketed antihistamine.

(e.g., emission, photoacoustic).[21, 22] The final conclusion that can be drawn is that it takes both an educated decision, as well as a trial-and-error process, to determine the best sampling method to answer a question. More sampling applications are needed in the literature to assist in determining these analytical procedures. A compendium of best methods based on past experiences would be of great value. This publication and its preceding symposium represent steps that need to be taken to approach the concern for proper sample handling with the instrumentation of today.

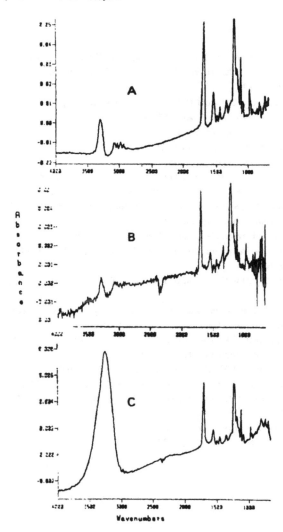

FIGURE 11. Infrared spectra of amphetamine obtained by cryogenic GC/FT-IR: 3 ng of material obtained in real time (A); 600 pg of material obtained in real time (B); and 600 pg of material signal averaged for 200 scans (C).

REFERENCES

1. Gilhooley, R.A., Richards, C.P., Warren, J., and Well, K.S., *Anal. Chem.*, 55, 639A, 1983.
2. Hopkins, J.L., Cohen, K.A., Hatch, F.W., Pitner, T.P., Stevenson, J.M., and Hess, F.K., *Anal. Chem.*, 59, 784A, 1987.
3. Brown, D.J., Schneider, L.F., and Howell, J.A., *Anal.Chem.*, 60, 1005A, 1988.
4. Kalasinsky, K.S., *Trends Anal. Chem.*, 9(3), 83, 1990.
5. Grasselli, J.G., *Chem. Times Trends*, October, 26, 1978.
6. Pattachi, S.C. and Porro, T.J., *Am. Lab.*, 20(8), 24, 1988.
7. Harrick, N.J., *Am. Lab.*, 20(10), 99, 1988.
8. Shearer, J.C., Peters, D.C., Hoepfner, G., and Newton, T., *Anal. Chem.*, 55, 874A, 1983.
9. Reffner, J.A., Coates, J.P., and Messerschmidt, R.D., *Am. Lab.*, 19(4), 86, 1987.
10. Lang, P.L., Katon, J.E., and Bonanno, A.S., *Applied Spectrosc.*, 42, 313, 1988.
11. Young, P.H., *Spectroscopy*, 3(9), 24, 1988.
12. Harthcock, M.A. and Atkin, S.C., *Applied Spectrosc.*, 42, 449, 1988.
13. *Infrared Microscopy Theory and Applications*, Messerschmidt, R.G. and Harthcock, M.A., Eds., Marcel Dekker, New York, 1988.
14. Centeno, J.A., Kalasinsky, V.F., Johnson F.B., Vinh, T.N., and O'Leary, T.J., *Lab. Invest.*, 66, 123, 1992.
15. *Chromatography/Fourier Transform Infrared Spectroscopy and Its Applications*, White, R., Ed., Marcel Dekker, New York, 1990.
16. McGorrin, R.J., Pofhal, T.R., and Croasmun, W.R., *Anal. Chem.*, 59, 1109A, 1987.
17. Kempfert, K., *Applied Spectrosc.*, 42, 845, 1988.
18. Duncan, W. and Soine, W.H., *J. Chromatogr. Sci.*, 26, 521, 1988.
19. Levine, B.S., Smith, M.L., and Froede, R.C., *Clin. Lab. Med.*, 10(3), 571, 1990.
20. Smith, M.L., Bronner, W.E., Shimomura, E.T., Levine, B.S., and Froede, R.C., *Clin. Lab. Med.*, 10(3), 503, 1990.
21. Chase, D.B., *Appl. Spectrosc.*, 35, 77, 1981.
22. Harris, T.D., *Anal. Chem.*, 54, 741A, 1982.

Chapter 11

INFRARED SAMPLING TECHNIQUES TRAINING

Thomas J. Porro and Silvio C. Pattacini

CONTENTS

I. Introduction 290
II. Annual Training Courses 291
III. Scheduled Courses in FT-IR Provided by Instrument Manufacturers 292
IV. Other Courses Providing IR Sampling Techniques Training 293

0-8493-4203-1/93/$0.00+$.50
© 1993 by CRC Press Inc.

I. INTRODUCTION

This chapter has been added to this basic text on infrared (IR) sampling techniques as a necessary compliment and follow up to the verbal descriptions of sample preparations to provide the reader with a listing of sites offering some form of "hands-on" training on a regular basis. This training in sample handling techniques is generally provided as a part of a broader offering of IR spectroscopic courses.

The intent here is to be as comprehensive as is practical for the situation as it exists today. However, the reader is advised that there will be changes with time, and inquiries should be made sufficiently in advance before making firm commitments.

The current state of the art of infrared sampling techniques differs widely from what it was some 15 or 20 years ago or about the time that the first Fourier Transform instruments came on the market. In those days, much of the analytical infrared spectroscopic work was done in central applied-research laboratories of large corporations, whether it was research, analytical services or support for quality control analyses.

Small laboratories either did not have infrared instrumentation or they used analytical service labs until they could justify obtaining their own equipment. Large universities and colleges then benefitted from relatively large government funding of research instrumentation which to a lesser extent continues today. Large and small teaching schools obtained equipment commensurate with their funding capability for their organic and physical chemistry undergraduate courses. And there were a plethora of schools in those days taking advantage of the Federal GI Bill largess.

Forensic, as well as life science, laboratories did not use infrared then to the extent they do today. In addition, there were fewer infrared technicians requiring training in the art of sample handling, although the number of institutions providing such training were about what they are today. What has changed?

With the advent of FT-IR instrumentation 20 years ago, society was already in the process of increasing its mobility. There was a greater and greater turnover of laboratory personnel. Additionally, business was good generally, and the advantages of FT-IR instrumentation broadened the infrared market so that the need for analytical equipment was expanding greatly. Also occurring during those years was a decentralization as well as broadening of the analytical laboratories of the larger corporations recognizing the need to provide more and more quality control to better assure lower manufacturing costs in the face of greater competition. These forces operated so as to produce a tremendous need for training more and more personnel. Though the established infrared courses have been able to adequately train technicians in basic theory and instrumentation, they were and are not sufficiently geared to provide adequate training in sample handling techniques. Thus, this book has been written to help fulfill the training requisites of infrared sampling techniques for those many people needing them.

Following are outlines of various courses in Infrared Spectroscopy given in this country (U.S.) which include some aspect of sample handling techniques. They have been divided into three main categories: (1) annual IR training courses, (2) scheduled IR courses by instrument manufacturers, and (3) other courses providing IR sampling techniques training.

The annual courses listed in Section II.A. (Bowdoin College), Section II.B. (Vanderbilt University) and Section II.C. (Arizona State University) provide treatments of FT-IR and Raman theory and instrumentation, interpretation of IR and Raman Spectra, applications of IR and Raman spectroscopy, and sample handling techniques, but as seen below the mix is different in each case. For example, the Bowdoin Course provides a full week of spectral interpretation and the better part of another week on "hands-on" laboratory exercises demonstrating various sampling techniques.

The Fisk course at Vanderbilt University and that at Arizona State University cover much the same ground but are less comprehensive because they are shorter courses. The Miami University and Weslyan University courses both emphasize IR microspectroscopy with the former giving a number of sampling experiments each by different instrument vendors, while the latter concentrates heavily on "hands-on" microsampling preparation training.

The "other courses" (Section III) category includes one presented by The Center for Professional Advancement in New Jersey which offers a course similar to those in the annual type except for less interpretation of spectra and more treatment of hyphenated techniques, but its frequency and location can vary considerably.

Most manufacturers of infrared equipment provide various user training courses for much of their instrumentation; however, the content and schedules for them are not as publicized as the others. In most cases, the user of particular equipment will take advantage of the specific vendor for training. However, included here is a listing of the various vendors of infrared instrumentation with their addresses and phone numbers so that one can obtain directly from any company any course information available.

II. ANNUAL IR TRAINING COURSES

A. Bowdoin College IR Course Brunswick, ME
   Fifteen Days (Three Weeks) Course in:
   1. Infrared Spectroscopy I. Interpretation of Spectra
   2. Infrared Spectroscopy II. Instrumentation, Polymer Spectra, Sample Handling, and Computer Assisted Spectroscopy
   3. Infrared Spectroscopy III. Workshop on FT-IR

   Contact: Professor Dana W. Mayo, Director
              Department of Chemistry
              Bowdoin College
              Brunswick, ME
              Tel. (207)725-3218 or 725-3601

B. Fisk Infrared Institute Vanderbilt University, Nashville,TN
   A Six Day Short Course in:
   1. Infrared and Raman Spectra Interpretation
   2. FT/IR Lectures
   3. Hands-on FT/IR Labs

   Contact: Dr. Clara Craver, Director
              Tel. (314) 962-5752
              Dr. Enrique Silberman, Director
              Fax. (615) 329-8634
              Box 15, The Fisk University, Nashville, TN 37208

C. Arizona State University Applied Molecular Spectroscopy, Tempe, AZ
   Five Day Course in:
   1. Infrared FT-IR
   2. Use of Mini and Micro Computers for Data Processing
   3. Search and Retrieval

    4. Quantitative Analysis
    5. Raman

       Contact: Dr. Jacob Fuchs, Director
               Chemistry Dept. Applied Molecular Spectroscopy
               Arizona State University, Tempe, AZ 85287
               Tel. (602) 965-4496 or (602) 965-3461

D. Molecular Microspectroscopy Workshop Miami University, Oxford,OH
    Three Day Course in:
    1. Theory and Instrumentation of Optical Microscopy, Infrared Microspectroscopy, and
       Raman Microspectroscopy
    2. Microspectroscopy Sampling Handling Techniques
    3. Instrument Manufacturers Demonstrations

       Contact: Dr. J.E. Katon, Director
               Molecular Microspectroscopy Laboratory
               Department of Chemistry
               Miami University, Oxford, OH 45056
               Tel. (513) 529-2874

E. Weslyan University Infrared Microscopy Workshop, Middletown, CT
    Three Day Course in:
    1. Theory and Instrumentation of FT-IR and Optical Microscopy
    2. "Hands-On" Laboratory Sessions on Sample Preparation and Operation of FT-IR
       Microscopes
    3. Experiments Using Microscopic Samples of Many Types

       Contact: Professor Wallace Pringle, Director
               Hall-Atwater Laboratories
               Department of Chemistry
               Weslyan University
               Middletown, CT 0649
               Tel. (203) 347-9411 Ext. 2361 or 2791

F. Eastern Analytical Symposium (EAS) IR Workshops
    Every year during the symposium, a number of workshops covering FT-IR and Raman
Spectroscopy topics are presented by instrument manufacturer technical specialists at least
one of which deals directly with sample handling techniques. These are tutorial in nature
and provide useful information. For additional information contact the EAS HOTLINE at
(302) 453-0785 or the EAS FAXLINE at (302) 738-5275

III. SCHEDULED COURSES IN FT-IR PROVIDED BY INSTRUMENT MANUFACTURERS

    1. The Perkin-Elmer Corp
       761 Main Ave.
       Norwalk, CT, 06859-0918
       Tel. (203) 762-1000

2. Nicolet Analytical Instruments
   5225 Verona Road
   Madison, WI 53711-4495
   Tel. (608) 271-3333
   Fax. (608) 273-5046
3. Bio-Rad, Digilab Division
   237 Putnam Avenue
   Cambridge, MA 02139
   Tel. (617) 868-4330
   Fax. (617) 499-4519
4. Mattson Instruments, Inc.
   1001 Fourier Court
   Madison, WI 53717
   Tel. (608) 831-5515
   Fax. (608) 831-2093
5. KVB Analect
   17819 Gillette Avenue
   Irvine, CA 92714
   Tel. (714) 660-8801
6. Bomem, USA
   7780 Quincey Street
   Willowbrook, IL 60521
   Tel. (800) 888-FTIR
   (708) 323-2373
   Fax. (708) 986-1091
7. Bruker Instruments, Inc.
   Manning Park
   19 Fortune Drive
   Billerica, MA 01821
   Tel. (508) 667-9580
8. Spectra-Tech, Inc.
   652 Glenbrook Road
   Stamford, CT 06906
   Tel. (203) 357-7055
   Fax. (203) 357-0609
9. Graseby Specac Inc.
   301 Commerce Drive
   Fairfield, CT 06430
   Tel. (800) 447-2558
   Fax. (203) 384-1524

IV. OTHER COURSES PROVIDING IR SAMPLING TECHNIQUES TRAINING

A. The Center For Professional Advancement
   Box H
   East Brunswick, NJ 08816-0257
   Tel. (201) 238-1600

Telex 139303 (cenproebrw)
Fax. (201) 238-9113

Fourier Transform Infrared Spectroscopy (FT-IR)
A Three-Day Intensive Course on
  1. Principles Techniques and Applications
  2. Sampling and Sample Preparation
  3. Selecting an Instrument
  4. Hyphenated Systems

B. SAS Short Course on Fourier Transform Infrared Spectrometry

Two day course given twice a year prior to the Pittsburg Conference and FACSS meeting given by Drs. Peter Griffiths, University of Idaho, Moscow, ID, Tel.(208) 885-6552 and James deHaseth, University of Georgia, Athens, GA, Tel. (404) 542-1968 and containing:

1. FT-IR Theory
2. Instrumentation Design
3. Infrared Microspectrometry
4. Reflection Techniques
    a. Internal Reflection Spectroscopy
    b. External Reflection Spectroscopy
    c. Photoacoustical Spectroscopy
5. FT Raman Spectroscopy

To register or obtain more information contact The Society of Applied Spectroscopy (SAS), 198 Thomas Johnson Drive, S-2, Frederick, MD 21702, Tel. (301) 694-8122.

INDEX

Symbols

% Transmittance 246
100% line 71–72, 241

A

abrading samples 98–101 *See also* silicon
    carbide
Absorbance 246
absorbance linearity 178
absorption 146
absorption coefficient 247
absorptivity 222, 247
accessory alignment
    DRIFT 95–96
    IRA 68–70
accessory dependence 228
additives 229–230
adsorption trap 174, 194
aggregated domains 234
air dryer; importance for purge 271
alignment of FT-IR 242–243
alkali halide disks 235 *See also* KBr pellets,
    pellets
alkali halide salt 228
analytical method; for quantitative analysis
    221–225
angle of incidence 37, 57–61, 89, 239–240
annual IR training courses 291–294
antistat 80 *See also* polymers
apodization functions 244, 245
appropriate standards 250
aqueous 141
aqueous media 172
aqueous solutions 26–27
    by internal reflection 69
arbitrary bias 246
artifact component 195
artifacts; in GC/FT-IR 207–211
artificial sweetener 202
aspartane 202-210
ASTM Standard E1421-91 270
asymmetric (single-sided) data collect 244
atmospheric interferences 245
ATR 44–46, 56, 108, 141, 258, 279–281
attenuated total reflection 44–46, 56
attenuated total reflection objective 161
attenuation 242
automated headspace analysis 185–187

automated thermal desorption system 181
automobile tire 138–139
average error 223

B

back-calculated values 223
back-calculations 223
backflush 180
background reference spectrum 89–92
background selection; for IRS 89–92
background spectrum 94
baseline correction 247–248
beam condenser 48–49
beam size 235
Beer-Bouguer-Lambert law 231 *See also*
    Beer's law
Beer's law 41, 47, 178, 223, 231, 246
bending vibration 6–10
bias 24
blackbody curve 241
blinds 222
block polymer 229–230
Boltzmann distribution 6, 231
bonded phase 177
bottoming out 12–14, 31, 35, 267
boxcar apodization 244
broadening 232
buffer system 226

C

C–H stretching 101–103
cadmium telluride 63, 64–69
caffeine 199, 203
calcination 228
calibration blends 222
calibration matrix 223
calibration plots 247
calibration references 221–222
calibration set 221–222
calibration standards 221–225, 251
carbon black 35, 36, 113, 114, 137–138
carbon dioxide 171–173
carbon disulfide 23–25
carbon tetrachloride 23–25
carbon-filled materials 45, 83–85, 137–138
carry-over 234
Carver press 35 *See also* hydraulic press
cast films 33–35, 233–234
cast-film artifacts 264

catalysts 229–230
cell thickness 22–23
centerburst 243–244
charcoal trap 196
chemical compatibility 228
chemical interactions 225
chemical reactions 228, 232
chlorinated pesticides 173
chlorinated-PVC 268
choosing cell thickness 23–25
chromatographic coupling 161–162
chromatographic separation 170
classical theory 2–6
cleaning 238
cleaning; in DRIFT 96
CO_2 modifier combinations 173
coal analysis 36, 125–129, 130–132
coatings 133–134, 140
cold finger 174–176
cold on-column injection 178–179
collection time 243, 248
collection vial 196
collinated beam 178
column bleed 208–209
combined analytical techniques 166–211
combustion gas 179
composites 138
condensed-phase 180–181
constructive interference 236
contact; inadequate 85
contamination 237
continuous liquid-liquid extraction 174-176
control chart 103
conventional distillation 185
convergent 178
cooled melts 259–261
copolymer 129–130
copolymerization 229–230
courses by instrument manufacturers 292–293
critical angle 57
critical point 171
critical pressure 171–173
critical temperature 171–173
cross linking 229–230
cross-linked, bonded stationary phases 209
cryofocusing 177
cryogrind 238
cryopump 210
cryotrapping 202
crystallinity 229–230, 233, 261
curved baseline 248
cushioning pad 67
cushioning pressure pads 66–68

D

data enhancement techniques 246–247
data processing 14–17

data treatment 246
deactivated, fused silica 179
decompression 173
degassing 232
degradation 234
degree of contact of IRE 87
degrees of freedom 7–10
demountable liquid cells 21-23, 236–237
dense fluid-phase chromatographies 167–168
deposition methods 169
depth of penetration 57–61
depth profiling; by PAS 110–113, 132–137
depth profiling of solid samples 99
derivative spectra 249
derivatized homologue 191–193
desorption 194, 207
destructive interference 236
detector 245 See also DTGS and MCT
detector icing 208–211
detector linearity 242
detector/optics cutoff 241
deuterated triglycine sulfate detector 20, 245
diamond anvil cell 49, 154–155, 158, 261-264, 266
diffraction 146, 155-157
diffuse reflectance 39-44, 93-104, 147, 149-150, 153, 154, 240, 246–247 See also DRIFTS
dilute gas streams 174
direct headspace 179–181, 189-194
directional orientation 76
dissolution 230
distillation 169–170
distillation flasks 174–176
distillation temperature 186
DRIFTS 41–44, 103, 108, 134-135, 137, 143, 240 See also diffuse reflectance
dry box 28, 31, 235
dry distillation 185–187
DTGS detector 20, 21, 245
duplication 227–228
dynamic headspace 181, 194-198

E

effect of particle size; in DRIFT 96–97
effective pathlength 69
EGA 167
electrolytic reactions 64
electromagnetic radiation 2–6
electronic filtering; incorrect 271
elemental analysis 203
emission 161
emulsions 80–92
energy 3–6
energy throughput for KBr powder 96–97
environmental 276–288

environmental analyses 20–21
EPA Methods (3560, 8015, 8410, 8440) 211–212
epoxycarvone 197–198, 200, 201
equilibrium concentrations 195
error sources 225
errors 233, 246
evaporation 227
event broadening 167
evolved gas analysis 167
external GC detectors 186
extraction 169–170
extrapolation 221, 222
extreme concentrations 221

F

far infrared 4, 25
fiber 75, 137-138
    by PAS 121–125
    use of DRIFTS for 98
filling thin cells 237
film orientation 76
film preparation technique 264
films 76, 233–234
    nondestructive DRIFTS analysis for 98–99
first derivative 202
flame ionization detector 179
flash vaporization 178–179
flow-through method 236–237
Fluorolube 28, 238
focused beam 178
fogged beamsplitter 270–271
fogging of windows 22, 24, 238
folding 247
forensic 47, 121, 124, 275-288
Fourier high-pass filter 249
Fourier Self-Deconvolution 248
Fourier transform 243–244
Fourier-space filter 248
fringe pattern 79–80, 86
fringe removal 139, 249
FT-IR hardware-induced artifacts 268–271
FT-IR/PAS 108–143 See also PAS
full saturation 110
fundamental mode 6 See also normal mode
fused silica capillary 173

G

gas chromatography/Fourier-Transform-Infrared 166
gas evolution 202
gas phase samples 17-21, 179, 194, 230–232
GC/FT-IR 166, 168–169, 283
GC/IR/MS 212, 283
GC/MS 180, 187
gels 141
germanium 45–46, 63-64

glass liners 170
Golay-Savitzky algorithm 15–17, 248
gold-coated lightpipe 168–169, 211
good quantitative analysis; rules for 251
GPC/FT-IR 167–168
Gram-Schmidt chromatogram 177
graphitic carbon 83–85
grazing angle objective 161
grinding 28, 30, 230
group frequencies 7–10
group theory 8

H

halide salt abrasion 77
halide salts 228
"hands-on" training 290–291
harmonic oscillator 4–6
headspace direct derivatization 193
headspace sampling 191–194
headspace vials 185, 193
heat capacity 37
helium 36, 109–110, 112–114, 122–126, 128
Helmholtz resonance 112
heterogeneous samples 267
high pressure liquid chromatography 167–168
high resolution GC 169
high selectivity 196
high-pass filter 248
homogeneity 146, 159, 228, 233, 235
homopolymers 229–230
Hooke's law 4–6
horizontal IRAs 73–74, 88-89
HPLC 167–168
HPLC/FT-IR 167–168
HRGC/FT-IR 169
hybridization 196
hydraulic press 235
hygroscopic 228
hyphenated techniques 165-212, 276

I

icing of MCT detector 241
immobilized phase 179
importance of diagnostic maintenance 270
importance of purge 270–271
improper reference standards 267
impurities 226, 233–234
inadequate contact to IRE 85–86
incidence angle 59–61
incompatibility 226
incubation temperatures 193
index of refraction 38
indirect headspace 181, 194-198
infrared courses 290–291
infrared microspectroscopy 145-162
injector 169–170
injector barrel 173

injector broadening 169–170
instrument line-shape function 244
interference fringes 25, 139, 233, 236
interferogram 243–244
intermolecular associations 225
intermolecular interactions 247
internal reference material; in DRIFTS 103–
    104
internal reflectance element 239
internal reflection 27, 44-46, 57-61
internal reflection attachment 56
internal reflection element 44-46, 56, 62-68
internal reflection spectroscopy 44-46, 56–92,
    239
internal standard 235, 238
interpolation 249–250
ion exchange 33, 230, 235
IR microscope 146 See also infrared
    microspectroscopy
IR spectroscopic courses 290–291
IRA alignment 68-70
IRA solid sampling accessory 68
IRAs 68
IRE 44–46, 239
IRE cleaning 65
IRE contamination 65–66
IRE crystals 63–68
IRE holder 68, 88
IRE properties 62–68
IRE selection 88
IRS 44–46, 56-92, 239

J

Joule-Thompson cooling 173

K

K-M 41–44
KBr crystal 260
KBr overlayer 98
KBr pellet 28, 30–33, 203, 260
kinetic experiments 83
Kramers-Kronig transform 38, 151-153
KRS-5 45–46, 62–63, 271
Kubelka-Munk 41-44, 135, 151, 240, 246–247
Kubelka-Munk theory 94, 150
Kuderna-Danish 174–176

L

large or irregular shape samples for IRS 75
least-squares regression 222
library searching 61, 118, 169
light scattering 229–230
lightpipe 168-169, 211
linear calibration plot of silane-treated fillers
    99–103
linear dependence 221-222
linear format 246, 251

linear regions 247
linearly independent concentration data 251
liquid cells 21-23, 236–238
liquid IRAs 64
liquid IRS 68-72, 85
liquid phase samples 169–170
liquid solvent extraction 188–189
liquids 21–27, 98, 115, 141
long path cells 20–21

M

macro-liquid cells 236
MAGIC 168
magnesium perchlorate 113, 117
mass discrimination 177
maximum entropy 248
MCT detector 20, 21, 122, 168, 210
measurement conditions defined; for quantita-
    tive analysis 221–225
memory effects 72, 228, 231, 232
mercury cadmium telluride detector 20, 122,
    168
micellar behavior 80–92
micro KBr pellets 48
micro-cavity cells 236–237
micro-scale preparations 174–176
micro-SDE apparatus 174–176
microdistillation apparatus 185
microparticle analysis 125–129, 281–282
microsample analysis 121–125
microsampling 47–49
microspectroscopy 49, 146, 281–282
migration 80–92
mild conditions 199
miniature diamond anvil cell 49 See also
    diamond anvil cell
misalignment in quantitative analysis 241—242
moderate conditions 199
modifier agents; in SFE 172
modulation frequency 37, 108–114, 116
moisture 227
moisture sensitive materials 235
molecular mixture analysis 166
monomer units 229–230
morphology 230
moving smoothing function 248
mulls 27–28, 238–239
multicomponent analysis 224
multihyphenation 212
multilayer film 76, 81
multiple examinations 228
multiple headspace extraction 180
multiple internal reflections 249

N

near-infrared 108
nitrogen oxides 179

nitrogen purged 101
nondestructive method 239
nonlinearity 241
nonstoichiometric loss 202
nonuniformity 233
nonvolatiles 169–170, 181–187
normalization 235
normal node 6
Nujol 28, 238

O

off-line collection 173
on-line collection 173
opaque materials 108, 137–138, 234
operator bias 246–247
operator-dependent bias 246
optical configurations 178
ordinate axis 246–247
organic chemical preparations 169–170
organic solvents 169–170
orientation 78, 234, 259-264
oriented films 139–140
out-of-tune beamsplitter 268–270
overdetermination 222–223
oxidative decarboxylation 194

P

P branch 17–20
packed columns 179
partial IRE coverage 86
partial pressures 231
particle size 96, 230
particulates 97–98, 234
partitioning 179–181
PAS 36–37, 246
pathlength of a cell 236–237
peak carry-over 208, 211
pellets 30–33
penetration depth 37, 57–61, 63
phase diagram boundary 247
photoacoustic signal generation 108–110
photoacoustic spectroscopy 36–37, 107–143
photoacoustic units 246
piecewise fit 247
pinholes/too-thick samples; analysis of 266
Planck's constant 2–6
plasticizers 237 See also polymers
polar materials 172, 199
polarizer 140
polyethylene 120–121, 129–130, 139
polymer analysis; by PAS 118–121
polymer fibers, films 139–140
polymers 33–36, 45, 85, 228-230, 238, 281–282
polymorphism 33, 230
polystyrene 135–138
polytetrafluoroethylene 66–68

polyvinyl chloride 179, 257–258, 261-267
poor contact with IRE 86
powders 77, 137–138
precision 224
pressed films 233
pressure 231–232
pressure effects 233, 261-264
pressure broadening 232
pressure dependence 198
pressure plates 70
pressure threshold 198
pressurization 180
pretailing 177
prism cell 88–89
process waste-gas analysis 179
purge and trap 181, 185
purge variation 208–209
purge-gas flow rate 186
pyrolysis 35–36, 173–174
pyrolysis probes 173–174

Q

Q branch 17–20
qualitative analysis 120–121, 124, 139, 180
   by PAS 118–121
quantitative analysis 23, 41, 139, 180, 217-252, 277-279
quantitative applications
   DRIFTS 99–104
quantitative method 224
quantum mechanical theory 2–6

R

R branch 17–20
R-A See refelection-absorption
Raman 4
random errors 223, 227–228, 246
reaction mechanism 229–230
reaction monitoring 83
reference standards 224
reflectance 146, 247, 279-281
reflectance spectrum 94–95
reflection techniques 37–46
reflection-absorption 38, 147–149, 153, 154
refraction 146, 159
refractive index 38, 58
repetitive analysis 81
reproducibility 225, 226, 227
   in DRIFTS 101—102
   of IRS 67, 239
residual monomer 179
residual solvent 227
residue 226
residue spectrum 203
resolution enhancement 248, 249
resolution selection 243–244
restrictor plugging 173

Reststrahlen 240
retardation 244
rolled films 234
rotational isomers 232
routine performance diagnostics 270
rubber 83–85

S

safety 23, 33, 45, 173
sample capacities 176
sample manipulations 225
sample pathlength; IRS 86
sample purity 146, 159
sample thickness 249
sample vapor 176
sampling depth 110-113, 132-137
sampling valves 186
sampling 180, 227–228
sample-preparation artifacts 264–267
saturation 241
Savitzky-Golay smoothing 15-17, 248
scale variations in FT-IR/PAS 129
scattering 158, 234
scattering coefficient 247
sealed liquid cells 21-23, 236
secondary decomposition 207
sediment aggregates 80–92
selection of standards 222
semisolids 79, 115, 141
separation efficiencies 176
separation zone 174–176
separatory funnel 174–176
septum 173
septum-capped vial 185–187
setting down criteria for quantitative analysis
        219–221
SFC/FT-IR 167–168
SFE 170–173, 198–199
short path cells 17–20
short pathlength IRAs 74
Si-Carb sampling technique for non-powdered
        sample 98–101 See also silicon carbide
signal-to-noise ratio 14, 113, 243
significant figures 224
silica filler systems 98
silicon 63–64
silicon carbide 240–241
silicone contamination 17, 257–259
simultaneous distillation extraction 174–176
single beam spectrum 241, 243–245
single particle analysis 125–129
single-sided data collection 244
"SMART" analysis for quantitation 219–252
SMATCH/FT-IR 167
smoothing 15-17, 248
Snell's Law 63

SNR 14, 26, 38, 44, 47-49,
        113, 114, 127, 243, 248
software compatibility 245–250
software-induced artifacts 267–268
solid sampling techniques 27–46
    cast films 33-35
    for quantitative analysis 228-230
    IRS 44-47, 73-80
    mulls 27-29
    nondestructive DRIFTS analysis of 98–99
    PAS 36-37, 115-118
    pellets 30-33
    pyrolysis 35-36
    specular reflectance 37-38
solid sorbent 181
solute identification 80–81
solution cells See liquid cells
solutions 21–27
solvent effects 225–227
solvent evaporation 227
solvent extraction 171
solvent retention 234
solvent volatilization 234
sources of error 227
Soxhlet extraction 170, 172
spectral fringes 249
spectral library 118
spectral manipulations 251
spectral noise 243–244
spectral smoothing 248
spectral subtraction 24, 72, 89, 211
    using PAS 120–125
spectrophotometer performance 241
specular reflectance 37-38, 140, 147, 151-154
split injection 176–177
split ratios 176
split solution 23–25
split-mull 28
split/splitless GC injector 173, 176-179
stabilizers 76, 226
stacked plot 207
standard mixtures 223
standard reference samples 229–230
standards preparation for quantitative analysis
        221–222
standby 180
static headspace 179–181, 189-194
stationary phase 169–170, 177
stationary phase breakdown products 209
steam distillation 195
steel flow-restrictor 198
step-scan FT-IR 108, 110, 135
stereochemistry 229–230
stray light 68
strenuous conditions 199
stretching vibration 6–10

structural damage 230, 235
subambient operating accessories 186
sulfur oxides 179
supercritical fluid chromatography 167–168
supercritical fluid extraction 170–173, 198–199
supercritical state 171–173
supercritical water 172
surface examination
 by IRS 80–92
 by PAS 135–138
surface treatments 77
symmetric, double-sided data collect 244
syringe 22
syringe needle 194

T

temperature effects 231–232
temperature programmed run 173–174, 211
Tenax-TA 174, 181, 196, 207
TG/FT-IR 166–167, 202–207
TG/GC/FT-IR 174, 202–207
TG/GC/FT-IR/MS 174
TGA 166–167, 173–174
TGA-IR 258–259 See also TG/FT-IR
thallium iodide-thallium bromide 63
theory of infrared spectroscopy 1–10
thermal conductivity 37
thermal conductivity detectors 179
thermal desorption 181
thermal diffusivity 110–111
thermal expansion 110
thermal-wave decay coefficients 110-111
thermal-waves 109–110
thermochemical properties 202–207
thermogravimetric analysis 166–167, 173-174
thermogravimetric FT-IR 166–167, 202-207, 258-259
thermolysis 193, 195
thick films 177
thickness 233
thin layer chromatography 167–168
three-factor, fully nested design 102
tilt baseline correction 247–248
TLC 167–168
TLC/FT-IR 167–168
total absorption 86–89
total ion chromatogram 177
total oxidation 202
total petroleum hydrocarbons 173
trace additives 88–89
trace analysis 20–21
trading rules 243
transfer lines 186
transmission 141, 143, 147-148, 153, 154, 276-279

trapped air bubbles 85
trapping effect 177
trapping temperature 181
tungsten needle 125–126

U

undersampling 244
unheated injector 207
uniformity 230, 233
unsaturate level 81
unwanted optical effects 153–159
use of DRIFTS to study coupling agents on fibers/fillers 97, 102

V

vacuum distillation 176
validity 224
vapor phase analysis 191, 222, 232
vapor pressure 191
variable angle IRAs 59–61
variable-temperature liquid cells 236–237
verification 224
vibrating ball mill 31
vibration-rotation energy levels 230, 231
volatiles 169–170

W

weight-loss events 173–174, 202–206
wetting agent 235
window material 23-25, 112

X

XAD-2 174, 181
xenon 167

Z

zero crossing 244
zero path difference 269
zero-filling 244, 250
zinc selenide 63-69
zinc sulfide 63
ZPD 269, 270, 271